£2.90
HPD

**Questions and
Problems in
Pre-University Physics**

Questions and Problems in Pre-University Physics

P M Whelan MA
Sherborne School

M J Hodgson MA
King's School, Canterbury

Illustrated by **T Robinson**

John Murray   Albemarle Street   London

Filmset in Belgium by Grafikon Ltd
Printed and Bound in Great Britain by
Cox & Wyman Limited, London, Fakenham and Reading

0 7195 2585 3

Preface

Aims

Any school physics book published nowadays needs to justify itself on at least two counts. First, it must use SI units exclusively, and second, it must offer something different from the great number of books already available. The majority of problems to be found at the ends of chapters in textbooks, and indeed in many books devoted exclusively to questions, are collected from GCE examination papers. Such questions are written specifically to test an examination candidate's understanding at the end of his two year course, and therefore are not necessarily suitable as vehicles for introducing a new topic to the inexperienced student.

The questions in this book have been written with the express aim of developing confidence, understanding and interest. We have tried to make many of them simple to answer by structuring them in such a way that the student is led easily from one step to the next. Where a more difficult idea is introduced we have sometimes given a hint for the solution. It must be emphasized that this book is *not* linked in any way to examination *technique,* but we hope that its effect will be to improve examination performance by developing a real *understanding* of the ideas of physics. We hope that this understanding will give the student the necessary ability to tackle a university course with confidence.

While writing the questions we have put emphasis on the desirability of the student acquiring (*a*) a feeling for *orders of magnitude,* (*b*) a realization of the importance of *energy* as a linking concept, and (*c*) a familiarity with microscopic ideas and the behaviour of *electrons, atoms* and *molecules.*

Content

For convenience we have planned the organization of this book to match that of our A/S level textbook *Essential Principles of Physics,* but we have been careful to ensure that it can be used independently. We have included few questions of the type set in examinations because these are becoming available in ever-increasing quantity from the various examining boards in book form. Similarly there are few full scholarship problems. We have tried to make a systematic collection in which no topic of importance has been omitted.

We have separated the questions in nearly every chapter into two groups, *Qualitative* and *Quantitative.* The former are intended for discussion in class: seldom do they ask a question which can be answered by appealing directly to the concise text that we visualize many students will have. They do not require him to reproduce the sort of description that he will find in a detailed text, but more often they ask the question *'Why?'* Not all of them have clear-cut answers. Some are intended to develop the basic bookwork beyond that which he need *learn,* and are presented so that they involve useful application of those ideas with which he must become familiar. Many will require him to make use of more detailed monographs in the library. Others are written expressly to develop creative thinking, especially with regard to orders of magnitude.

It is regrettable that much of a science student's work at school is analytical rather than creative. For this reason we have included an extensive chapter on **Essays**. While some topics are historical and biographical, most of them have been chosen to promote reading and thought in a subject area of interest both to the student and to the future development of physics.

Details

(*a*) *Classification of questions.* The subject matter has been broken down into detailed chapters, and further subdivided within a chapter where it was thought to be helpful. Where a question aims to develop a particular point this has been indicated, and an index has been included to assist the location of such questions. Questions marked † are particularly easy, whereas those marked * either deal with a simple topic in an advanced way, or treat what is usually regarded as a scholarship topic. Questions marked **O-M** (standing for *order of magnitude*) are discussed on page 22.

(*b*) Material in *italic* is of two kinds: (*i*) *hints* for solving more sophisticated problems, and (*ii*) *teaching information* to point out the relevance of a problem, its connection with other topics, or perhaps the significance of the order of magnitude of the answer.

(*c*) In many minor ways we have tried to keep the working of the problems simple. For the most part information is presented to two significant figures, and where possible in round numbers. We have quoted volumes and areas rather than radii, weights rather than the ubiquitous (mass $\times g_0$) etc., since many students lose sight of the physical principles when they are faced by a lengthy numerical computation.

(*d*) *Answers.* Where a question has been broken down into several steps, we have sometimes given selected answers only. If the method of solution is so simple that it is indicated by the answer, then that answer is omitted.

Acknowledgements

(*a*) All the questions were written expressly for this book, but naturally few of them are original in concept. We have referred widely to current textbooks and examination papers, and we have collected from them the ideas on which these questions have been based. We are not aware that any question here duplicates that in any other publication, but freely acknowledge our indebtedness to the books listed in the Bibliography of *Essential Principles of Physics*. In particular we would like to pay tribute to *Physics*, by Halliday and Resnick, from which we derived the idea of the short but penetrating qualitative questions.

(*b*) Mrs E. G. Hodgson again earns our sincere thanks for taking on the demanding job of preparing the typescript, a task which she performed with notable accuracy.

(*c*) We would like to thank Kenneth Pinnock and Howard Jay of John Murray who advised us at all stages of the production of both this book and its companion, and who were most cooperative in following our wishes on details of layout and illustration.

(*d*) Dr J. W. Warren of Brunel University kindly undertook the uninviting task of scrutinizing the questions for errors. This attention to detail has been very helpful in reducing minor mistakes and obscurities, and we are most grateful.

(*e*) Once again we thank Tim Robinson for the first class job that he has made of the diagrams.

(*f*) One of us (PMW) would like to thank the Warden and Fellows of Merton College, Oxford, for their generous hospitality during his term as a schoolmaster student at the College; the Headmaster of Sherborne School for allowing him leave of absence; and his wife Sue for enduring for some years now what W. M. Gibson has delightfully called a 'severe case of authorship'.

A book of this nature is more than usually prone to minor mistakes, both numerical and otherwise. We encourage readers, students particularly, to write and tell us about those that they discover.

August, 1972

P.M.W.
Sherborne, Dorset

M.J.H.
Canterbury, Kent

Contents

Subject Matter Listed by Chapters

I General Introduction

ERRORS

DIMENSIONS

GRAPHS

Introduction

About this Book

Before reading this section you are advised to refer to the *Preface,* in which we discuss the purpose of the book, and the methods that we adopt to achieve it. In particular you should refer to the ways in which we have classified the different types of questions.

Many of the questions require numerical values for the fundamental constants, or for particular physical properties, and in these problems we have indicated at the end which are to be used. For convenience the values of these constants are printed on the last two pages of the book. No *other* information should be assumed in answering these problems. For example the quantities R, N_A, k and the ideal-gas molar volume at s.t.p. are all closely related, and to assume constants other than those indicated could destroy the point of the question.

Some of the questions are marked **O-M**, which is an abbreviation for *order of magnitude*. For these questions you will have to make your own assumptions. More details about them, and a worked example, are given on p. 22.

If you make reference to older literature you will find information expressed in non-SI units. Conversion factors will be found in *Tables of Physical and Chemical Constants*, by **Kaye** and **Laby** (*Longman*).

The Meanings of Words in Questions

Although this book is not concerned with examination questions, you will note the recurrent use of words habitually used in such questions, and you are advised to clarify in your own mind exactly what they mean. Here are some examples.

(*a*) **Define** means 'Give a short but exact formal statement'. Thus the definition of a physical quantity is best given by an algebraic equation. The symbols used in the equation must first be defined. Definitions are discussed further on p. 4.

(*b*) **Explain** (or *'What is meant by...?'*) requires much more detail. There would be a considerable difference between a formal statement *defining* pressure, and a microscopic *explanation* as to the nature and causes of the pressure at a point in a fluid. (Nevertheless it is often helpful if this explanation is developed from the definition.)

(*c*) The word **physical** in a question usually implies that the answer must include an explanation of the *mechanism* by which some process occurs. Very often this will include a description involving the imagined behaviour (according to some particular model) of submicroscopic particles such as electrons or molecules.

Suppose, for example, that we are asked to give a physical explanation for the reason why a metal generally conducts more heat under specified conditions than a non-metal. It would be no answer at all to quote the defining equation for thermal conductivity λ, and then to point out that the value of λ for a metal is generally larger than that for a non-metal. We would need to go into the microscopic description of the physical processes by which the two types of substance transport energy through themselves.

(*d*) Other words which occur frequently in questions are *Describe, Discuss, Compare, Contrast,* etc., and you must always check, before answering such questions, that you are doing just what is asked. For the latter two it is not sufficient to draw up a pair of corresponding lists which leave the reader to make the necessary comparisons.

Prefixes for SI Units

Certain prefixes have been agreed for use with the basic and derived SI units. All those listed below are of the form 10^{3n}, where n is a whole positive or negative number. Not all these prefixes have been used widely in the past, but it seems probable that they will be in the future. For that reason we have used them extensively in the questions of this book. It is essential that you become familiar with them as soon as possible, and to help you a number of concrete examples are given in the table below. The values of the prefixes will also be found on the front and back end-papers.

Prefix	Symbol	Meaning	Common example	
atto	a	10^{-18}	the electronic charge	$e = -0.16 \text{ aC}$
femto	f	10^{-15}	energy of a typical X-ray photon	$W = 1 \text{ fJ}$
pico	p	10^{-12}	capacitance of a mica tuning capacitor	$C = 6.0 \text{ pF}$
nano	n	10^{-9}	the wavelength of sodium light	$\lambda = 589 \text{ nm}$
micro	μ	10^{-6}	the leakage current in a junction diode	$I = 3.5 \text{ μA}$
milli	m	10^{-3}	inductance of an air-cored solenoid	$L = 5.7 \text{ mH}$
kilo	k	10^{3}	power of an electric heater	$P = 2.5 \text{ kW}$
mega	M	10^{6}	standard pressure	$p_0 = 0.10 \text{ MPa}$
giga	G	10^{9}	the half-life of radium decay	$T_{\frac{1}{2}} = 51 \text{ Gs}$
tera	T	10^{12}	typical frequency of middle i.r. waves	$f = 10 \text{ THz}$

Equations and Quantities

The Nature and Status of Physical Equations

The Order in which Physical Quantities are Defined

The Nature and Status of Physical Equations

The solution of physics problems is too often regarded as a question of selecting an appropriate formula into which numbers can be substituted. While this may *sometimes* be a quick method of arriving at an answer, it is no way to learn the concepts of physics. Nevertheless physics is primarily an exact (quantitative) discipline, and this means that much of our work is concerned with equations.

It is the aim of this paragraph to indicate the relative importance of equations, so that you can decide for yourself what has to be *learned*. Not every equation can be fitted into a neat classification, but we can distinguish the following.

(a) **Defining equations.** These equations summarize the series of operations by which we relate a new physical quantity to other pre-existing quantities, and thus by implication to the seven fundamental quantities. For example we define electric potential V by relating it to the electric p.e. W of a test charge Q_0 through the defining equation $V = W/Q_0$.

In one sense a defining equation is a convention, and, as such, cannot be wrong: nevertheless private conventions are less useful than universally observed conventions, and much confusion is avoided if we all use the same convention. Therefore *defining equations must be learned.*

(b) **Laws.** A law is a statement that summarizes, with great precision and simplicity, ideas of fundamental importance. It is usually expressed in the form of an equation relating symbols that represent experimental observations, or, more simply, physical quantities. Thus the equation $F = Gm_1m_2/r^2$, once we have defined explicitly all the symbols involved, represents *Newton's* law of universal gravitation. *Equations representing laws must be learned,* or at least a physicist must be able to write down the equation that symbolizes a verbal statement of a law.

(c) **Principles and useful results.** There are many equations in physics that are the consequence of applying laws and defining equations to situations which are frequently encountered. These results are then used so often that much time is saved if they are committed to memory. The following well-known equations are a few examples selected at random:

$$Fs = \tfrac{1}{2}mv^2 - \tfrac{1}{2}mu^2$$

$$p + \tfrac{1}{2}\rho v^2 + h\rho g = \text{constant}$$

$$p = \tfrac{1}{3}\rho \overline{c^2}$$

$$n = \frac{\sin[(A + D_{\min})/2]}{\sin A/2}$$

$$V = \left(\frac{1}{4\pi\varepsilon_0}\right)\frac{Q}{r}$$

$$E = \sigma/\varepsilon_0$$

$$\mathscr{E} = I(R + r).$$

It should be stressed that it is not strictly *necessary* to learn any of these equations, but the derivation of (e.g.) $p = \tfrac{1}{3}\rho\overline{c^2}$ is an exercise which many beginning students find difficult and time consuming. Thus *it is useful to learn these equations*, provided that one remembers that the ideas behind the derivation of the equations are what really matter.

If you are uncertain whether you have remembered a derived equation correctly, you can always do a quick mental check that your version is at least dimensionally consistent.

Example: suppose you thought that the energy stored by an inductance L was of the form $\tfrac{1}{2}LI$. This has the unit $(\text{V s A}^{-1}) \times (\text{A}) = \text{V s} = \text{J s C}^{-1}$. To obtain the unit J we must multiply by C s^{-1}, i.e. A. We deduce that the equation is more likely to be of the form

$$W = \tfrac{1}{2}LI^2.$$

(We should immediately recognize that stored energy can be represented by an expression in which a variable is *squared*.)

(d) **Special results.** There are occasions when you may read in a textbook an analysis whose end-product is an equation applicable to one special situation only. For example it can be shown that the loss of electrical p.e. ΔW when two capacitors are joined is given by the equation

$$\Delta W = \frac{\tfrac{1}{2}(C_2Q_1 - C_1Q_2)^2}{C_1C_2(C_1 + C_2)}.$$

The effort spent in learning such an equation is better used in mastering the principles used to derive it. It is a common fallacy to imagine that the learning of such equations is equivalent to an understanding of the ideas behind them, and very often they are misquoted, and used even in circumstances to which they do not apply. *These equations should not be learned.*

Of course there are some results which a specialist will use a large number of times. For example he may want to use the equation

$$\lambda = \frac{h}{\surd(2m_eeV)} = \frac{1.23 \text{ nm}}{\surd(V/\text{volts})}.$$

He would not *learn* this equation (although no doubt he would become familiar with it by repetitive use), but rather he would write it down and refer to it when necessary.

Summary

When you first meet a particular equation, ask yourself whether it is

(a) a definition,

(b) a law

(c) an important *idea* which you will want to use time and time again, or

(d) applicable to one situation only.

Do not learn equations of type (d).

The Order in which Physical Quantities are Defined

The logical structure of physics, in some subject areas particularly, is based on a distinct hierarchy, and it is important to appreciate the order in which quantities have been defined. For example if we choose to define magnetic flux Φ by the scalar product $\Phi = \mathbf{B} \cdot \mathbf{A}$, then we cannot later define magnetic flux density $\mathbf{B}$ by the quotient $B = \Phi/A$. The latter equation can be regarded as a satisfactory definition of $\mathbf{B}$ only if both Φ and $\mathbf{A}$ have already been defined by some other means. (There exists an alternative approach to electromagnetism in which one defines Φ from $\mathscr{E} = -\mathrm{d}\Phi/\mathrm{d}t$, and then $B = \Phi/A$ becomes acceptable as the formal definition of $\mathbf{B}$.)

The schemes which follow show a logical order of defining physical quantities in five branches of physics. Every equation shown is a *definition*, and no attempt has been made to relate quantities by other equations. For simplicity the definitions are given in an elementary form. The boxes of the fundamental quantities (the logical starting points) are shown in grey.

Linear Dynamics (see scheme 1 below)

Notes

(*a*) No arrows are shown for mass ratio. The equation $\mathbf{F} = m\mathbf{a}$ defines the measure of *force*, and is also used to measure *inertial mass* through $\mathbf{F} = m_1\mathbf{a}_1 = m_2\mathbf{a}_2$.

(*b*) The definitions of rotational dynamics are not included—they follow a similar scheme.

Thermal Properties of Matter

(See scheme 2 on page 6.)

Electrostatics (see scheme 3 on page 6)

Notes

(*a*) In this scheme quantities are introduced from mechanics only where they make a significant contribution to the understanding of the whole.

(*b*) Note carefully that electric charge is not a fundamental quantity, since it is derived from electric current through $Q = \int I\mathrm{d}t$.

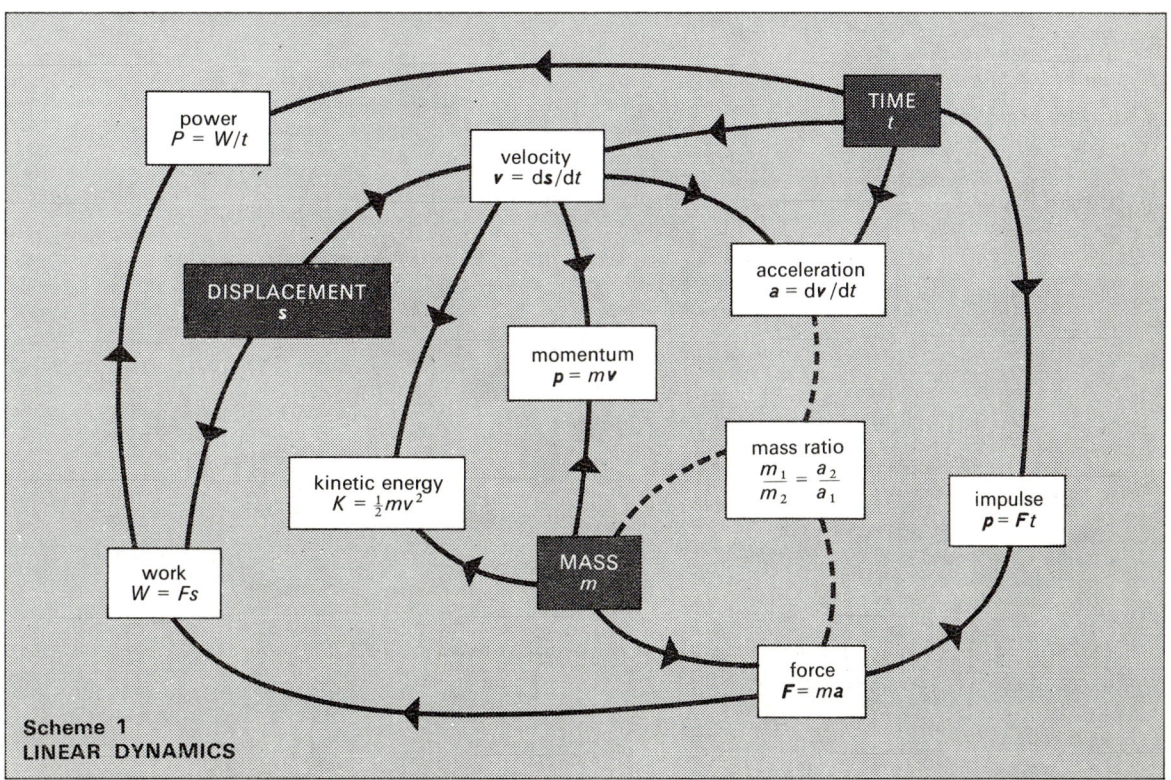

Scheme 1
LINEAR DYNAMICS

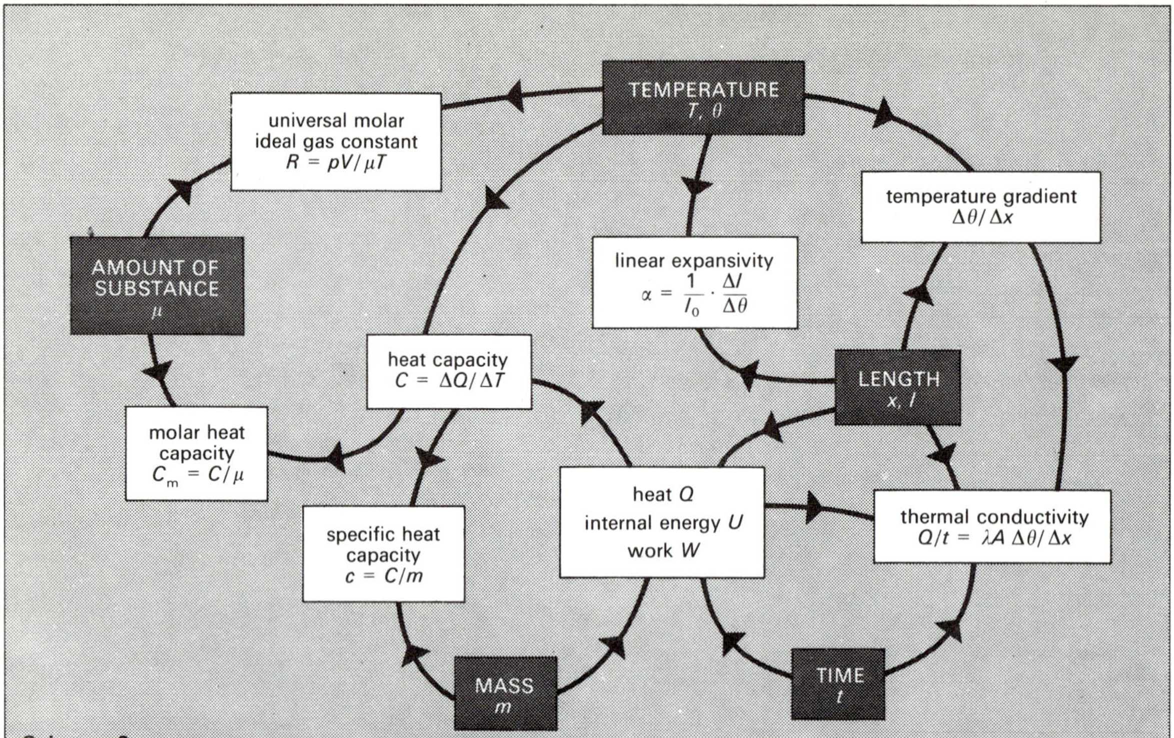

Scheme 2
THERMAL PROPERTIES OF MATTER

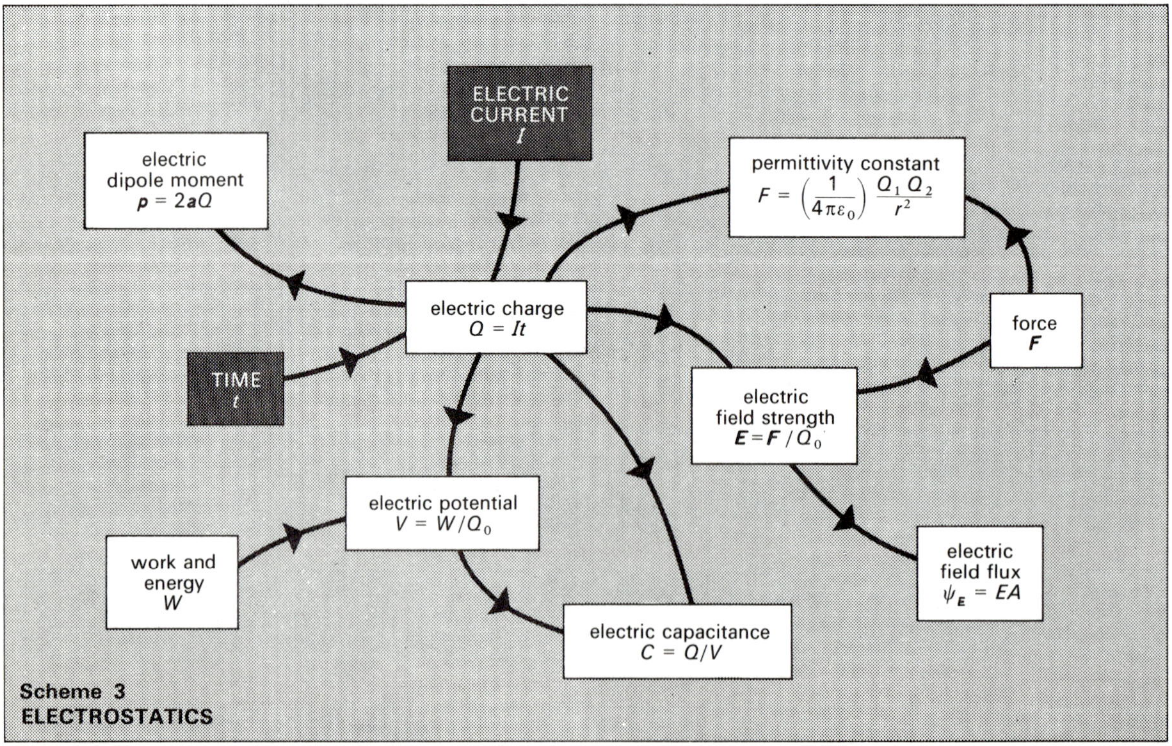

In this scheme the following boxes and relations appear:

TEMPERATURE T, θ

universal molar ideal gas constant $R = pV/\mu T$

temperature gradient $\Delta\theta/\Delta x$

AMOUNT OF SUBSTANCE μ

linear expansivity $\alpha = \dfrac{1}{l_0} \cdot \dfrac{\Delta l}{\Delta\theta}$

heat capacity $C = \Delta Q/\Delta T$

LENGTH x, l

molar heat capacity $C_m = C/\mu$

specific heat capacity $c = C/m$

heat Q internal energy U work W

thermal conductivity $Q/t = \lambda A\, \Delta\theta/\Delta x$

MASS m

TIME t

Scheme 3
ELECTROSTATICS

ELECTRIC CURRENT I

electric dipole moment $p = 2aQ$

permittivity constant $F = \left(\dfrac{1}{4\pi\varepsilon_0}\right)\dfrac{Q_1 Q_2}{r^2}$

electric charge $Q = It$

TIME t

force F

electric field strength $E = F/Q_0$

electric potential $V = W/Q_0$

work and energy W

electric capacitance $C = Q/V$

electric field flux $\psi_E = EA$

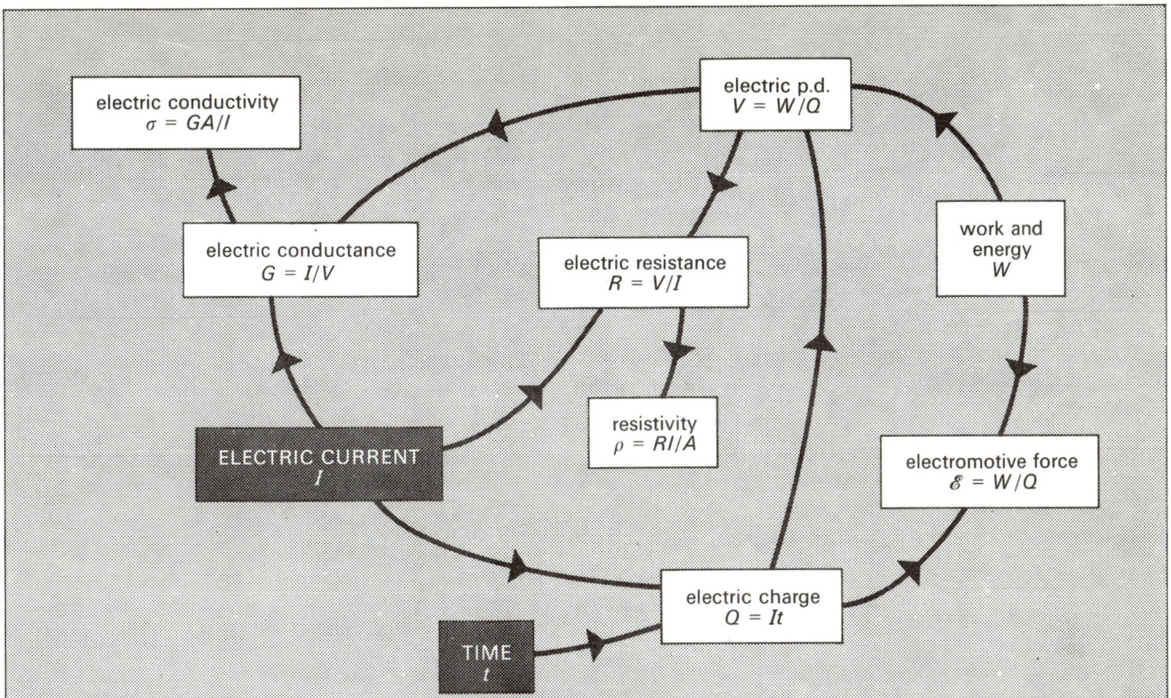

Scheme 4
CURRENT ELECTRICITY

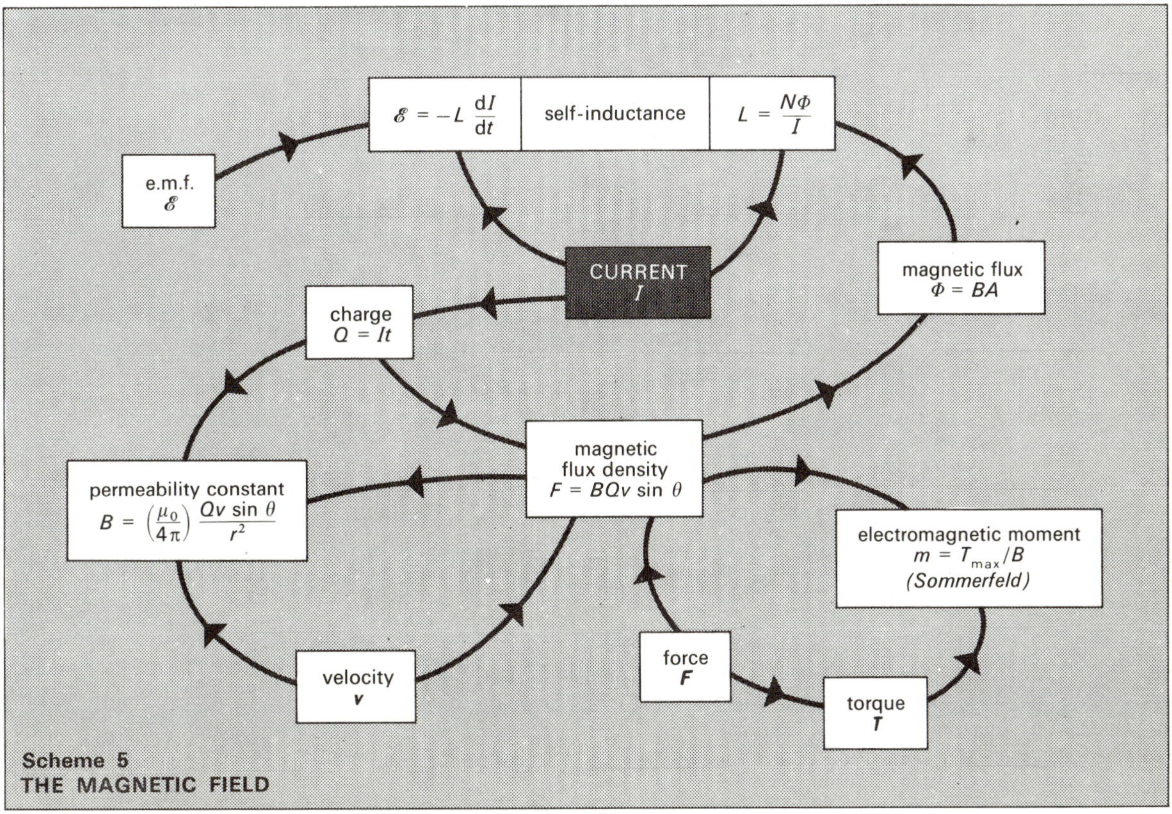

Boxes and equations in Scheme 4:

electric conductivity $\sigma = GA/l$

electric conductance $G = I/V$

ELECTRIC CURRENT I

electric p.d. $V = W/Q$

electric resistance $R = V/I$

resistivity $\rho = RI/A$

work and energy W

electromotive force $\mathscr{E} = W/Q$

electric charge $Q = It$

TIME t

Scheme 5
THE MAGNETIC FIELD

Boxes and equations in Scheme 5:

$\mathscr{E} = -L \dfrac{dI}{dt}$ | self-inductance | $L = \dfrac{N\Phi}{I}$

e.m.f. $\mathscr{E}$

CURRENT I

magnetic flux $\Phi = BA$

charge $Q = It$

permeability constant $B = \left(\dfrac{\mu_0}{4\pi}\right)\dfrac{Qv\sin\theta}{r^2}$

magnetic flux density $F = BQv\sin\theta$

electromagnetic moment $m = T_{max}/B$ (Sommerfeld)

velocity v

force F

torque T

The following notes refer to the schemes on page 7.

Current Electricity

Notes

(*a*) Mechanical quantities are introduced as in the previous scheme.

(*b*) The defining equations for $\mathscr{E}$ and V are symbolically identical, and it is important to identify carefully the quantity represented by W in each definition.

The Magnetic Field

Notes

(*a*) There are several alternative approaches to electromagnetism. The one used here is that recommended by the *Association for Science Education.*

(*b*) The equations $F = BI \, l \sin \theta$ and

$$\delta B = \left(\frac{\mu_0}{4\pi}\right) \frac{I\delta l \sin \theta}{r^2}$$ are exactly equivalent to the two

used above to define $\boldsymbol{B}$ and μ_0 respectively.

Mathematics

The Slide Rule

Significant Figures

Approximations

Some Useful Mathematics

Differential Equations in Physics

The Slide Rule

A slide rule should be regarded as a vital piece of equipment for everyone who studies physics above 'O' level. For school use it is not necessary to invest in a particularly complicated one and a small pocket-size slide rule may be sufficient. Whatever type you have you should always ensure that it is clean, and that the slide and cursor are neither too tight nor too loose in their movement. Having bought a slide rule, it is essential to spend some time reading the accompanying instruction booklet. This is usually comprehensive and, apart from explaining the special markings on the cursor and scales for commonly recurring operations, it will also discuss the wide variety of calculations that can be carried out. The booklet should contain all the detailed information you require, and only a few general points will be emphasized here.

(1) Be realistic in your use of the slide rule and be aware that certain calculations require more significant figures than it can offer. Thus tables will sometimes be essential and you must be able to use them.

(2) Become familiar with the basic scales before attempting to use the more advanced ones. Ensure that the slide rule is always perpendicular to your line of vision so that you avoid parallax errors.

(3) The slide rule deals only with significant figures in a calculation and so the position of the decimal point must be established by making a rough calculation. These mental checks are also most important for deciding which section of the rule should be used for finding square roots and cube roots. Remember that the results of all calculations must be adjusted to the number of significant figures appropriate to the original information.

(4) Each slide movement may introduce a small error. Therefore (*i*) simplify the expression by mental arithmetic as far as possible, (*ii*) spend a short time planning compound multiplication and division, (*iii*) make full use of the reciprocal scale which can be used to cut down the number of slide movements.

(5) The more practice and experience you have, the more efficient you will become in getting the best from your slide rule.

Significant Figures

No physical measurement is exact, and if a reported figure is to make an appropriate impression, then its uncertainty must also be given. Thus a mass might be quoted as $m = (1.25 \pm 0.02)$ kg. This implies that the best estimate of the third significant figure is 5, but that it is also recognized that during the measurement of this mass there was an experimental uncertainty of 0.02 kg, and that the true value of m could be as high as 1.27 kg or as low as 1.23 kg. On this basis, if we used the quoted value of m for calculation purposes, we would not be confident that the result was accurate to better than about 1 part in 60. When the results of experimental work are to be published, it is essential that the extent of the experimental uncertainty should be quoted explicitly if the results are to be of value to others.

A way of *implying* the precision and accuracy of a measurement is simply to quote a sensible number of significant figures. If we write $m = 1.25$ kg, then there are two implications. One is that 5 is our *best estimate* of the third significant figure. The other is that we have not ruled out the possibility that the third figure might be (say) 1. We are not, by writing three significant figures, claiming an uncertainty of ± 0.01 kg. When we use this value of m for calculation purposes, it is pointless to express further results to more than three significant figures (except in special circumstances). There exist extensive formal rules for analysing probable errors, and the technique of their application is an important weapon of the physicist's armoury. Nevertheless there is no point in subjecting every calculation in a book of this type to such an analysis.

Accordingly in this book we adopt the following procedure. With few exceptions the numerical *information* in the questions is given to *two* significant figures (to save time being spent on laborious calculation). You are recommended to work each stage of your calculation as carefully as possible by slide rule (to about *three* significant figures), and then to present the final conclusion

rounded back to *two* significant figures. It is recognized that on occasion this procedure is not fully justified, but it is felt that there is little to be gained from extreme rigour, and much time will be saved.

You should be on the look-out for problems in which the nature of the calculation reduces considerably the number of figures which should be quoted in the answer. Consider as an extreme example the calculation of the difference of two close quantities, as in

$$\Delta \lambda = \lambda_2 - \lambda_1 = 589.59 \text{ nm} - 589.00 \text{ nm}$$
$$\underline{\Delta l = 0.59 \text{ nm.}}$$

Five-figure information has resulted in a quantity whose uncertainty is at least 1 part in 60.

Approximations

Very often a result can be obtained far more easily, or with greater accuracy, by making simplifying approximations in the course of a calculation. No hard and fast rules can be given because every situation must be treated on its own merits, and experience enables one to judge what is justifiable in any particular problem.

The examples which follow are intended to guide you in your thinking. They show that the obvious way of dealing with a problem, involving perhaps a fair amount of four-figure arithmetic, sometimes leads to an answer which is less reliable than that obtained by careful thinking and a slide rule.

Example 1–1   A small length change.

Refer to the diagram. A load has been added at P, the mid-point of a wire APB, and has depressed it by 10 mm. What is the extension of half the wire?

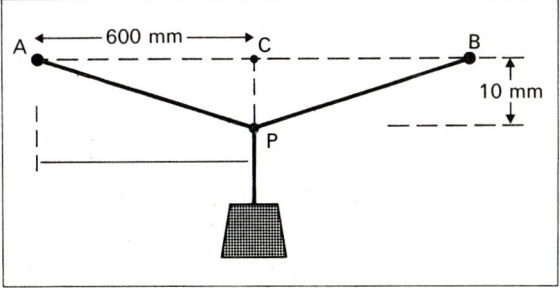

(*a*) Suppose the new length AP $= l$. Then by *Pythagoras*
$$l^2 = (600 \text{ mm})^2 + (10 \text{ mm})^2$$
$$= 36.01 \times 10^4 \text{ mm}^2.$$

In four-figure tables we find $l = 600.1$ mm.
The extension of half the wire is 0.1 mm.

(b) Let Δx be the extension of half the wire, so that AP = (600 mm + Δx). Then

$$(600 \text{ mm} + \Delta x)^2 = (600 \text{ mm})^2 + (10 \text{ mm})^2$$
$$1200 \,\Delta x \text{ mm} + \Delta x^2 = 100 \text{ mm}^2.$$

At once the 10 mm shows itself as the important quantity which determines the size of Δx. Since $\Delta x \sim 0.1$ mm, it follows that $\Delta x^2 \sim 0.01 \text{ mm}^2$, and we may neglect 10^{-2} in comparison with 10^2.

$$\therefore\ 1200 \,\Delta x \approx 100 \text{ mm},$$

giving
$$\Delta x \approx 0.083 \text{ mm}.$$

Notes

(i) Unintelligent use of the four-figure tables gave us an answer whose percentage error was $\sim 20\%$.

(ii) The idea used in (b) is similar to that which enables us to write (for small α)

$$\sqrt{(1 + 2\alpha)} \approx 1 + \alpha.$$

Example 1–2   The difference between two nearly equal quantities.

The quantity $(1/R_1 + 1/R_2)$ for a thin lens is 1.00×10^{-2} mm^{-1}. Calculate the difference between its focal lengths for light of the C (red) and F (blue) Fraunhöfer lines, for which $n_C = 1.514$ and $n_F = 1.524$.

(a) We can calculate the focal lengths as follows:

$$1/f_C = (n_C - 1)\,(1/R_1 + 1/R_2)$$
$$= (0.514) \times (10^{-2} \text{ mm}^{-1})$$
$$f_C = 19(5) \text{ mm by slide rule.}$$
Similarly $\qquad f_F = 19(1)$ mm.

Hence the difference between the focal lengths

$$\Delta f = f_C - f_F = 4 \text{ mm.}$$

Note that there is an uncertainty in Δf of about 2 mm, which is a percentage error of about 50%. Careless work with the slide rule could easily have led to answers of 2 mm or 6 mm.

(b) A better procedure is to calculate the *difference* between the focal lengths *directly*. Since

$$1/f = k(n - 1), \text{ where } k = 1.00 \times 10^{-2} \text{ mm}^{-1},$$

differentiating, we have

$$-\delta f/f^2 = k\delta n$$

so $\qquad\qquad \Delta f \approx -kf^2\,\Delta n.$

The minus sign tells us that f increases as n decreases (as is seen from the reciprocal). Using the *mean* value for f of 193 mm (above)

$$\Delta f \approx -(1.00 \times 10^{-2} \text{ mm}^{-1})\,(3.72 \times 10^4 \text{ mm}^2)\,(1.514 - 1.524)$$

from which $\qquad\qquad \Delta f = 3.7 \text{ mm.}$

Careless work with the slide rule in method (b) would have led to an error less than the uncertainty implied by the fact that Δn is known to only one part in 10.

Some Useful Mathematics

A number of results are collected here for reference.

Algebra

Binomial Theorem

$$(1 + x)^n = 1 + nx + \frac{n(n - 1)}{1 \times 2}x^2 + \frac{n(n - 1)\,(n - 2)}{1 \times 2 \times 3}x^3 + \ldots$$

If $x \ll 1$ $\qquad (1 + x)^n \approx 1 + nx$
$$(1 + x)^{-n} \approx 1 - nx.$$

These useful approximations are valid when x^2 is negligible.

Quadratic Equations

The equation $\qquad ax^2 + bx + c = 0$

has the solutions $\qquad x = \dfrac{-b \pm \sqrt{b^2 - 4ac}}{2a}.$

Logarithms

Common logarithms are those to base 10.
Thus if $\qquad\qquad\qquad x = 10^y$
$$y = \lg x$$
where $\lg x$ implies $\log_{10} x$.

Natural logarithms are those to base e.
Thus if $\qquad\qquad\qquad x = e^y$
then $\qquad\qquad\qquad\quad y = \ln x$
where $\ln x$ implies $\log_e x$.   e is defined by

$$e = \lim_{n \to \infty} \left(1 + \frac{1}{n}\right)^n = 2.718 \ldots$$

$$\ln(1 + x) = x - \frac{x^2}{2} + \frac{x^3}{3} - \ldots$$

Natural and common logarithms are related by
$$\ln x = (2.303)\,\lg x, \text{ and}$$
$$\lg x = (0.434)\,\ln x.$$

Trigonometry

The angle θ subtended by a length s of arc of a circle of radius r is given by
$$\theta[\text{rad}] = s/r,$$
from which it follows that
$$1 \text{ rad} = 180°/\pi \approx 57.3°.$$
If $x = \sin\theta$, then we say 'θ is the angle whose sine is x', and we write
$$\theta = \arcsin x.$$

Useful Relationships

Signs: particular trig. functions are positive as follows:

first quadrant, all: second, sine: third, tangent: fourth, cosine.

$$\tan\theta = \sin\theta/\cos\theta$$
$$\sin^2\theta + \cos^2\theta = 1$$
$$\sin A + \sin B = 2\sin\left(\frac{A + B}{2}\right)\cos\left(\frac{A - B}{2}\right)$$

Cosine law: for any triangle

$$a^2 = b^2 + c^2 - 2bc \cos A$$

Since

$$\sin \theta = \theta - \frac{\theta^3}{3!} + \frac{\theta^5}{5!} - \dots \quad , \text{ and}$$

$$\tan \theta = \theta + \frac{\theta^3}{3!} + \frac{\theta^5}{5!} + \dots$$

it follows that when $\theta \ll 1$ (say ~ 0.1 rad)

$$\sin \theta \approx \tan \theta \approx \theta \text{ , but } \cos \theta \approx 1.$$

Calculus

Derivatives and Integrals

$y = \text{f}(x)$	$\dfrac{dy}{dx} = \text{f}'(x)$	$\int y \, dx$
$y = x^n$	$n\,x^{n-1}$	$\dfrac{x^{n+1}}{n+1} + C \, (n \neq -1)$
$y = x^{-1}$	$-x^{-2}$	$\ln x + C$
$y = \ln x$	x^{-1}	$x \ln x - x + C$
$y = \sin x$	$\cos x$	$-\cos x + C$
$y = \cos x$	$-\sin x$	$\sin x + C$

C is the constant of integration, whose value is determined by the limits of integration.

Average Value of a Function

The **mean** or **average value** $<y>$ of $y = \text{f}(x)$ over the interval $x = a$ to $x = b$ is given by

$$<y> = \frac{1}{b-a} \int_a^b y \, dx.$$

Similarly

$$<y^2> = \frac{1}{b-a} \int_a^b y^2 \, dx.$$

The **root mean square** (r.m.s.) value of y is given by

$$y_{\text{r.m.s.}} = \sqrt{<y^2>} \ .$$

For a periodic function the interval $(b - a)$ is understood to be taken over an integral number of periods or half periods.

*Differential Equations in Physics

Their Origin

Differential equations arise in many branches of physics, as the following examples show.

(*a*) **Mechanics.** The solution of many problems in dynamics depends upon the equation

$$\boldsymbol{F} = m\boldsymbol{a} = m(\text{d}^2\boldsymbol{x}/\text{d}t^2).$$

$\boldsymbol{F}$ may be a function of t, $\boldsymbol{x}$ and/or $\dot{\boldsymbol{x}}$, and so we might have to integrate to find $\boldsymbol{x}$ as a function of t.

(*b*) **Oscillations and waves.** When $\boldsymbol{F}$ can be written in the form $-k\boldsymbol{x}$, the equation becomes

$$m\,(\text{d}^2\boldsymbol{x}/\text{d}t^2) + k\boldsymbol{x} = 0,$$

which is referred to as the **differential equation of s.h.m.**

Its solution is discussed below. Wave motions have a similar important differential equation: in one dimension

$$\frac{\partial^2 y}{\partial x^2} = \frac{1}{c^2} \frac{\partial^2 y}{\partial t^2} .$$

(*c*) **Heat flow.** In one dimension the rate of heat flow through a material is described by the differential equation

$$\text{d}Q/\text{d}t = -\lambda A \,(\text{d}\theta/\text{d}x).$$

If, in a given situation, $\text{d}Q/\text{d}t$ is constant, and both λ and A depend upon x, then we would have to integrate to find the variation of θ with x.

(*d*) **D.C. circuits.** The growth of current in a d.c. LR circuit is described by the equation

$$\mathscr{E} = L(\text{d}I/\text{d}t) + IR.$$

The decay of charge on the plates of a capacitor in a d.c. CR circuit is described by the equation

$$IR + Q/C = 0$$

$$R(\text{d}Q/\text{d}t) + Q/C = 0.$$

(*e*) **A.C. circuits.** The application of the law of energy conservation in an a.c. circuit leads to the differential equation

$$\mathscr{E}_0 \cos \omega t = L(\text{d}I/\text{d}t) + Q/C + IR.$$

Since $I = \text{d}Q/\text{d}t$, we may write

$$\mathscr{E}_0 \cos \omega t = L\frac{\text{d}^2 Q}{\text{d}t^2} + R\frac{\text{d}Q}{\text{d}t} + \frac{1}{C}Q,$$

a differential equation which can be solved to find Q (and I) in terms of t.

(*f*) **Radioactive decay.** The disintegration of a radioactive nucleus is an entirely *random* process. This means that $-\text{d}N/\text{d}t$, the rate of decay of a particular sample, is proportional to the (large) number N of active nuclei that it contains. We write

$$\text{d}N/\text{d}t = -\lambda N.$$

If a radioisotope is being created in a nuclear reactor at the constant rate C, but at the same time the new nuclide disintegrates at the rate $-\lambda N$, then we can calculate the net rate of increase of nuclei from

$$\text{d}N/\text{d}t = C - \lambda N.$$

The number of nuclei present after a given time interval must be found by solving this differential equation. (Note the formal similarity between

$$\text{d}N/\text{d}t + \quad \lambda N - \quad C = 0 \text{ , and}$$

$$\text{d}I/\text{d}t + (R/L)I - (\mathscr{E}/L) = 0 \quad \text{(above)}.$$

Similarities of this kind enable us to solve many differential equations by comparing them with others whose solutions are already known.)

The Solution of Differential Equations

There exists a considerable scheme for classifying differential equations as to their order and degree, and also a formal set of rules for their solution. Most of the differential equations which we meet can be solved by one of two methods.

(*a*) **By separation of variables.** If the variables can be separated onto the two sides of the equation, then the solution will be obtained by direct integration. Several examples are given below.

(*b*) **By inspection.** Because the equations that we deal with tend to treat a small number of distinct physical situations, previous experience and physical intuition will often enable us to *guess the form of the solution*. We can then test our proposed solution by substituting it into the original differential equation.

Example: the LC circuit. Experience teaches us that an equation of the form

$$\frac{d^2Q}{dt^2} = -\left(\frac{1}{LC}\right)Q$$

describes an oscillatory situation, since it closely resembles

$$\frac{d^2x}{dt^2} = -\left(\frac{k}{m}\right)x.$$

Suppose we write $(1/LC) = \omega^2$, and then suggest

$$Q = A \sin \omega t + B \cos \omega t$$

as our solution of

$$d^2Q/dt^2 = -\omega^2 Q.$$

A and B are arbitrary constants. Differentiating twice, we have

(1) $dQ/dt \quad = A\omega \cos \omega t - B\omega \sin \omega t$
(2) $d^2Q/dt^2 = -A\omega^2 \sin \omega t - B\omega^2 \cos \omega t$
$\qquad\qquad = -\omega^2 Q.$

Since this is the original equation, we have verified that

$$Q = A \sin \omega t + B \cos \omega t$$

is a solution.

Constants

The above solution can also be expressed in the form

$$Q = C \sin (\omega t + \delta),$$

where C and δ are two further arbitrary constants related to A and B. To eliminate *two* such constants, and hence arrive back at the original equation, it was necessary to differentiate *twice*. This illustrates a general rule:

*The **general solution** of an n th order equation (involving* $d^n y/dx^n$*) will have n constants.*

A **particular solution** is one in which the arbitrary constants have been given specified values. Thus to obtain the *particular* solution of a differential equation describing s.h.m. we need to have *two* separate items of information, or **boundary conditions**. For example we might be told that $\dot{x} = 0$ and $x = +a$ when $t = 0$, and this would enable us to write a particular solution involving no arbitrary constants.

Worked Examples

Example 1–3   Vertical motion under gravity.
A body is projected vertically upwards from the Earth's surface at a speed u. What will be its speed v at a distance r from the centre of the Earth? Neglect air resistance and the Earth's rotation.

Let the Earth have radius R, and let the acceleration at the surface caused by gravity be g_0 (downwards). Considering the upward direction as positive, we have

$$\frac{dv}{dt} = a$$

$$\frac{dv}{dr}\frac{dr}{dt} = v\frac{dv}{dr} = -g_0\left(\frac{R}{r}\right)^2$$

(We have here used the fact that $g_1 r_1^2 = g_2 r_2^2$.)

Separating the variables, and inserting the limits

$$\int_u^v v \ dv = -g_0 R^2 \int_R^r \frac{dr}{r^2}$$

$$\therefore \left[\tfrac{1}{2}v^2\right]_u^v = g_0 R^2 \left[\frac{1}{r}\right]_R^r.$$

Hence the speed at distance r is given by
$$v = \sqrt{[u^2 - 2g_0 R(1 - R/r)]}.$$

Example 1–4   Centrifuge problem.
A cylindrical tube of length h is full of an incompressible liquid of density ρ. It is rotated in a horizontal plane at an angular speed ω about an axis through the open end of the tube (i.e. through the free liquid surface).
(a) What is the pressure difference between the two ends of the tube?
(b) What angular speed would produce a pressure difference of 1.0 kPa (one hundredth of standard pressure) in a tube of water 0.10 m long?

(a) Refer to the diagram. Applying $F = ma$ to the element of liquid shown

$$\delta p A = (A\rho\delta r)\omega^2 r.$$

Integrating and inserting limits

$$\int_{p_x}^{p_y} dp = \rho\omega^2 \int_0^h r \ dr$$

$$(p_y - p_x) \equiv \Delta p = \tfrac{1}{2}\rho\omega^2 \left[r^2\right]_0^h$$

$$= \tfrac{1}{2}\rho\omega^2 h^2.$$

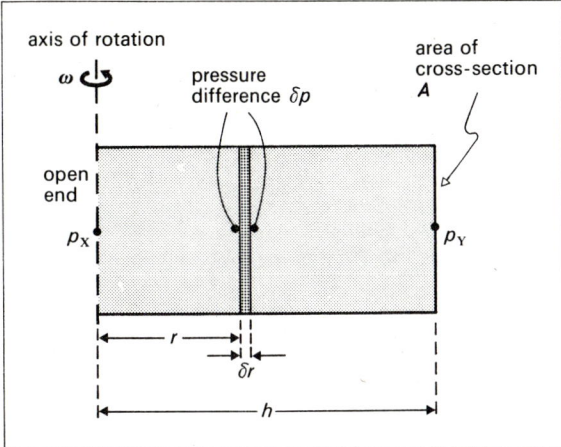

axis of rotation

ω

pressure difference δp

area of cross-section **A**

open end

p_X

p_Y

r

δr

h

For Example **1–4**

The pressure difference between the ends of the tube is

$$\Delta p = \tfrac{1}{2}\rho\omega^2 h^2 .$$

(b)
$$\omega^2 = \frac{2\,\Delta p}{\rho h^2}$$

$$= \frac{2 \times (10^3 \text{ N m}^{-2})}{(10^3 \text{ kg m}^{-3}) \times (10^{-2} \text{ m}^2)}$$

$$= 2.0 \times 10^2 \text{ rad}^2 \text{ s}^{-2}.$$

An angular speed of 14 rad s^{-1} would produce a pressure difference of 1.0 kPa.

Example 1–5 *Variation of pressure with altitude.*

Make the assumption that the density of the atmosphere is proportional to its pressure (i.e. neglect variations in temperature), and determine how atmospheric pressure depends upon height h above the Earth's surface.

Consider the equilibrium of a horizontal layer of air of thickness δy and area of cross-section A. Measure y vertically upwards. Then

upward force on lower surface = pA
downward force on upper surface = $(p + \delta p)A$
downward pull of Earth = $(\rho A\,\delta y)g_0$

For equilibrium

$$pA = (p + \delta p)A + (\rho A\,\delta y)g_0$$

$$\frac{\mathrm{d}p}{\mathrm{d}y} = -\rho g_0 \qquad \dots\dots (1)$$

The negative sign tells us that p decreases in the direction in which h increases. Suppose that **g** takes the constant value g_0 over the range of y with which we are concerned. At sea level $y = 0$, let $p = p_0$ and $\rho = \rho_0$. At the height $y = h$, the pressure and density are p and ρ.

Using $p_0/\rho_0 = p/\rho$, equation (1) becomes

$$\frac{\mathrm{d}p}{p} = -\left(\frac{g_0\rho_0}{p_0}\right)\mathrm{d}y.$$

$$= -k\,\mathrm{d}y$$

in which we have put $k \equiv \left(\dfrac{g_0\rho_0}{p_0}\right)$.

$$\therefore \int_{p_0}^{p}\frac{\mathrm{d}p}{p} = -k \int_{0}^{h}\mathrm{d}y.$$

Integrating

$$\Big[\ln p\Big]_{p_0}^{p} = k\Big[-y\Big]_{0}^{h}$$

$$\therefore \ln(p/p_0) = -kh$$

$$\therefore \qquad p = p_0\,\mathrm{e}^{-kh}.$$

The pressure at height h under these conditions is given by

$$p = p_0\,\mathrm{e}^{-kh},$$

in which $k = \dfrac{g_0\rho_0}{p_0} \approx \dfrac{(9.8 \text{ m/s}^2) \times (1.3 \text{ kg/m}^3)}{(1.0 \times 10^5 \text{ N/m}^2)}$

$$= 1.3 \times 10^{-4} \text{ m}^{-1}.$$

Note that the constant k may also be expressed as $g_0 M_m/RT$, in which M_m is the molar mass.

Example 1–6 *The Clausius-Clapeyron equation.*

The laws of thermodynamics lead to the equation

$$\frac{\mathrm{d}p}{\mathrm{d}T} = \frac{L_m}{T(V_f - V_i)_m}.$$

L_m *is a molar heat of transformation accompanying a phase change in which a substance has initial and final molar volumes* V_i *and* V_f. *p is the equilibrium vapour pressure, and T the thermodynamic temperature. Use this equation to find the dependence of vapour pressure on temperature.*

Let us make the following assumptions.

(a) The molar heat of vaporization L_m is independent of temperature (this is never exactly true, but is to a first approximation).
(b) The molar volume of the vapour is much greater than that of the liquid, $V_f \gg V_i$ (this is very nearly exact, except near critical conditions).
(c) The vapour shows ideal-gas behaviour, so that

$$p\,V_f = RT.$$

(This follows from (b).)

Using $V_f = RT/p$, the equation

$$\frac{\mathrm{d}p}{\mathrm{d}T} = \frac{L_m}{T(V_f - V_i)_m}$$

becomes
$$\frac{1}{p}\frac{\mathrm{d}p}{\mathrm{d}T} = \frac{L_m}{RT^2}.$$

Integrating $\quad \ln p = -\dfrac{L_m}{RT} + \text{constant}$

or $\quad \ln \dfrac{p}{p_0} = -\dfrac{L_m}{RT}.$

The vapour pressure varies with temperature according to

$$p = p_0\,\mathrm{e}^{-L_m/RT}.$$

p_0 is some (unknown) constant of proportionality. (During the integration we assumed that L_m is independent of temperature. At low temperatures, where $L_m \gg RT$, the exponential result is reasonably accurate.)

Example 1–7   Acceleration with variable mass.
A raindrop falls from rest, and its mass increases at a rate proportional to its surface area. If at time $t = 0$ its size is negligible, what will be its velocity after it has fallen for 10 s? Ignore air resistance, and take $g_0 = 10$ m s^{-2}.

The mass increases at a rate $\propto 4\pi r^2$.

$$dm/dt = k\,4\pi r^2$$

But since $m = \frac{4}{3}\pi\rho r^3$, we also have

$$dm/dt = 4\pi\rho r^2 \frac{dr}{dt}$$

from which

$$\frac{dr}{dt} = \frac{k}{\rho}. \qquad \dots\dots (1)$$

Integrating, and putting $r = 0$ when $t = 0$, we have

$$\int_0^r dr = \frac{k}{\rho}\int_0^t dt$$

$$\therefore\ r = \left(\frac{k}{\rho}\right)t. \qquad \dots\dots (2)$$

The equation of motion for this situation must be written

$$F = \frac{d}{dt}(mv)$$

$$mg_0 = m\frac{dv}{dt} + v\frac{dm}{dt}$$

$$\therefore\ \frac{4}{3}\pi r^3\rho g_0 = \frac{4}{3}\pi r^3\rho\frac{dv}{dt} + v\,4\pi r^2\rho\frac{dr}{dt}$$

$$\therefore\ g_0 = \frac{dv}{dt} + 3\frac{v}{r}\frac{dr}{dt}$$

$$= \frac{dv}{dt} + \frac{3v}{t}$$

in which we have used (1) and (2). To find the velocity as a function of time, we must integrate, but this cannot be done directly. If we multiply both sides by t^3, which is called an **integrating factor**, we have

$$g_0 t^3 = t^3\frac{dv}{dt} + 3t^2 v.$$

We *can* integrate this, since we have made the right hand side a **perfect differential**.

$$\left[\frac{g_0 t^4}{4}\right] = \left[t^3 v\right]$$

$$\therefore\ \left[\frac{g_0 t^4}{4}\right]_{t=0}^{t=10s} = \left[v\right]_{v=0}^{v=v_f}.$$

The velocity acquired is that of a body of constant mass falling at an acceleration $g_0/4$. It is less than g_0 because of the momentum acquired by still water vapour which it collects as it falls.

After 10 s the raindrop has a downward velocity of 25 m s^{-1}.

Physics Problems

Problem Solving	Worked Examples
Discussion of Some Common Mistakes	

Problem Solving

How to Approach a Problem

(1) Read through the question rapidly to centre your thoughts on the subject area concerned. Determine, for example, whether the question is concerned with particle dynamics, wave superposition, etc.

(2) Now read through the question more thoroughly to establish *exactly* what it is that has to be calculated.

(3) The next step is the difficult one, since it requires creative thinking. Ask yourself what needs to be known in order to evaluate the end-product. The technique to be learned is that of establishing priorities, and then asking the right questions in the right order. There can be no set rules for developing such a technique, but listed below are some of the questions that might be asked:

(*a*) To what other quantities is the end-product related? What general principles relate it to these other quantities or similar quantities which can be calculated directly from the information given in the question? Under what conditions can these principles be applied?

(*b*) What assumptions can be made about the conditions of this problem? For example, are there quantities which can be treated as constants? Can we make simplifying approximations?

(*c*) Are the required conditions observed in this particular situation?

What you are trying to do here is to connect together your different ideas, which are based on first principles and *general* concepts, and to relate them directly to this *particular* problem. (At the same time you must judge which of your original ideas are relevant to this problem, and which should be put aside.) The ability to do this can be initiated by reading through worked examples in textbooks, but there is no real substitute for the first-hand experience gained by working through problems on your own, and then discussing their solution with a supervisor.

Many of the problems in this book have been broken down into simple step-by-step questions. It is hoped that these will help you to develop a feel for their solution. You must make a conscious effort, as you work at a given problem, to note the relationship between the different questions, and the order in which they have been asked.

Presenting a Solution

(1) While there is little point in repeating the question, there is a lot to be gained from *summarizing the information*. It will often help you considerably in your identification of the problem if you draw a diagram on which you have marked the data. It is also helpful to express numerical data in terms of the coherent SI units.

(2) You should *state clearly the principle* and, where relevant, the assumptions on which your solution depends. For elementary problems this may be simply the statement of a defining equation (p. 4), but in general it is likely to be a combination of principles and defining equations, or perhaps a well-known derived result. Nevertheless as a general rule your solution should rest as far as possible on first principles rather than on derived equations.

(3) Do the *numerical work* in such a way that somebody else can find your mistakes. For example, cancellations (in pencil perhaps) should be distinguished clearly from deletions (in ink).

(4) *Units.* There are two ways of using units in numerical solutions.

(*a*) Replace each symbol of an equation by its numerical measure only (expressed in the basic SI unit), and *assume* that the unit of what you have calculated is

what you intended. Since the SI is a coherent system this unit will be unambiguous.

(*b*) Alternatively you can replace each symbol for a physical quantity by both its numerical measure *and* the symbol for the unit. Treat the unit symbols algebraically, and *calculate* the unit of the answer.

While method (*a*) is used almost invariably by practising physicists, there is no doubt that method (*b*) is far more beneficial to the student, since apart from the valuable familiarity it gives with derived units, it also checks automatically whether the equation being used is dimensionally correct.

(5) *State your answer* with a complete sentence. Make sure that it is of a sensible order of magnitude (p. 19), or comment appropriately if you suspect that it is not.

Discussion of Some Common Mistakes

(*a*) *Not reading the question carefully.* Of all the reasons for inadequate answers, this is probably the most common, and fortunately the one most easily put right.

(*b*) *Using equations in situations to which they do not apply.*

(*i*) **Defining equations,** in their most general form, are by their nature always true. Nevertheless to achieve simplicity new quantities are often defined by equations which apply in special cases only. This need not matter provided that you are aware that this has been done. Thus we can define the magnetic flux through a surface by any one of the following equations:

$$\Phi = BA \ , \ \Phi = (B \cos \theta)A$$
$$\Phi = \Sigma B \cos \theta \Delta A, \text{ and } \Phi = \int B \cos \theta dA.$$

The practising physicist will choose the simplest equation which *does* apply in the situation he is treating at the time. The beginning student must realize whether the equation he wants to use is applicable.

(*ii*) In the same way some of the statements of the **laws of physics** are not of general application. For example

(1) the equation $F = Gm_1m_2/r^2$ is a correct statement of the law of gravitation, but cannot be applied *directly* to (say) two cubic masses

(2) the equation $F = ma$ is a simplified statement of *Newton's Second Law* which may not be applied to a rocket of variable mass.

Similarly it is known that there are violations of *Newton's Third Law*, whereas the law of conservation of momentum is always observed.

(*iii*) Perhaps the most common mistake of this type is to attempt to use **derived equations** in situations other than those for which they were derived. Consider the scheme below, in which is indicated a method for classifying different types of motion. Each type of motion can be described by a set of appropriate equations, and it is vital to choose the right ones. One may *not*, for example, apply $v^2 = u^2 + 2as$ to any part of a simple harmonic oscillation.

Our general conclusion is this:

If one memorizes an equation, it is equally important to memorize the conditions under which it holds true.

(*c*) *Assuming that variable things are constant.* In elementary work we often, out of necessity, make the implicit assumption that a particular quantity is held constant. (We have to do this to reduce the mathematical difficulties.) It is a good principle to make these assumptions *explicit*, so that we can at a later stage recognize situations where they are not justified—even if this does mean that no further progress is possible on that problem.

Example: Suppose that we write the defining equation for the work ΔW done by a force F, which moves its point of application a displacement Δs, in the forms

(i) $\Delta W = F\Delta s$, and

(ii) $\Delta W = \int_0^{\Delta s} F \cos \theta \, ds$.

Then if a body of mass m is lifted *vertically* upwards through a *small* height h from the surface of the Earth, we would calculate the increase of gravitational energy of the system from the simple equation

$$W = mg_0 h,$$

corresponding to $W = F\Delta s$.

If, on the other hand, the change of height h was such that it caused F to vary, then we would have to find ΔW from

$$\Delta W = \int_R^{R+h} -\frac{GmM}{r^2} \, dr.$$

Once again the important thing is to determine whether the situation with which you are dealing is covered by the equation you propose to use.

(*d*) *Units.* In this book you will not have to convert information from one set of units to another, but there is the possibility that you may forget to convert from a fraction or multiple of a unit into the unit itself. This omission should always reveal itself by giving an answer of the wrong order of magnitude. The factor will usually be 10^3, $\sqrt{10^3}$, etc.

(*e*) *Miscellaneous.* The following points are mentioned here (in spite of their apparent triviality) because they are very common causes of mistakes. Do not confuse *radii* and *diameters*. Distinguish carefully between *weight* (measured in newtons) and *mass* (measured in kilograms). Time taken over careful arithmetic is seldom wasted: beware especially of taking incorrect *square roots* (e.g. $\sqrt{8.1 \times 10^4} \neq 900$), and forgetting to convert a *reciprocal*. (Careless thought might lead us to write

$$1/R = \tfrac{1}{3}\,\Omega^{-1} + \tfrac{1}{2}\,\Omega^{-1} = 5/6\;\Omega^{-1}$$

$$\therefore R = 5/6\;\Omega.)$$

Worked Examples

The purpose of working these examples here is not so much to help you with particular physical concepts, but rather to illustrate the points made in the preceding paragraphs. For obvious reasons the information of the question is not invariably written out in the solution, and the solutions should not be treated as model answers.

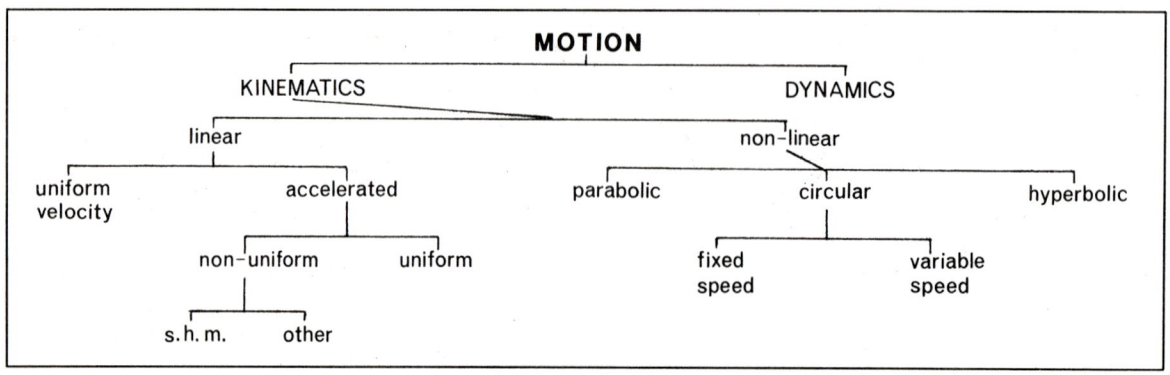

Example 1–8

A man of mass 80 kg and weight 0.80 kN climbs into a car, and this causes the centre of mass to be depressed by 4.0 mm. The car has a mass of 8.4×10^2 kg. What is its natural period of vibration when the driver carries one passenger?

To find the time period of an oscillation we need to calculate the inertial and elastic characteristics of the oscillating system.

Inertial. We make two assumptions: (i) that the mass of the passenger will equal that of the driver (80 kg), and (ii) that the whole car will oscillate. (In practice of course the car has a certain amount of what is referred to as 'unsprung weight'.) Then the total mass to be set oscillating is given by

$$m = (2 \times 80 + 840) \text{ kg} = 1.0 \times 10^3 \text{ kg}.$$

Elastic. We wish to calculate the spring constant k from

$$F = kx$$
$$\therefore k = F/x = 800 \text{N}/(4 \times 10^{-3} \text{ m})$$
$$= 2.0 \times 10^5 \text{ N m}^{-1}$$

Combining these, the time period T of oscillation is given by

$$T = 2\pi \sqrt{\left(\frac{\text{inertial characteristic}}{\text{elastic characteristic}}\right)} = 2\pi \sqrt{\frac{m}{k}}$$

$$= 2\pi \sqrt{\left(\frac{1.0 \times 10^3 \text{ kg}}{2.0 \times 10^5 \text{ N/m}}\right)}$$

$$= \left(\frac{2\pi}{10\sqrt{2}}\right) \times \sqrt{\left(\frac{\text{kg m}}{\text{kg m/s}^2}\right)}$$

$$= 0.44 \text{ s}.$$

When the driver carries one passenger the natural period of oscillation is 0.44 s.

This answer is consistent, within an order of magnitude, with observed periods of oscillation.

Example 1–9

The observed or apparent acceleration caused by gravity is greater at the North pole than it is at the equator. Assume that the difference between them (say 50 mm s^{-2}) results from the rotation of the Earth, and use this figure to calculate a value for the radius of the Earth.

Let us choose the following symbols:
ω = angular speed of Earth's spin
R = Earth's radius
g_0 = true acceleration caused by gravity
g_p and g_e the observed values at the *pole* and *equator* respectively.
Then $g_0 = g_p$
and $g_0 = g_e + \omega^2 R$, since $\omega^2 R$ is the centripetal acceleration of a point on the equator.
Equating the values for g_0, we see that

$$\Delta g = (g_p - g_e) = \omega^2 R.$$

The Earth completes one rotation in a time
$$T = 24 \times 60 \times 60 \text{ seconds} = 8.64 \times 10^4 \text{ s}.$$
$$\therefore \omega = 2\pi/T = 7.27 \times 10^{-5} \text{ rad s}^{-1}$$
$$R = \frac{\Delta g}{\omega^2} = \frac{5 \times 10^{-2} \text{ m/s}^2}{52.9 \times 10^{-10} \text{ rad}^2/\text{s}^2}$$
$$= 9.5 \times 10^6 \text{m}.$$

The assumptions made in arriving at this answer do not enable us to justify two significant figures.

The radius of the Earth (calculated from this method) is 1×10^7 m, or about 10 Mm.

In fact the radius of the Earth is 6.4 Mm, so our answer agrees within an order of magnitude. Since there are factors (other than the Earth's rotation) which we have ignored, the apparent discrepancy does not make us feel that there is a mistake in the solution.

Example 1–10

Green light of wavelength 550 nm illuminates a single slit, the light from which is incident on two very narrow slits separated by 0.40 mm. Where should a viewing screen be placed if a demonstrator wishes to show an interference pattern in which the fringe width is of the order 5 to 6 mm?

There are four quantities to be related, w the fringe width, s the double-slit separation, D the slit-screen distance, and λ the wavelength of the light. Let us suppose that you do not remember the exact details of the equation relating the variables, and that you do not have time to derive it analytically. We will derive it by an intuitive argument.

Common sense and reference to simple interference patterns in the ripple tank suggest that

(*a*) $w \propto D$, and
(*b*) $w \propto \lambda$.

A dimensional argument shows that (provided no other variables are concerned) w must be proportional to s^{-1}, otherwise the dimensions of our answer could not be L^1. No information is available about the dimensionless constant, but the student is likely to remember that it is equal to 1.

$$\therefore w = D\lambda/s.$$

Using the figures given
$$D = sw/\lambda$$
$$= \frac{(0.4 \times 10^{-3} \text{ m}) \times (5 \times 10^{-3} \text{ m})}{(550 \times 10^{-9} \text{ m})}$$
$$\approx 4 \text{ m}.$$

Since $w \sim 5$ to 6 mm, there is no point in using two significant figures.

The screen should be placed 4 m away.

*Example 1–11

Sound waves of frequency 0.50 kHz are passing through air of density 1.3 kg m^{-3}. They cause the air molecules to undergo s.h.m. of amplitude 8.0 µm. What is the energy density of the wave motion?

A body of mass m undergoing an s.h.m. of amplitude a at a pulsatance ω has a maximum k.e. K, where

$$K = \tfrac{1}{2}mv_{max}^2 = \tfrac{1}{2}m\omega^2 a^2,$$

and this also equals the total energy W of the motion.

Consider an element of air, of cross-sectional area A and length $c\,\Delta t$. It has a mass $m = V\rho = (Ac\,\Delta t)\rho$.

Its energy density is the energy per unit volume, (W/V), which is given by

$$\frac{W}{V} = \frac{\frac{1}{2}(Ac\,\Delta t)\,\rho\omega^2 a^2}{(Ac\,\Delta t)}$$

$$= \frac{1}{2}\rho\omega^2 a^2 \,.$$

Comparison with $K = \frac{1}{2}m\omega^2 a^2$ suggests that the dimensions are correct. (We can visualize the energy density as being the energy carried by the wave through a transverse area 1.0 m^2 as the wavefront advances by 1.0 m.)

For this problem $\omega = 2\pi f = 3.14 \times 10^3$ s^{-1}

$$a = 8.0 \times 10^{-6} \text{ m}$$
$$\rho = 1.3 \text{ kg m}^{-3}.$$

Substituting into the equation which we have derived, we find that the energy density

$W/V = \frac{1}{2}(1.3 \text{ kg/m}^3) \times (3.14 \times 10^3 \text{ s}^{-1})^2 \times (8.0 \times 10^{-6} \text{ m})^2$

$= (4.2 \times 10^{-4}) \times \left(\dfrac{\text{kg m}^2}{\text{s}^2}\right) \times \left(\dfrac{1}{\text{m}^3}\right)$

$= 0.42 \text{ mJ m}^{-3}.$

The energy density of the air is 0.42 mJ m^{-3}.

It is unlikely that you will have the experience to judge whether the order of magnitude is sensible. In fact this corresponds to a very loud sound.

Example 1–12

Two spherical lumps of lead of mass 6.0 kg lie with their centres 1.0 m apart. They are each given a charge Q (which may be assumed to be distributed with spherical symmetry) such that each exerts a zero resultant force on the other.

  (a) What is their mutual gravitational attraction?
  (b) What is the value of Q?
  (c) How many electrons should be removed from each sphere to achieve this result?
Assume G, $1/4\pi\varepsilon_0$ and e

(a) Using *Newton's Law of Gravitation*

$$F = Gm_1m_2/r^2$$

(which *can* be applied here because the masses have spherical symmetry), we have

$$F = \frac{(6.7 \times 10^{-11}\,\text{N m}^2/\text{kg}^2) \times (6.0 \text{ kg})^2}{(1.0 \text{ m})^2}$$

$$= 2.4 \times 10^{-9} \text{ N}.$$

Each attracts the other with a gravitational force 2.4 nN.

(b) We can use *Coulomb's Law* because the electric charges have spherical symmetry. The size of the electric force equals that of the gravitational force.

$$\therefore F = \left(\frac{1}{4\pi\varepsilon_0}\right)\frac{Q_1Q_2}{r^2} = 2.4 \times 10^{-9} \text{ N}$$

$$\therefore Q^2 = \frac{(2.4 \times 10^{-9} \text{ N}) \times (1.0 \text{ m}^2)}{(9.0 \times 10^9 \text{ N m}^2/\text{C}^2)}$$

$$= 26.7 \times 10^{-20} \text{ C}^2$$

$$\therefore Q = \pm 5.2 \times 10^{-10} \text{ C}.$$

The charges have the same sign (minus or plus) and a size 0.52 nC.

(c) A body can acquire a positive charge Q by losing N electrons, each of charge magnitude e, where

$$Q = Ne.$$

Thus $\quad N = \dfrac{5.2 \times 10^{-10} \text{ C}}{1.6 \times 10^{-19} \text{ C/electron}}$

$$= 3.2 \times 10^9 \text{ electrons.}$$

This could be achieved by removing 3.2×10^9 electrons.

Note (i) the relative numerical values of the mass and charge, and

  (ii) 6 kg of lead make up about 30 moles of lead atoms. If we remember that each mole contains 6×10^{23} atoms, it will be seen that 10^9 is a minute fraction of the total number of available electrons.

Example 1–13

A point charge 40 pC is placed at the centre of a sphere of radius 0.50 m. Assume the truth of Coulomb's Law, and use this situation to show that Gauss's Law is consistent with Coulomb's Law. Assume ε_0, $1/4\pi\varepsilon_0$

Place a test charge Q_2 at the surface of the sphere. Then it experiences a force F where

$$F = \left(\frac{1}{4\pi\varepsilon_0}\right)\frac{Q_1Q_2}{r^2}. \qquad \text{(Coulomb's Law)}$$

The value of the field E at the surface is given by

$$E = F/Q_2 = \left(\frac{1}{4\pi\varepsilon_0}\right)\frac{Q_1}{r^2}$$

$$= \frac{(9.0 \times 10^9\,\text{N m}^2/\text{C}^2) \times (40 \times 10^{-12}\text{ C})}{0.25 \text{ m}^2}$$

$$= 1.44 \text{ N C}^{-1}.$$

Over the surface of the sphere this field is everywhere of the same size, and everywhere cuts the surface normally (since the field lines are radial).

Thus $\quad \psi_E = EA$

$$= E4\pi r^2$$
$$= (1.44 \text{ N/C}) \times (4\pi \times 0.25 \text{ m}^2)$$
$$= 4.5 \text{ N m}^2 \text{ C}^{-1}.$$

The quantity $\varepsilon_0\psi_E$ evaluated over the closed surface has the value $\quad (8.85 \times 10^{-12} \text{ C}^2/\text{N m}^2) \times (4.5 \text{ N m}^2/\text{C})$

$$= 40 \text{ pC}.$$

Now the net charge enclosed by the closed surface is 40 pC.

Thus we have shown that, for this special case, $\varepsilon_0\psi_E = \sum Q$.

Example 1–14   Radiocarbon dating.

Investigation of the wood from the case of an Egyptian mummy showed a specific activity of 1.2×10^2 s^{-1} kg^{-1}. Comparable living wood gave a value 2.0×10^2 s^{-1} kg^{-1}. The half-life of carbon-14 is 5.7×10^3 years. What is the approximate time interval since the burial? (Give the answer in years.)

The rate of disintegration is given by

$$-\frac{dN}{dt} = \lambda N, \qquad \text{from which}$$

$$\cdot N = N_0\, e^{-\lambda t}\,.$$

To solve for the time interval t in terms of the half-life $T_{1/2}$, we make two substitutions:

(a) $\dfrac{N}{N_0} = \dfrac{1.2 \times 10^2}{2.0 \times 10^2} = 0.60$

(since the activity of a sample is proportional to the number of active nuclei it contains), and

(b) $\lambda = 0.693/T_{1/2}$

a result which can be quoted without proof, or derived very simply from $N = N_0\, e^{-\lambda t}$.

Then $\qquad 0.60 = \exp(-0.693t/T_{1/2})$

$$\therefore\ 0.693\, t = (T_{1/2}) \times \ln(1/0.6)$$
$$= (5.7 \times 10^3 \text{ years}) \times (0.510).$$

So $\qquad\qquad t = 4.2 \times 10^3$ years.

$\underline{4.2 \times 10^3 \text{ years have elapsed since the burial.}}$

(Strictly this is the time interval since the heartwood of the tree was formed.)

Orders of Magnitude

Developing a Feeling for Orders of Magnitude

Orders-of-Magnitude Tables

Mental Arithmetic

Orders-of-Magnitude Problems

Developing a Feeling for Orders of Magnitude

In physics the phrase 'correct to an order of magnitude' means that the value quoted is reliable to within a factor of ten or so, and for many purposes the information may be sufficiently precise in this form. Suppose you are designing an experiment to demonstrate two-source interference of light waves, and you want to know a suitable separation for the twin slits. You may be told that, to an order of magnitude, 0.5 mm gives good results. This does not mean that 0.4 mm is unsuitable, but it does imply that 0.05 mm or 5 mm would (for different reasons) give a less satisfactory effect.

It is most important for a physicist to acquire a working knowledge of the size of typical quantities. This enables him to do two things:
(a) to judge the plausibility of any quantity that he has calculated, or which is presented to him, and
(b) to estimate the possible sizes of further quantities (p. 22).

Even the best physicists (including *Newton*) make elementary errors in their working, but only a bad physicist is happy to present a solution which is patently absurd. Suppose the result of a calculation suggests that the speed of sound through air at s.t.p. is 0.11 km s^{-1}. It is at once apparent that there is a mistake which derives from a factor of about 3.1, and it should be noted that this is a factor often introduced by looking up a square root incorrectly.

If the conclusion of a lengthy calculation is absurd, then you should always say that it is, stating the order of magnitude that you had estimated to be appropriate, and the reasons for your estimation. A comparison of the calculated and estimated values may, as above, indicate the sort of factor which has caused the discrepancy, and thereby lead to the mistake.

For discussion. Explain, *with reasons,* whether or not each of the following statements could be plausible as an answer to a calculation:

(a) There are 10^{13} molecules in a drop of water.
(b) The random speed of electrons within a hot tungsten filament is 2 Gm s^{-1}.

(c) The angular speed of the Moon in its motion round the Earth is 4 rad s^{-1}.

(d) The amount of air in a room is 13 mol.

(e) The two spheres of a teaching laboratory *Van de Graaff* machine acquire charges of ± 0.3 C.

(f) The energy released by the fission of a uranium nucleus is 2 mJ.

(g) The rate of emission of electrons from the cathode of a c.r.o. is 10^{-6} s^{-1}.

Orders-of-Magnitude Tables

Energy

One of the principal interests of physicists is energy, and its conversion from one form to another. One reason for the concept of energy holding such an important place is

that we can often use it to decide whether a particular physical change is likely to occur. A physical system achieves equilibrium by making its p.e. a minimum. Thus a liquid surface contracts, an electron in an atom moves as close as possible to the nucleus, and a radioactive nucleus disintegrates, each making an attempt to reduce the p.e. of the system.

In the SI *all* energies are expressed in *joules*, and this has the enormous advantage of revealing the true comparative values of energies previously expressed relative to other quantities (such as the electronvolt). The table below should be studied closely, since many useful conclusions may be drawn from it. For example one can predict that photons of violet light will cause photoelectric emission from caesium, but not from tungsten; or that air is ionized by photons from the middle ultraviolet, but not by those of lower frequency.

Description of energy W	W/J	Description of energy W	W/J
binding energy of Earth-Sun system	10^{33}	rest-mass-energy of 1 unified atomic mass unit	1.6×10^{-10}
translational k.e. of Moon	10^{28}		
energy radiated by the Sun in 1 second	10^{26}	energy released by fission of 1 uranium nucleus	3×10^{-11}
energy received per day on Earth from Sun	10^{22}		
estimated Human energy requirement per year (1950)	10^{20}	typical binding energy per nucleon	1.3×10^{-12}
			1 pJ
energy associated with a strong earthquake	10^{20}	minimum energy of γ-ray photon for pair-production	1.6×10^{-13}
energy released by annihilation of 1 kg of matter	9×10^{16}		
		rest-mass-energy of 1 electron	8×10^{-14}
energy released by fission of 1 mol of ^{235}U	10^{13}	photon energy of typical X-ray	1 fJ
	1 TJ	photon energy of middle u.v.	10^{-17}
energy dissipated by a lightning discharge	10^{10}	average energy to create 1 ion-pair in air	5×10^{-18}
	1 GJ	ionization energy of hydrogen atom	2×10^{-18}
energy released by combustion of 1 kg of petrol	10^{8}	width of forbidden region between energy bands in diamond crystal	
		change of binding energy per atom in a typical chemical change	1 aJ
energy converted by a 1 kW heater operating for 1 hour	3.6×10^{6}		
energy required to charge a car battery	1 MJ	work function energy of tungsten	7×10^{-19}
energy provided by a slice of bread	10^{5}	photon energy of visible light (violet)	5×10^{-19}
k.e. of a bowled cricket ball	10^{1}	work function energy of caesium	3×10^{-19}
energy stored in the magnetic field of a 2 H inductor carrying a current of 1 A	1	photon energy of visible light (red)	2.5×10^{-19}
energy stored in the electric field of a 1 μF capacitor charged to 200 V	2×10^{-2}	width of forbidden region between energy level bands in germanium crystal	1×10^{-19}
		photon energy of middle i.r.	2×10^{-20}
	1 mJ	translational k.e. of ideal-gas molecule at 300 K	6×10^{-21}
recommended maximum ionizing radiation dose to be absorbed per year by 1 kg of living tissue	4×10^{-4}		
		photon energy of microwaves	10^{-23}
		photon energy of medium band broadcast radio waves	10^{-27}
	1 μJ		
maximum energy of proton in CERN synchrotron	10^{-8}		
	1 nJ		

Some Miscellaneous Tables

The tables on this page list, over a wide range, some representative values of four common physical quantities.

Earth-pull on	F/N
fly's wing	10^{-6}
postage stamp	10^{-4}
coin	10^{-1}
apple	$10^0 = 1$
large man	10^3
oil tanker	10^9
Moon	10^{20}
Sun	10^{22}

Temperature description	T/K
produced by adiabatic demagnetization of nuclear alignment	10^{-6}
produced by adiabatic demagnetization of paramagnetic salts	10^{-3}
b.p. of hydrogen	2×10^1
m.p. of water	3×10^2
m.p. of tungsten	4×10^3
produced by shock wave in air ($Ma = 20$)	2×10^4
interior of Sun	10^7
helium thermonuclear reaction	10^8

Electric charge description	Q/C
elementary charge e	10^{-19}
positive charge on gold nucleus	10^{-17}
maximum charge which can be held by a sphere of radius 1 mm in air at s.t.p.	10^{-10}
typical charge produced by rubbing	10^{-8}
typical charge measured by ballistic galvanometer	10^{-6}
charge transferred by lightning flash	10^1
negative charge passing in ten minutes through a torch bulb	10^2
positive charge on the nuclei of a copper penny	10^5

Electric field strength description	$E/(\text{N C}^{-1})$ or (V m^{-1})
field in which an electron experiences a force of the same size as the Earth-pull on it	10^{-10}
$E_{\max}$ in electromagnetic wave near a radiating 100 W lamp	10^1
between deflecting plates of a c.r.o.	10^4
breakdown field for air at standard pressure	10^6
field which produces field emission from surface of a typical metal *in vacuo*	10^8
field at the most probable location of an orbital electron in the hydrogen atom in ground state	10^{12}
field at the surface of a gold nucleus	10^{21}

You should try constructing similar tables for yourself—the exercise would be both instructive and rewarding. The following quantities are suggested as some of those likely to prove particularly interesting: mass, length, time, linear and angular speed, energy density, amount of substance, mean free path, number density, electric current and current density, electric potential, electrical and thermal conductivity, and magnetic flux density (the B-field).

Facts about Topics

It is often helpful to group together facts about certain topics. Much can be learned, for example, by listing typical values of current and p.d. for semiconducting devices such as junction diodes and transistors, and comparing them with those for their vacuum-tube counterparts.

You will find it a useful revision exercise to do this for (say) the electromagnetic spectrum, photoelectric emission, elementary particles and radioactivity. It would be sensible to record these facts about topics together with your order-of-magnitude tables in a special notebook kept for the purpose.

Mental Arithmetic

Most of the quantitative problems in this book can be worked by slide rule (p. 8), and a mental estimate will then be needed to see where to place the decimal point. Such mental arithmetic will also be useful for calculating answers to the order-of-magnitude problems discussed in the next paragraph. Experience and practice will enable you to judge what approximations are justified, and to develop your own techniques. For the inexperienced the following method is recommended:

(1) Write all the numbers in standard form to one significant figure. If one figure is rounded down, it is usually possible to compensate by rounding (up or down) another so as to give a balancing effect.

(2) Group together and evaluate the powers of 10.

(3) Cancel where possible, and do the now simple computation.

(4) Before taking an nth root, express the number in the form $N \times 10^{nx}$, where x is an integer.

Worked example. Estimate a value for T in the equation

$$3.94 \times 10^{26} = (5.67 \times 10^{-8}) \times [4\pi \times (7.0 \times 10^8)^2] \times T^4$$

$$\therefore \quad T^4 \approx \frac{4 \times 10^{26+8-16}}{6 \times 4\pi \times 49}$$

$$\approx \frac{10^{18}}{20 \times 50} \qquad \text{(putting } 6\pi = 20\text{)}$$

$$\approx 10^3 \times 10^{12}.$$

$$\therefore \quad T \approx \sqrt[4]{1000} \times 10^3 \approx \sqrt[2]{30} \times 10^3$$

$$\approx 5 \text{ or } 6 \times 10^3.$$

More accurate working with a slide rule yields the figures 580, and the true figure is 5.8×10^3.

Orders-of-Magnitude Problems

Under the heading *Questions for Discussion* in the following chapters you will find many problems marked **O-M**. These are intended to encourage creative thinking, a feeling for realistic values of the sizes of physical quantities, and an awareness of reliable sources from which such information can be obtained.

Here is a suggested method of solving such problems:

(1) There may be several alternative methods of approaching the same problem, and you could usefully discuss their relative merits. Choose the one that seems most promising. Survey the physical principles needed for your method, and use them as a guide for selecting those factors which seem to be relevant to the situation, and for dismissing those that are not.

(2) Express the principles that you are using in the form of a quantitative relationship from which you can derive an equation for the quantity being estimated.

(3) Look up in a reference book, or make a reasoned estimate for, any quantities whose numerical values are not known. (It is here that commonsense and experience play a large part.) Insert these values, and use either mental arithmetic or a slide rule (according to the

number of significant figures that you need) to evaluate the answer.

(4) Indicate which of the estimated quantities has most bearing in determining the uncertainty of the final answer, and discuss briefly how reliable you think this answer is.

Worked Example 1–15

Of what size are the *Amperian* currents that we imagine to flow round the surface of a typical laboratory bar magnet?

(1) *Principles.* We can approach this problem by drawing a quantitative conclusion from the qualitative similarity between the external magnetic fields of a solenoid and bar magnet. This similarity is emphasized in the diagram opposite ((a) and (b)).

We imagine the magnetic field established by the magnet to be the result of currents flowing round its surface (part (c) of the diagram). The size of the field established by a solenoid at a point P on its axis is given by

$$B = \tfrac{1}{2}\mu_0 nI (\cos \theta_2 - \cos \theta_1),$$

in which θ_1 and θ_2 are defined in part (d) of the diagram, and n is the number of turns per unit length.

(2) *Quantitative relationship.* For our cylindrical magnet we wish to calculate the value of (nI), where

$$nI = 2B/\mu_0 (\cos \theta_2 - \cos \theta_1).$$

(3) *Estimation.* We need to know the defined value of μ_0, and to estimate a set of corresponding values for B, θ_1 and θ_2. Now a typical magnet oriented with its axis pointing N→S gives a neutral point in the Earth's field where r (in part (d)) is about 0.1 m. Consider, for simplicity, a cylindrical magnet (and hence its corresponding solenoid) to have length 0.1 m and radius 10 mm.

Then for the solenoid equation we have

$$\cos \theta_1 = 10/\sqrt{101} = 1/\sqrt{1.01}$$

$$\cos \theta_2 = 20/\sqrt{401} = 1/\sqrt{1.0025}$$

$$\cos \theta_2 - \cos \theta_1 = (1 + 0.0025)^{-\frac{1}{2}} - (1 + 0.01)^{-\frac{1}{2}}$$
$$= [1 - \tfrac{1}{2}(0.0025) + \ldots] - [1 - \tfrac{1}{2}(0.01) + \ldots]$$
$$\approx 0.00375 \approx 4 \times 10^{-3}.$$

The horizontal resolved part of the Earth's field is about 20 μT.

So
$$nI \sim \frac{2 \times 20 \times 10^{-6} \text{ Wb m}^{-2}}{(4\pi \times 10^{-7} \text{ Wb A}^{-1} \text{ m}^{-1}) \times (4 \times 10^{-3})}$$

$$\sim \frac{10 \times 10^3}{4\pi} \text{ A m}^{-1}$$

$$\sim 10^3 \text{ A m}^{-1}.$$

(4) *Reliability.* The value quoted for μ_0 is absolute, and that for B is accurate to better than 20%. The main source of uncertainty is the estimated value of $(\cos \theta_2 - \cos \theta_1)$, but this is not likely to be in error by a factor of more than 10. Since the properties of permanent magnets vary considerably from specimen to specimen, there is little point in trying to be more precise than this.

Conclusion. The *Amperian* currents associated with teaching laboratory magnets are likely to lie in the range 10^2 to 10^4 A m^{-1}.

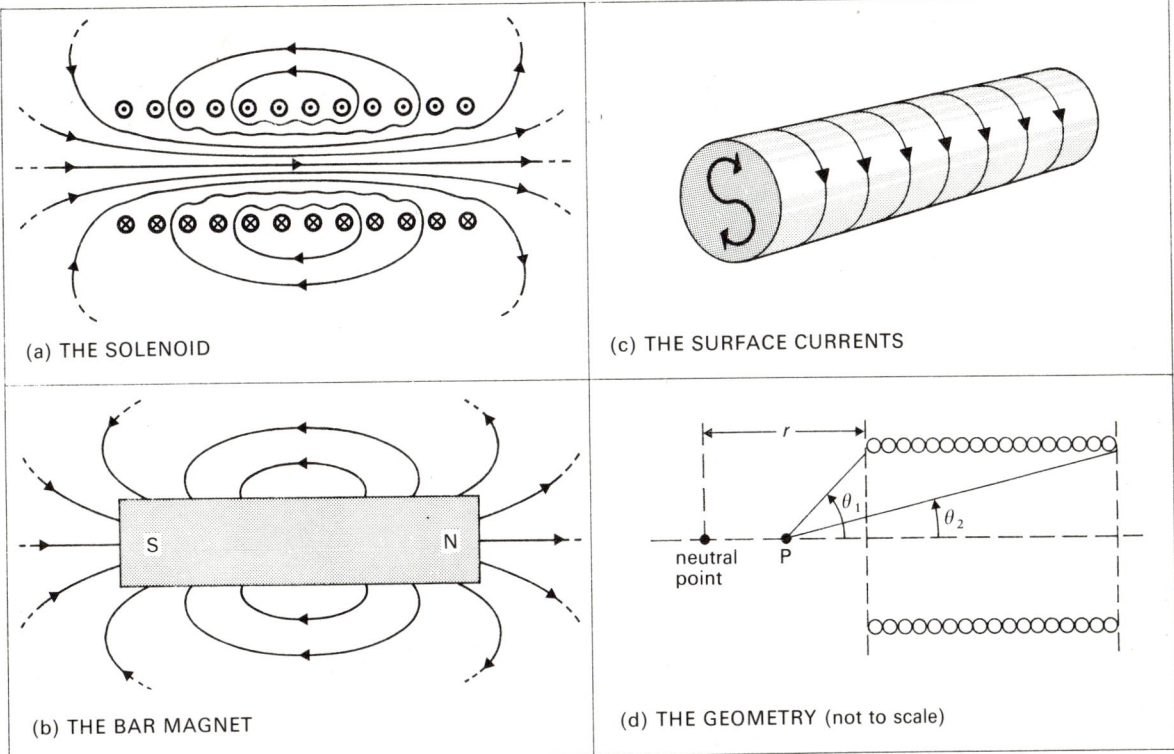

(a) THE SOLENOID

(c) THE SURFACE CURRENTS

(b) THE BAR MAGNET

(d) THE GEOMETRY (not to scale)

For Example **1–15**

Questions

Errors

1–1 Explain the difference between *systematic* and *random* errors, using particular operations (such as the use of a clock or a ruler) to illustrate your answer. Which type of error is likely to be the more misleading?

1–2 Give two examples of
(*a*) an accurate measurement of low precision
(*b*) an inaccurate precise measurement.

1–3 Select three well-known experiments in which the aim is to measure some physical quantity. List the measurements that would be made in the course of the experiments, and indicate in each case the measurement over which you would take the most trouble. Say why you would choose that measurement.

1–4 Suppose you were doing an experiment in order to plot a force-extension graph to find the force constant k of a helical spring. Would there be any advantage in adding the loads in a *random* order (rather than **monotonically**)?

1–5 Use a sine table to find the largest angle θ ($|\theta| < \pi/2$) for which $\sin \theta = \theta$ to an accuracy of 1%.

1–6 In each of the following examples evaluate v/c, where c is the speed of light, and discuss whether relativistic corrections would be necessary in kinematic calculations.

(a) v is the speed of the Earth in its orbit round the Sun. Take the distance from Earth to Sun as 0.15 Tm.

(b) v is the speed of electrons in a c.r.o. after being accelerated through a p.d. of 5.0 kV.

Assume $c, e/m_e$ [(a) 10^{-4} (b) 10^{-1}]

1–7 What *precise* meaning can you attach to the statement $r = (24.0 \pm 0.3)$ mm, where r is the radius of a tube?

1–8 *Mean value.* If several readings of a single physical quantity are all taken under the same experimental conditions, then the mean is the best estimate of that quantity. Using a mean value helps to reduce the effect of the *random* errors. Calculate the mean of the following measurements for the resistance R of an electrical device: $R/\Omega = 3.19, 3.15, 3.20, 3.23, 3.17, 3.25, 3.22, 3.20, 3.21$ and 3.18.

1–9 *Errors from graphs.* Use the following results to plot points on a graph to show how the p.d. across an ohmic device varies with the current flowing through it.

V/V	3.0	3.6	4.6	5.0	6.2	6.6	7.0	8.4
I/A	0.50	1.0	1.3	1.5	2.0	2.3	2.7	3.0

V/V	9.4	10.0	10.4	12.4
I/A	3.5	4.0	4.6	4.8

(a) Draw the best straight line through the points and deduce the resistance of the device from the slope of the graph.

(b) Join points 1–7, 2–8, 3–9, 4–10, 5–11, 6–12 and measure these six slopes. Calculate the mean value of the slopes, and again deduce the resistance of the device.

(c) Why is result (b) more reliable?

[(b) 2.1 Ω]

1–10 *Combination of errors.* Calculate the value of the quantity X and its maximum error ΔX from the given measured values of $(a \pm \Delta a)$, $(b \pm \Delta b)$, $(c \pm \Delta c)$, in each of the following examples.

(i) $X = 6a + 4b$ $a = 40 \pm 2$, $b = 20 \pm 2$

(ii) $X = a^3$ $a = 10.0 \pm 0.3$

(iii) $X = a/b$ $a = 100 \pm 4$, $b = 50 \pm 1$

(iv) $X = ab^2/c$ $a = 0.200 \pm 0.004$
                             $b = 0.100 \pm 0.003$
                             $c = 0.050 \pm 0.002$

[(i) 320 ± 20 (ii) 1000 ± 90 (iii) 2.0 ± 0.1 (iv) 0.040 ± 0.005]

†1–11 Two forces have magnitudes (50 ± 3) N and (25 ± 2) N. Calculate the absolute and percentage maximum uncertainty in

(a) the sum of the forces,

(b) the difference of the forces.

[(a) 5 N, 7%   (b) 5 N, 20%]

†1–12 The force exerted by a gas on a piston of radius (8.0 ± 0.4) mm is measured to be (30 ± 2) N. Calculate

(a) the uncertainty in the area of the piston, and

(b) the maximum percentage uncertainty in the calculated value of the gas pressure in the cylinder.

[(a) 20 mm^2 (b) 17%]

1–13 If the maximum error in measuring the period of a simple pendulum is 0.05% and in measuring the length is 0.6%, what is the maximum percentage error in the calculated value of the gravitational acceleration?

[0.7%]

†1–14 Calculate the maximum percentage uncertainty in the following example: $X = a - 2b$ where $a = 50 \pm 1$ m and $b = 24 \pm 0.5$ m. (*The answer illustrates the need for careful measurement when dealing with the difference between two almost equal quantities.*)

[100%]

1–15 A rectangular block has a mass of (1.5 ± 0.1) kg, and a volume that can be calculated from the following dimensions: (80 ± 2) mm, (50 ± 1) mm and (30 ± 1) mm. Assuming that all the errors are independent, calculate the maximum uncertainty in the measured values of

(a) the volume of the block

(b) the density of the block.

[(a) 9.4×10^{-6} m^3 (b) 1.8×10^3 kg m^{-3}]

*1–16 *Standard deviation.* This gives a measure of the spread or scatter of a number of observations. It is defined as the r.m.s. value of the deviations from the mean. The following readings are for a particular time interval Δt:

$\Delta t/$s $= 190, 186, 181, 195, 191, 189, 196$ and 184.

Calculate

(a) the mean of the observations

(b) the deviations from the mean

(c) the sum of the squares of the deviations from the mean

(d) the mean square deviation

(e) the standard deviation.

[(a) 189 s   (e) 4.8 s]

Dimensions

1–17 What is meant by a *dimensionless* quantity? Give as many examples as you can from various branches of physics. Do you think that the quantity *angle* is dimensionless?

*1–18 *The fine structure constant.* The ratio of the radius of an electron estimated from scattering experiments to its value estimated from pair production and annihilation data is known as the **fine structure constant** α, and is given by $\alpha = e^2/4\pi\varepsilon_0 \hbar c$. Establish the dimensions of this constant, and express its numerical value as a fractional quantity. ($\hbar = h/2\pi$)

Assume $e, \varepsilon_0, \hbar, c$ [1/137]

1–19 Compare the dimensions of (a) linear and angular acceleration (b) linear and angular momentum (c) force and moment (torque) of a couple.

1–20 It is suggested that the speed c of a wave across a liquid surface is given by $c = \sqrt{gd}$, where g is the gravitational acceleration and d is the mean depth of the liquid. Discuss whether this relationship is possible, and if so the conditions under which it is likely to be applicable.

1–21 The speed v of a projectile which was launched with initial velocity u at an angle α to the horizontal is thought to be given by $v^2 = u^2 - 5ugt \sin \alpha + 2g^2t^2$, where t is the time interval since launching. Discuss the possibility of this relationship being correct.

1–22 Find, by dimensional analysis, how the period T of a satellite orbiting a planet of mass M depends upon the gravitational constant G, the radius of circular orbit R, and M. Do you think we are justified in omitting the mass of the satellite?

$$[T = k(R^3/MG)^{\frac{1}{2}}]$$

1–23 Use dimensional analysis to find expressions for the speed c of plane surface waves across a liquid in terms of their wavelength λ. Consider two situations (a) deep water waves, and (b) surface ripples, and discuss which of the following quantities are relevant in each case: the liquid viscosity η, density ρ, surface tension γ, and the gravitational field strength g.

1–24 Explain why, in extreme circumstances, the viscous force F on a sphere travelling through a liquid depends only on its radius a, its velocity v and the viscosity η of the liquid, whereas in other circumstances F depends only on a, v and ρ the density of the liquid. Use dimensional analysis to derive the exact relationships.

*1–25 The time period T of a simple pendulum is thought to depend upon the length l, the mass m of the bob, the gravitational field strength g and the arc s of swing. Use dimensional analysis to find what you can about the relation between T and these quantities.

$$[T = k(l/g)^{\frac{1}{2}} f(s/l)]$$

1–26 If frequency f, momentum p and velocity v were chosen as fundamental physical quantities, what would be the dimensions of mass, length, time and work?

$$[[PV^{-1}], [VF^{-1}], [F^{-1}] \text{ and } [PV]]$$

*1–27 Suppose that a system of units were developed which had G, g_0 and the speed of light c as the standards of its fundamental physical quantities. Derive the dimensions of mass, length and time in terms of the above fundamental quantities.

1–28 *Dimensions in electricity.* In addition to mass [M], length [L] and time [T], current [I] is introduced as a further funda-

mental quantity. Derive the dimensions of the following physical quantities in terms of [M], [L], [T] and [I]: (a) charge (b) power (c) e.m.f. (d) resistance (e) capacitance (f) electric field strength (g) magnetic flux (h) magnetic flux density (i) electromagnetic moment (j) inductance.

$$[(d) \ [ML^2 \ I^{-2} \ T^{-3}] \quad (g) \ [ML^2 \ I^{-1} \ T^{-2}] \quad (j) \ [ML^2 \ I^{-2} \ T^{-2}]]$$

1–29 The *ampere* is the unit of one of the fundamental quantities of the SI. Can one therefore obtain valid results in dimensional analysis by using the *coulomb*, the unit of a derived quantity?

1–30 (a) Use *Coulomb's Law* expressed for a vacuum, $F = \left(\dfrac{1}{4\pi\varepsilon_0}\right)Q_1Q_2/r^2$, to derive the dimensions of ε_0 in terms of [M], [L], [T] and [I].

(b) Use the expression for the force between two parallel current-carrying conductors *in vacuo*, $F = \left(\dfrac{\mu_0}{2\pi}\right)\dfrac{I_1I_2 \, l}{a}$ to derive the dimensions of μ_0.

(c) What are the dimensions of $(\mu_0\varepsilon_0)^{-1/2}$? Evaluate the numerical value of this quantity and comment on your answer. *Assume* ε_0, μ_0

$$[(b) \ [ML \ T^{-2} \ I^{-2}]]$$

1–31 *Dimensions in heat and thermodynamics.* In addition to mass [M], length [L] and time [T], temperature [Θ] is introduced as a further fundamental quantity. Derive the dimensions of the following physical quantities in terms of [M], [L], [T] and [Θ]: (a) specific heat capacity (b) specific latent heat (c) expansivity (d) thermal conductivity (e) the *Stefan-Boltzmann* constant (f) spectral radiant emittance.

$$[(a) \ [L^2 \ T^{-2} \ \Theta^{-1}] \quad (d) \ [MLT^{-3} \ \Theta^{-1}]]$$

1–32 When dealing with the dimensions of physical quantities which involve an *amount of substance* [N] (for which the SI unit is the mole), this quantity is treated as being fundamental. Derive the dimensions of the following: (a) molar latent heat (b) molar volume (c) molar heat capacity (d) the *van der Waals* pressure correction coefficient a.

$$[(c) \ [ML^2 \ T^{-2} \ \Theta^{-1} \ N^{-1}] \quad (d) \ [ML^5 \ T^{-2} \ N^{-2}]]$$

Graphs

1–33 What uses have graphs in experimental physics?

1–34 How could you use a plane mirror to assist in finding an accurate value for the slope of a curve at a particular point?

1–35 Given that y and x are the variables in the following equations, how would you describe the graph that each represents?

(a) $y = -8x + 5$

(b) $y = 4x^2 - 3x + 2$

(c) $y = 25 \exp(-8x)$

(d) $y = (2x + 7)^2$

(e) $xy = 6$

(f) $y = 10^{-2} x^3$.

1–36 In the following relationships, two quantities are variables and any others are constants. In each case

(a) state the graph that you would plot to find the constants, and

(b) describe how to evaluate them.

Relationship	Variables	Required constant(s)
$m + 1 = v/f$	m , v	f
$T = kl^x$	T , l	x
$xy = f^m$	x , y	m
$T = 2\pi((h^2 + k^2)/gh)^{1/2}$	T , h	k
$d = kWx^3/4ab^3$	d , x	k
$T^2 = \alpha\,(l + \beta)^3$	T , l	α , β
$r = R\left(\dfrac{\mathscr{E} - V}{V}\right)$	V , R	r
$T = 2\pi\left(\dfrac{m + \frac{5}{3}\,\mathrm{kg}}{gk}\right)^{1/2}$	m , T	gk
$N = N_0 \exp(-\lambda t)$	N , t	λ , N_0
$Q = Q_0 \exp(-t/RC)$	Q , t	C

1–37 The surface tension γ of a molten metal can be measured by making observations on ripples or capillary waves. The speed c of the waves is given by $c = \sqrt{2\pi\gamma/\rho\lambda}$, where ρ is the density of the metal and λ is the wavelength of the waves. How would you find γ from a graph of experimental results?

*1–38 The amplitude θ_n of the nth swing of a vertical cylinder undergoing torsional oscillations with a length l immersed in a liquid is said to be given by $\theta_n = \theta_0 \exp(-nAl/T)$, where T is the period of oscillation and A is a constant.

(a) What do you think θ_0 represents?

(b) How would you investigate graphically whether the above expression does in fact describe the motion?

1–39 The effective resistance R of a loop of wire is measured by a metre bridge circuit for various values of l, where l is the length of that part of the loop between the terminals which does not contain the join. (R is thus the combined resistance of the two parts of the loop connected in parallel.) A graph is plotted with R as ordinate and l as abscissa. Sketch the graph that you would expect to obtain, and explain how you would use it to calculate the resistance per unit length of the wire.

1–40 *Object-image distances.* Mark on a large sheet of graph paper the principal axis, lens position and principal foci of a converging lens of 100 mm focal length. Draw an object of

appropriate size, and find the position of its image over a range of values for u. Draw up a table of values for u and $(u + v)$. Plot a graph of $(u + v)$ (y-axis) against u (x-axis), and use it to evaluate

(a) the smallest distance between the object and its real image

(b) the value of u when this occurs

(c) the value of $(u + v)$ as u tends to f.

1–41 Draw a suitable graph from which you can find the radius of curvature of a concave mirror given the following measurements of object distance u and corresponding image distance v.

u/mm	204	250	345	455	714
v/mm	1000	526	333	270	222

[0.34 m]

1–42 The periodic time T for a certain type of oscillation is related to the length l of the suspension by the equation $T = kl^m$. Use the following measurements to plot a graph from which you can find values for k and m.

l/mm	224	282	398	501	847
T/s	0.944	1.06	1.26	1.41	1.85

$[k = 2\text{ s m}^{-\frac{1}{2}},\, m = \frac{1}{2}]$

1–43 The table of results indicates how the coefficient of viscosity of gaseous carbon dioxide varies with temperature.

T/K	225	289	335	502	615
η/µPa s	11.4	14.8	17.0	24.0	28.2

Plot a graph of η against $T^{1/2}$ and say what you can deduce from it. (*Kinetic theory predicts that $\eta \propto T^{1/2}$.*)

1–44 The table of results indicates how the coefficient of viscosity of liquid acetone varies with temperature.

T/K	215	232	252	269	294
η/mPa s	0.885	0.659	0.505	0.414	0.319

(a) Plot a graph of ln η against $1/T$, and say what you can deduce from it.

(b) *Sketch* a graph of η against T.

II Principles of Mechanics

2 Kinematics

Questions for Discussion

2–1 Answer these questions for one-dimensional and two-dimensional motion.

(*a*) Is it possible for a body to be accelerating while travelling at constant speed?
(*b*) Can a body have a constant velocity and a varying speed?
(*c*) Can the direction of a body's velocity change when it has constant acceleration?
(*d*) If a body has zero velocity, can it be accelerating?
(*e*) Can a body be accelerating in a direction opposite to that of its velocity?

2–2 Would it be medically safe to send an astronaut in a rocket which takes 100 km to reach its escape speed of 11 km s^{-1}?

2–3 A gun is aimed at a distant target. Describe what happens if the target is released in free fall at the same time as the bullet leaves the barrel of the gun.

2–4 A small body is released so that it falls to the floor of a lift which is descending at constant velocity. How would the *acceleration* of the body as observed by someone in the lift compare with that observed by a person stationary with respect to the lift shaft?

2–5 *Range of a projectile.* Show that the range R of a projectile launched at an angle α to the horizontal is given by $R = u^2 \sin 2\alpha/g$, where u is the initial velocity. Discuss the effects of air resistance and large altitude on the path of the projectile. What difference would the curvature and rotation of the Earth make?

2–6 What is the relationship between the velocity of A relative to B, and the velocity of B relative to A?

2–7 A ball is thrown vertically upwards from a trolley that is moving horizontally with constant velocity. Describe the path of the ball as seen by an observer (*a*) on the trolley, and (*b*) on the ground near the trolley.

2–8 Discuss the best method of moving from point A to point B in steady vertically falling rain so as to keep as dry as possible.

2–9 A spherical ball is rolled from one side of an empty lorry to the other while the lorry is in motion. Discuss the possible paths of the ball relative to an observer on the lorry and relate these paths to the motion of the lorry in each case.

2–10 (*a*) Sketch the velocity vectors (relative to the ground) of a number of points on a rolling wheel by considering the rotational and translational motions, and then using the principle of superposition.

(*b*) Sketch the path traced out by a point on the rim of a rolling wheel, and show on your diagram the instantaneous velocity and acceleration of the point at an arbitrary instant of time.

2–11 O-M What is the greatest height to which a man could throw a cricket ball on the Moon?

Quantitative Problems

Linear Kinematics

(*For the kinematics of s.h.m. see Chapter* **10**.)

2–12 *Average speed.* A man cycles to his destination with the wind at 8.0 m s^{-1}, and returns home against the wind at 4.0 m s^{-1}. What is his average speed?

[5.3 m s^{-1}]

†2–13 A research rocket is launched vertically and has a uniform acceleration of 0.40 km s^{-2} for 10 s until all the fuel has been burned. It then falls freely back to the ground.

(*a*) What is the maximum speed of the rocket?
(*b*) When does it reach its maximum height?
(*c*) What is this height?
(*d*) Sketch a_y-t and v_y-t graphs.
Ignore the Earth's rotation.

Assume g$_0$ [(*a*) 4.0 km s^{-1} (*b*) 0.42 ks   (*c*) 0.84 Mm]

†2–14 In the electric field of a cathode-ray tube an electron is accelerated along the axis from a velocity of 5.0 km s^{-1} to a velocity of 20 Mm s^{-1} over a distance of 10 mm. What is the acceleration?

[2.0 × 10^{16} m s^{-2}]

2–15 *Displacement-time graph.* Refer to the diagram. Describe the motion of the moving body between points A and H. Use the information to plot a graph of velocity against time.

2–16 *Velocity-time graph.* Refer to the diagram. Use the information given to plot a graph of displacement against time.

2–17 *Acceleration-time graph.* Refer to the diagram. Use the information given to plot a graph of velocity against time assuming that the body being considered starts from rest.

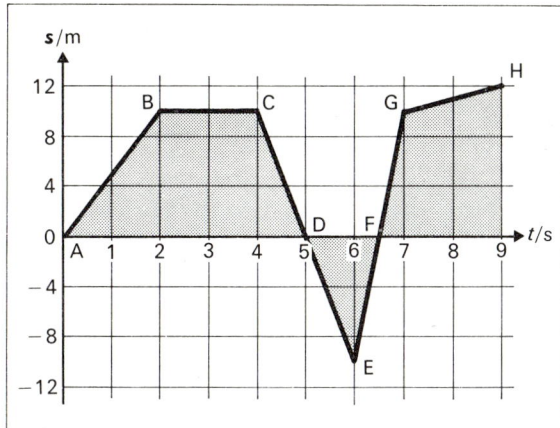

For Qu. **2–15**

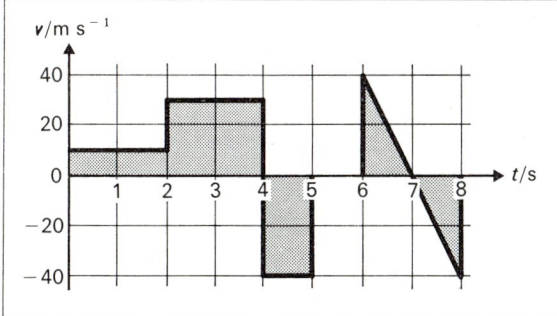

For Qu. **2–16**

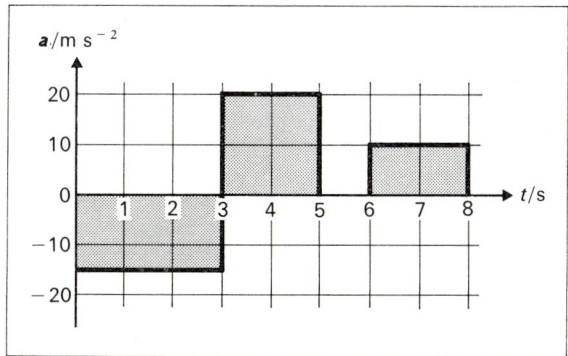

For Qu. **2–17**

2–18 (*a*) Calculate the velocity with which a bullet must be fired vertically to reach a height of 5.0 km assuming that air resistance can be neglected, and that *g* remains constant.

(*b*) How long after firing will the bullet arrive back at the level of firing?

(*c*) What is the difference between the total distance travelled by the bullet and its final displacement?

Assume g_0 [(*a*) 0.31 km s^{-1} (*b*) 64 s]

2–19 *A molecular velocity selector.* Two identical discs are mounted a distance *s* apart on the same horizontal axle, and each has a vertical slot through which particles can pass. The slot of the second disc is offset by an angle θ relative to that of the first. At what maximum speed should a molecule travel horizontally through the first slot if it is also to travel through the second when the axle is rotating at angular speed ω?

$[s\omega/\theta]$

2–20 A stone is dropped from a cliff of height *h* at the same moment as another stone is thrown vertically upward from the bottom of the cliff with initial velocity **u**. The stones are at the same horizontal level after a time *t*. Find an expression for *t* and the condition for the stones to have equal speeds at this level.

$[t = h/u,\ u = \sqrt{2gh}]$

Projectiles and Parabolic Motion

2–21 *Graphs for parabolic motion.* (*a*) A cricket ball is thrown horizontally at 20 m s^{-1} from the top of a cliff which is 300 m high. Draw accurately the following graphs: $a_x - t$, $v_x - t$, $x - t$; $a_y - t$, $v_y - t$, $y - t$.

(*b*) Draw the $v_y - t$ and $y - t$ graphs for a cricket ball thrown with the same speed at an angle of 30° to the horizontal. Take $g = 10$ m s^{-2}.

†**2–22** A gas molecule has an instantaneous horizontal velocity of 0.50 km s^{-1}. Through what vertical distance would it fall if it could cross a container of width 0.10 m? (*This would need a very low pressure.*)

Assume g_0 [0.20 μm]

2–23 A stone on the edge of a vertical cliff is kicked so that its initial velocity is 4.0 m s^{-1} horizontally. If the cliff is 0.20 km high, calculate

(*a*) the time taken for the stone to reach the beach
(*b*) how far from the cliff the stone will hit the beach
(*c*) the stone's velocity at this instant.
Ignore air resistance.

Assume g_0
     [(*a*) 6.4 s   (*b*) 26 m   (*c*) 63 m s^{-1}, 3.7° to the vertical]

2–24 A shell is fired from a gun with a velocity of 400 m s^{-1} at 25.0° to the horizontal. What is the range of the shell if it is fired on a horizontal surface? Neglect air resistance.

What is the minimum initial velocity for the shell to achieve this range?

Assume g_0 [12.5 km, 350 m s^{-1}]

Relative Velocity

†2–25   Rain is falling vertically at 8.0 m s^{-1} relative to the ground. The raindrops make tracks on the side window of a car at an angle of 30° below the horizontal. Calculate the speed of the car.

[14 m s^{-1}]

†2–26   A boat is being rowed across a river 1.0 km wide which is flowing at 0.60 m s^{-1}, and its heading is perpendicular to the banks. If the speed of the boat relative to the water is 0.80 m s^{-1}, calculate

(a) the velocity of the boat relative to the banks,
(b) the time taken to cross the river, and
(c) the lateral displacement when the boat reaches the far bank.

[(a) 1.0 m s^{-1} at 37° to heading   (b) 1.2 ks   (c) 0.75 km]

2–27   A plane X flies on a bearing of 330° with a groundspeed of 0.10 km s^{-1}, and a plane Y flies due North at 0.15 km s^{-1}. Calculate

(a) the velocity of X relative to Y, and
(b) that of Y relative to X.

(Hint: this can be solved quickly by scale drawing, since only two significant figures are involved.)

[(a) 81 m s^{-1} at 218°   (b) 81 m s^{-1} at 38°]

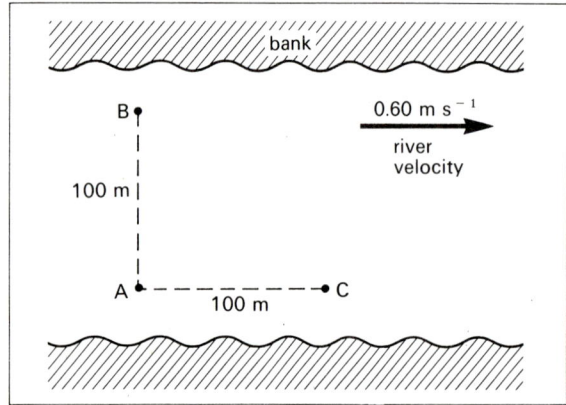

For Qu. **2–28**

2–28   *Kinematics of the Michelson-Morley experiment.* Refer to the diagram. A boat can be rowed in still water at 1.0 m s^{-1}. Calculate the times taken to traverse the paths (a) A B A and (b) A C A, each by the shortest route possible.

Could the oarsman distinguish the two routes entirely by the time taken?

(The aim of the Michelson-Morley experiment was to detect a medium analogous to the river (the ether). *The attempt failed.)*

[(a) 0.25 ks   (b) 0.31 ks]

Circular Motion

†2–29   Over a period of 5.0 s a point changes its velocity from 10 m s^{-1} at a bearing of 90° to 10 m s^{-1} at a bearing of 150°. Calculate

(a) the change of velocity, and
(b) the average acceleration.

[(b) 2.0 m s^{-2} at 210°]

†2–30   The orbit of an electron in a hydrogen atom may be considered to be a circle of radius about 50 pm and the period of motion is 0.15 fs. Calculate

(a) the frequency of rotation,
(b) the angular velocity,
(c) the linear speed, and
(d) the centripetal acceleration.

[(a) 6.7 × 10^{15} s^{-1} (b) 4.2 × 10^{16} rad s^{-1}
(c) 2.1 Mm s^{-1} (d) 8.8 × 10^{22} m s^{-2}]

†2–31   The Moon revolves round the Earth once in about 28 days with an average orbit radius of 0.38 Gm. Calculate

(a) the angular velocity,
(b) the linear speed, and
(c) the centripetal acceleration of the Moon.

[(a) 2.6 μrad s^{-1} (b) 0.99 km s^{-1} (c) 2.6 mm s^{-2}]

†2–32   Calculate the angular velocity of a car moving at 30 m s^{-1} round a bend of radius 0.45 km. What is the centripetal acceleration of the car?

[67 mrad s^{-1}, 2.0 m s^{-2}]

†2–33   A hump-back bridge has a radius of curvature of 40 m. Calculate the maximum speed at which a car could travel across the bridge without leaving the road at the top of the hump.

Assume g_0 [20 m s^{-1}]

†2–34   Calculate the speed at which a plane must be flying when looping-the-loop of radius 0.80 km so that the pilot feels no force from either his harness or his seat.

Assume g_0 [89 m s^{-1}]

2–35   A particle is moving in a circular path described by the equation $\theta = (3 \text{ rad s}^{-2}) t^2 + (2 \text{ rad s}^{-1}) t$. Calculate the angular velocity and the angular acceleration at $t = 6.0$ s.

[38 rad s^{-1}, 6.0 rad s^{-2}]

*2–36   A particle is slightly displaced from rest at the top of a fixed smooth sphere of radius r. Find the vertical height through which the particle descends before leaving the sphere.

[$r/3$]

3 Newton's Laws

Questions for Discussion

3-1 Explain what is meant by a *frame of reference*. Include some everyday situations as examples to help your explanation. Is the Earth an inertial frame of reference? Is there an *absolute* frame of reference?

3-2 Why is it necessary to define frames of reference when considering the motion of bodies? Discuss the meaning of any fictitious forces that may be introduced by the choice of a particular frame.

3-3 Suggest some situations in which it appears that a constant resultant force produces a constant velocity. Explain why this is, in fact, not so.

3-4 A body of large mass is suspended by means of a string A from the ceiling, and another similar string B is attached to the bottom of the body. Which string would break when string B is pulled (*a*) sharply, and (*b*) steadily? Why?

3-5 Explain why *Newton's Second Law* of motion can be applied even when frictional forces are present.

3-6 Discuss the possible physiological sensations experienced when moving in a lift. Sketch free-body diagrams for a passenger in different situations, and for each example that you give say what the effect would be on a balanced lever balance. What is the condition for a passenger to exert no force on the floor?

3-7 Two men want to break a cord. They first pull against each other. Then they tie one end to a wall, and pull together. Is either procedure better than the other?

3-8 Why is the following sentence misleading? 'A man standing on the ground experiences two forces, his weight acting downwards and an equal and opposite reaction from the ground upwards.'

3-9 Give four examples, with each pair taken from a different branch of physics, of the action-reaction pair forces referred to in *Newton's Third Law*.

3-10 How can a man move a box horizontally if, as *Newton's Third Law* predicts, 'the push of the man on the box equals the push of the box on the man'?

3-11 Does a centripetal force *always* have a centrifugal third-law-pair force?

*3-12 Give examples of situations where *Newton's Third Law* breaks down because the two forces concerned (*a*) do not

correspond at all instants of time (*b*) are not oppositely directed along the same straight line.

3-13 Discuss whether it would be more satisfactory to replace the prototype kilogram by some atomic standard. Why is the name *kilogram* an anomalous one for a basic standard?

3-14 Outline ways by which masses can be measured. Include a wide range of orders of magnitude in your considerations.

3-15 Discuss how you would measure the mechanical forces applied to a body (*a*) dynamically, and (*b*) statically.

3-16 How could you set up and calibrate a fixed mass and helical spring for use as an accelerometer in a rocket?

3-17 Why does a car with soft springing lean backwards while it is increasing its speed?

3-18 Describe how you would construct a simple instrument for measuring the acceleration of a train given a pendulum bob and some thread. What will happen to the bob when the train travels round a bend?

3-19 How is the velocity of an object changed when a force acts on it in a direction perpendicular to its path? Mention some examples of this situation.

3-20 What is the resultant force on an astronaut travelling in a space ship in a circular orbit round the Earth?

3-21 Under what conditions should a motor car turning a corner be considered as a rigid body, and when is it justifiable to treat it as a particle?

3-22 A small disc is placed on the turntable of a record player and the rate of rotation is increased. When will the disc leave the turntable?

3-23 An insect crawls from the edge of a rotating record directly towards the centre of the record. Describe the insect's motion as seen from a stationary point immediately above the record. Compare the observer's view with what the insect feels is happening in his accelerated frame of reference.

3-24 What are the desirable features in the design of a racing car? Explain the physical principles involved.

3-25 Discuss the importance of frictional forces in our everyday lives.

3-26 Comment on these two statements: (*a*) Frictional forces always oppose motion. (*b*) The motion of nearly all motor cars is caused by friction.

3-27 Describe, as fully as you can, what happens on a microscopic scale when one metal body is pulled across another. How can the frictional resistance be greatly reduced?

Quantitative Problems

Linear Motion

†**3-28** A resultant force of 1.0 N acts on an initially stationary mass of 1.0 kg for a time 1.0 s. How far will it move it?

[*Not* 1.0 m!]

†**3-29** Forces of 12 N and 16 N act on a body of mass 4.0 kg. If the forces are perpendicular to each other, calculate the size of the acceleration of the body.

[5.0 m s^{-2}]

†**3-30** The speed of a locust at take-off is 3.4 m s^{-1}. If the body is accelerated over a distance of 40 mm and the mass of the locust is 3.0 g, calculate the average force exerted by the ground on the hind legs of the locust.

[0.43 N]

†**3-31** (*a*) Calculate the maximum acceleration of a car of mass 1.6×10^3 kg on a horizontal sheet of ice if the wheels slip when the push of the wheels on the ice exceeds 0.40 kN.

(*b*) What would be the maximum acceleration of a car of smaller mass?

[(*a*) 0.25 m s^{-2}]

†**3-32** A lift has a mass of 1.2×10^3 kg. Calculate the tension in the supporting cable when the lift is

(*a*) descending at uniform velocity
(*b*) descending with downward acceleration 2.0 m s^{-2}
(*c*) at rest
(*d*) ascending with upward acceleration 2.0 m s^{-2}
(*e*) ascending with downward acceleration 2.0 m s^{-2}.

Assume g$_0$

†**3-33** An electron is accelerated from rest to a velocity of 25 Mm s^{-1} in the space of 12 mm between the cathode and anode of a thermionic vacuum tube. (*a*) What is the electric force on the electron? (*b*) How does this compare with the gravitational force exerted on it by the Earth?

Assume m$_e$, g$_0$ [(*a*) 2.4×10^{-14} N]

3-34 An aeroplane of mass 1.0×10^3 kg lands at a relative speed of 0.10 km s^{-1} on the deck of an aircraft carrier. The ship can provide a braking force of 40 kN. In what distance will the aeroplane be brought to rest?

[0.12 km]

3-35 *Atwood's machine.* Masses of 3.0 kg and 2.0 kg are connected by a string which passes over a frictionless and massless pulley. Calculate the tension forces exerted by the

string and the acceleration of the masses. Discuss the consequences of friction in the bearings and the finite mass of the pulley. (*Hint: draw separate free-body diagrams for the two masses.*)

Assume g$_0$ [24 N, 2.0 m s^{-2}]

3-36 A force of 30 N halves a body's velocity in 9.0 m. If the mass of the body is 5.0 kg, calculate the original velocity and the time for which the force acts.

[12 m s^{-1}, 1.0 s]

3-37 A spring balance carrying a mass of 20.0 kg in a lift registered 250 N. What was the acceleration of the lift? Draw free-body diagrams from which you can calculate the balance readings for

(*a*) free-fall, and
(*b*) movement at constant velocity.

Assume g$_0$ [2.7 m s^{-2} up (*a*) 0 (*b*) 0.20 kN]

3-38 A force of 8.0 N gives a mass m_1 an acceleration of 12 m s^{-2}, and a mass m_2 an acceleration of 48 m s^{-2}. Calculate the acceleration that this force would give the two masses when they are attached.

[9.6 m s^{-2}]

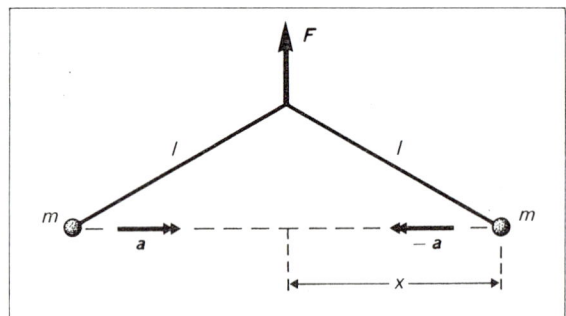

For Qu. **3-39**

3-39 Refer to the diagram. A constant force *F* is applied to the midpoint of the light string which joins two particles of mass *m*. Find an expression for the acceleration component *a* of one of the particles when it is a perpendicular distance *x* from the line of action of *F*. Sketch a graph to show the variation of *a* with *x*.

[$a = Fx/2m \, (l^2 - x^2)^{\frac{1}{2}}$]

*****3-40** A 5.0 kg mass moves on a smooth horizontal surface under the action of a horizontal force given by $F/N = 80 + t^2/s^2$. What is the velocity of the mass when $t = 3.0$ s if the body was at rest at the origin when $t = 0$ s? What is the displacement of the body when $t = 2.0$ s?

[+ 50 m s^{-1}, + 32 m]

*3–41 A particle of mass m is projected vertically upwards with initial velocity $\boldsymbol{u}$ and experiences a resisting force of kv^2 when it has velocity $\boldsymbol{v}$ (k is a constant). Find an expression for the height reached by the particle.

$$\left[\frac{m}{2k}\ln\left(1+\frac{ku^2}{mg_0}\right)\right]$$

Rotational Motion

3–42 A metal sphere of mass 0.10 kg moves in a horizontal circular path at one end of a string 0.80 m long, the other end of which is fixed. Calculate the tension in the string when the periodic time for the motion is 0.20 s.

In your calculation, you have probably assumed that the string is horizontal. Explain why this is not so and calculate the angle that the string would make with the horizontal. Would this make much difference to your original answer?

[79 N, 43′]

3–43 A person can be suspended against the inside of a hollow cylindrical room which is rotating about its central vertical axis even if there is no floor. If the radius of the cylinder is 2.0 m and the linear speed of the wall is 8.0 m s^{-1}, calculate the coefficient of friction which will prevent the person from falling.

Assume g_0   [$\mu_{\mathrm{s}} = 0.31$]

3–44 *Banking of a track*. A circular track of radius 0.36 km is banked at an angle θ. A high performance car is driven round the track at 60 m s^{-1}.

(*a*) Draw a free-body diagram for the car.
(*b*) Write down the condition for the resolved part of the weight of the car down the line of greatest slope to equal the resolved part of the required centripetal force in the same direction.
(*c*) Calculate the value of θ which avoids the need for a lateral frictional force.

Assume g_0   [(*c*) 46°]

3–45 *The conical pendulum*. A small massive body revolves in a horizontal circle at constant speed at the end of a string of length 1.2 m. As the body revolves, the string describes the surface of a right circular cone. If the semiangle of this cone is 30°, calculate the speed of the body and its periodic time.

Assume g_0   [1.8 m s^{-1}, 2.0 s]

3–46 A horizontal bar of length 8.0 m has a rope of length 2.0 m attached to each end, and each rope carries a metal sphere at its end. When the bar rotates about a vertical axis through its centre, the ropes are inclined to the vertical at an angle of $\pi/6$ rad. By considering the arrangement as part of a conical pendulum, calculate the periodic time for the rotating bar.

Assume g_0   [5.9 s]

3–47 *Tension in a rotating rope*. Consider a circular rope of radius r and linear mass density μ rotating at angular speed ω about its own centre of mass.

(*a*) Draw a free-body diagram for an element of rope which subtends a small angle $2\,\Delta\theta$ at the centre.
(*b*) Calculate, in terms of its tension T, the (centripetal) resultant force acting on the element.
(*c*) Use $\boldsymbol{F} = \boldsymbol{ma}$ to show that $T = \mu\omega^2 r^2$.

*3–48 A steel wheel with a thin aluminium rim is rotated about an axis through its centre and perpendicular to the plane of the wheel. Calculate the speed of the rim at which the breaking stress of the aluminium is reached. The density of aluminium is 2.7×10^3 kg m^{-3}, and its breaking stress is 0.12 GN m^{-2}.

[67 m s^{-1}]

Frictional Forces

3–49 *Toppling*. A railway truck has its centre of gravity 0.80 m above the rails, which are 1.2 m apart. At what maximum speed could the truck travel safely round an unbanked curve of radius 50 m? (*Hint: choose an accelerated frame of reference, and draw a free-body diagram for the truck in that frame of reference.*)

Assume g_0   [19 m s^{-1}]

3–50 *Skidding*. Calculate the minimum radius of a circle in which a car travelling at 20 m s^{-1} can turn if the coefficient of friction between the tyres and the road is 0.30. At what angle should this bend be banked to avoid the need for a centripetal frictional force at this speed? (*Assume the car will skid before it topples, and neglect the effect of air resistance.*)

Assume g_0   [0.14 km, 17°]

3–51 A box lies on the flat horizontal surface at the back of an estate car travelling at 30 m s^{-1}. Calculate the shortest distance in which the car can stop if the box is not to slip, given that the coefficient of friction between box and car floor is 0.50.

Assume g_0   [92 m]

3–52 A body is projected directly up a plane which is inclined at 30° to the horizontal. When it returns to its starting point, its speed is half its initial speed. Calculate the coefficient of dynamic friction between the body and the plane.

[0.35]

3–53 An inclined plane is adjusted so that a body just starts slipping when the angle of inclination with the horizontal is θ_{s}. What is the coefficient of static friction between block and plane? If the block were moving down the plane at constant velocity when the angle of inclination was θ_{k}, find an expression for the coefficient of dynamic friction.

*3–54   **Rope brake.** The diagram shows a rope wrapped round a rough post. The tensions marked are different because of the friction between the two surfaces. By drawing a free-body diagram for a small element of rope in contact with the post show that $F = F_0\, e^{\mu\theta}$, where μ is the coefficient of friction for the pair of surfaces. If $\mu = 0.20$ and the rope is wrapped round 5 times, calculate the value of F/F_0, and comment on your result.

[5.3×10^2]

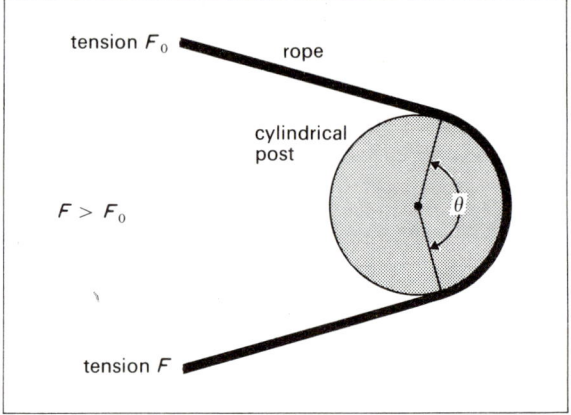

For Qu. **3–54**

*3–55   A cyclist, turning a corner on a horizontal surface on the Earth, has to lean over at an angle of $\pi/30$ rad to the vertical. At what angle would he have to lean, travelling at the same speed on a similar surface, on the Moon? Calculate the coefficient of friction between tyres and road if there is to be no slip in either case. (*Hint: use an accelerated frame of reference.*)

Assume $g_M = g_0/6$ [$32°$, $\mu_E = 0.11$, $\mu_M = 0.63$]

4   Work and Energy

Questions for Discussion

4–1   Describe some situations in which work is *not* done by a *particular* force that is applied to a body. How could you calculate the total work done by *all* the forces acting on the body?

4–2   **Work done by a variable force.** Consider a force that varies in magnitude only. How would you find the work done by the force from a graph of force F plotted against displacement x? Extend this idea to calculate the work done by the applied force F in stretching a spring from x_1 to x_2 given that $F = kx$ where k is the force constant of the spring. Outline the method of calculation if the direction of the force were also varying.

4–3   Explain why it is easy to become physically tired when pushing a car even when you fail to move it (and therefore fail to do any work on the car).

*4–4   **Work and k.e. in moving frames of reference.** Explain why two people in different frames of reference (e.g. one person on a moving ship and another at rest on the harbour wall) calculate different values for their measurement of the work done by a force acting on a particle initially at rest in one reference frame.

4–5   Is it possible to increase the k.e. of a system without applying external forces?

4–6   A ball is released from rest above a floor. Describe its motion and discuss the energy changes which take place. Would it bounce for ever in a vacuum? What would you deduce if the ball bounced higher than its original height?

4–7   Mechanical k.e. and p.e. are conserved only when conservative forces act. How then can total energy *always* be conserved?

4–8   Give one macroscopic and one microscopic example of each of the following types of collision: (a) inelastic (b) elastic (c) perfectly elastic (d) superelastic. In each example indicate any changes which occur in the k.e. or p.e. of the system you choose.

4–9   A compressed metal spring has its ends tied together and is dissolved in acid. What happens to its stored p.e.?

4–10   Explain carefully what is meant by potential energy. Why is it considered to be a property of a system rather than of a body?

4–11   What is the unit of a *potential energy gradient*? Is it the same for the p.e. resulting from all forms of interaction?

4–12   Discuss changes in the k.e. of a satellite describing an eccentric elliptical orbit, using the ideas of (a) force and work, and then (b) p.e. and energy conservation. Which method is simpler, and why?

4–13 A petrol tanker lorry is to be filled from an underground reservoir with the minimum expenditure of energy. Where would you fit the inflow pipe to the lorry, or is its position irrelevant?

4–14 Discuss the statement: 'The greater the efficiency of a machine, the more work it will save us.'

4–15 How does a frictional force do (*a*) negative work (*b*) positive work?

4–16 *Equivalence of mass and energy.* It can be shown from relativistic theory that, if the energy of a body increases, then so does its mass. If Δm is the mass increase corresponding to an energy increase ΔE, then $\Delta E = \Delta mc^2$, where c is the speed of light. Calculate the energy equivalent to a mass of 1.0 kg of sand. Discuss why this vast quantity of energy cannot be used.

*4–17 Compare the electrostatics equation $E = -\,\mathrm{d}V/\mathrm{d}x$ with $F = -\,\mathrm{d}E_\mathrm{p}/\mathrm{d}x$ in mechanics. Explain carefully what each symbol stands for, and indicate similarities and differences between the quantities they represent.

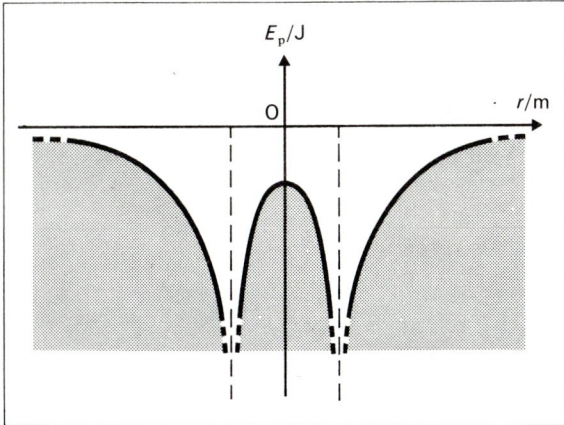

For Qu. **4–18**

*4–18 Refer to the diagram, which shows the p.e. of an electron in the vicinity of a pair of interacting atoms.

    (*a*) Discuss the size and direction of the force experienced by the electron as it moves from $r = -\infty$ m to $r = +\infty$ m.
    (*b*) Can the electron be in equilibrium near the two atoms? If so, what is the nature of the equilibrium?
    (*c*) What is the significance of the chosen zero of p.e.?
    (*d*) What interaction is responsible for this p.e.?

4–19 **O-M** If the non-tidal *Thames* could be harnessed, what would be the maximum power available?

4–20 **O-M** Estimate the power available from a windmill, taking care to discuss any assumptions you make.

*4–21 **O-M** What is the total translational k.e. of a mole of ideal gas at room temperature?

Quantitative Problems

Work

†4–22 Calculate the work done by a force of 15 N when its point of application moves through 8.0 m if the angle between the directions of the force and the displacement is

(*a*) 0 rad   (*b*) $\pi/6$ rad   (*c*) $\pi/2$ rad   (*d*) π rad.

[(*b*) 0.10 kJ]

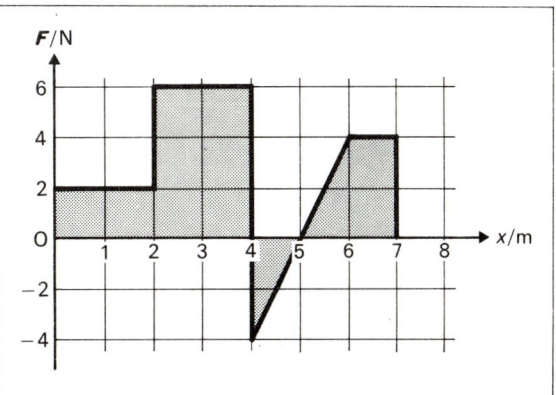

For Qu. **4–23**

†4–23 Refer to the diagram. Calculate the total work done by the force **F** on the body. What was the force's average rate of working if the work was done in 4.0 s?

[20 J, 5.0 W]

†4–24 How much work is done in stretching a spring of force constant 25 N m^{-1} from an extension of

(*a*) zero to 0.10 m, and
(*b*) 0.10 m to 0.20 m?

[(*b*) 0.38 J]

4–25 A resultant force of 30 N is applied for 4.0 s to a body of mass 10 kg which was originally at rest. Calculate

(*a*) the distance travelled,
(*b*) the work done on the body, and hence its final k.e., and
(*c*) the final velocity of the body.

[(*a*) 24 m   (*b*) 0.72 kJ   (*c*) 12 m s^{-1}]

4–26 A block of weight 0.15 kN is pulled 20 m along a horizontal surface at constant velocity. Calculate the work done by the pulling force if the coefficient of kinetic friction is 0.20 and the pulling force makes an angle of 60° with the vertical. Do any of the other forces acting on the block do work on it? What is the total work done on the block?

[0.54 kJ]

Kinetic and Potential Energy

†4–27   The time taken by a neutron to travel 4.0 m in a straight line is 0.20 ms. Calculate the k.e. of the neutron if it were travelling at constant speed.

Assume m_n [0.34 aJ]

†4–28   A flea has mass 0.50 mg and jumps so that it leaves the ground at 1.0 m s^{-1}. Calculate its initial k.e. Discuss ways in which this energy might be stored before jumping.

[0.25 µJ]

4–29   A simple pendulum of length 2.0 m is vibrating with an amplitude of $\pi/6$ rad. Calculate the speed of the pendulum bob at its lowest point.

State any assumptions you make in your calculation, and discuss the effect of using pendulum bobs of different masses.

Assume g_0 [2.3 m s^{-1}]

4–30   A cricket ball of mass 0.16 kg is thrown vertically upward with an initial velocity of 25 m s^{-1}. If the ball reaches a maximum vertical displacement of 20 m, what is the percentage loss of energy caused by air resistance?

Assume g_0 [37%]

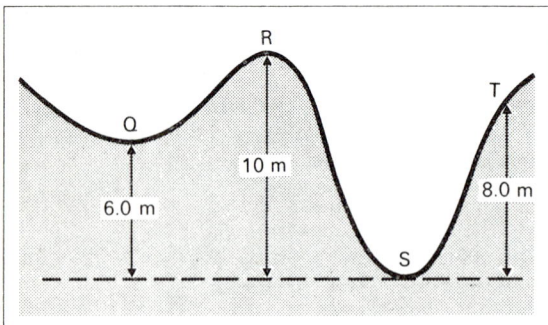

For Qu. **4–31**

4–31   Refer to the diagram. A small truck can be placed at various positions on the frictionless big dipper track and be given various initial speeds.

  (a) If the truck is released from R and allowed to move to the right, what would be its speed at S and at T?
  (b) With what speed should the truck be released from Q in order to reach S?

Assume g_0 [(a) 14 m s^{-1}, 6.3 m s^{-1} (b) 8.9 m s^{-1}]

4–32   A block of mass 0.40 kg is attached to a horizontal spring of force constant $k = 50$ N m^{-1}, and rests on a frictionless horizontal surface. If the spring is compressed 60 mm and then released, calculate

  (a) the maximum speed of the block, and
  (b) the work done by the spring force as the block moves from the position of maximum compression to maximum extension.

[(a) 0.67 m s^{-1}]

4–33   One end of a light inelastic string of length l is attached to a small body of mass m, and the other end is fixed. The body is released from a position where the string is taut and horizontal. Find the instantaneous values of the following when the string is vertical:

  (a) the body's k.e.,
  (b) its velocity,
  (c) its acceleration, and
  (d) the tension force exerted by the string.

(*Hint: for (d) draw a free-body diagram.*)

[(d) – 3 mg_0]

4–34   A satellite of mass 1.0×10^3 kg moves in a circular orbit of radius 7.0 Mm round the Earth. At this height $g = 8.2$ m s^{-2}. Calculate

  (a) the k.e. of the satellite,
  (b) its linear speed,
  (c) the work done by Earth-pull per revolution, and
  (d) the gain in speed per revolution.

[(a) 29 GJ   (b) 7.6 km s^{-1}]

4–35   *A model for atomic oscillation.* Refer to the diagram, which is drawn to scale. A mass of 1.0 kg is attached to a spring which does not obey *Hooke's Law*. The graph shows how the p.e. E_p of the spring varies with its extension x. The mass is set into oscillation so that the system has a constant total mechanical energy 10 J.

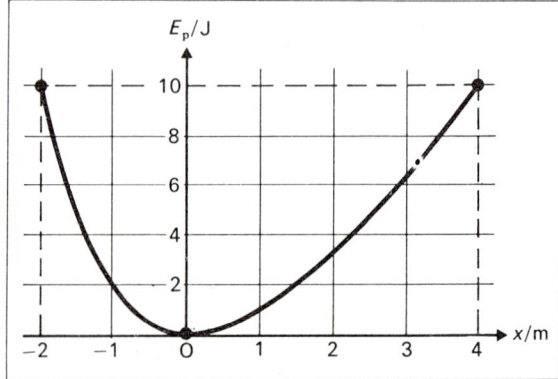

For Qu. **4–35**

  (a) In what position does the mass experience a force of maximum size, and what is its approximate value?
  (b) Where is its mean position?
  (c) What is the speed of the mass when its p.e. is 4.0 J?
  (d) By how much would the mean position change if the system's mechanical energy were halved?
  (e) Is this an *isochronous* system?

*4-36 *Interatomic p.e.* The p.e. E_p of two nitrogen atoms separated by a distance r can be expressed approximately as $E_p = a/r^{12} - b/r^6$, where $a = 4.8 \times 10^{-138}$ J m^{12},

and $b = 5.5 \times 10^{-78}$ J m^6.

(a) Sketch E_p as a function of r.
(b) At what values of r is E_p equal to zero?
(c) Calculate the force between the atoms as a function of r, and sketch F against r.
(d) What is the average separation between the atoms when they are combined to form a molecule?
(e) What energy must be given to the molecule to dissociate it into separate atoms?

[(b) ∞ m, 98 pm   (d) 0.11 nm   (e) 1.6 aJ]

*4-37 *Relativistic kinetic energy.* According to relativistic theory the k.e. K of an electron is given by

$$K = m_e c^2 \left[\left(1 - \frac{v^2}{c^2} \right)^{-\frac{1}{2}} - 1 \right]$$

where v is its speed, and m_e its rest mass.

(a) By expanding $\left(1 - \frac{v^2}{c^2} \right)^{-\frac{1}{2}}$ show that when $v^2 \ll c^2$ this expression reduces to $K = \frac{1}{2} m_e v^2$.
(b) Suppose an electron is accelerated from rest through a p.d. V which gives it a speed $0.60c$. Use both classical and relativistic methods to find values for V, and compare the two.
(c) Is any correction necessary for the average c.r.o.?

Assume e/m_e   [(b) 92 kV, 13(0) kV]

*4-38 A particle is moving in a circle of radius r under an attractive force F, given by $F = k/r^2$. Calculate

(a) the p.e. of the system,
(b) the speed of the particle, and
(c) the total energy of the system.

Assume that no k.e. is associated with the force centre.

[(a) $-k/2r$ (b) $(k/mr)^{\frac{1}{2}}$]

Power

†4-39 A man of weight 0.80 kN climbs 15 m up a vertical rope in 30 s. Calculate his average power during the climb.

[0.40 kW]

†4-40 A car is doing work at a rate of 80 kW. If it is moving at a constant velocity of 40 m s^{-1}, calculate the push of the wheels on the road. Explain why the car is not accelerating.

[− 2.0 kN]

†4-41 What is the power of an engine which drives a car of weight 15 kN up a slope of 1 in 10 at a steady speed of 12 m s^{-1}? Ignore dissipative effects.

[18 kW]

4-42 A belt drive transmits 11 kW to a pulley 0.70 m in diameter. If the pulley rotates 300 times every minute, what is the difference between the tensions in the two straight portions of the belt?

[1.0 kN]

4-43 A stone of mass 2.0 kg falls vertically from rest through a distance of 25 m under the influence of the Earth's gravitational field. Draw graphs of

(a) the work done by the gravitational force, and
(b) the power of the force,

both as functions of time. When could the power be negative? What happens to the p.e. when the power is negative? Take $g = 10$ m s^{-2}.

4-44 An upward force of 10 kN accelerates a lift from rest at 2.0 m s^{-2} for 5.0 s. Calculate

(a) the average power required during the whole time interval, and
(b) the instantaneous power after 1.0 s and 4.0 s.

[(a) 50 kW   (b) 20 kW, 80 kW]

5 Momentum

Questions for Discussion

5–1 Show that the translational k.e. K and linear momentum p of a particle of mass m are related by $K = p^2/2m$.

5–2 How can an engine be used to accelerate a car? Remember that only an *external* force can change the state of motion of the centre of mass of a body.

5–3 Discuss the design of a *recoilless* gun.

5–4 A baby is strapped in a pram which is resting on a horizontal surface. If the baby starts kicking, describe the motion of the pram with (*a*) the brake off, and (*b*) the brake on.

5–5 A moving particle makes a perfectly elastic collision with a stationary particle. After the collision, the second particle is moving faster than the first. What can you say about the relative masses of the two particles?

5–6 Two blocks of masses m_1 and m_2 are joined by a spring, and both rest on a horizontal frictionless table. The blocks are pulled apart and released. Describe as fully as you can the subsequent motion of the system.
   Compare this situation with that of a ball thrown up from the Earth by a person and then caught on its return.

5–7 A truck rests on a frictionless horizontal track. A man runs along the truck and then stops. Describe the motion of the truck by considering the motion of the centre of mass of the system. How would the situation be altered if the man jumped off at one end?

5–8 A shell explodes while in flight. Discuss the motion of the fragments and explain carefully any momentum and k.e. changes that take place.

5–9 What is meant by a *collision*? Discuss some examples from different branches of physics.

5–10 Discuss the conservation of momentum, kinetic energy and total energy in various different types of collision.

5–11 If, in a completely inelastic collision, the two bodies involved move off as one, what is the value of the coefficient of restitution?
   Discuss the physical interpretation of a collision in which the value of the coefficient of restitution is (*a*) negative, and (*b*) greater than one. Would these two values be possible for collisions between two solid spheres?

5–12 Describe carefully the measurements you would make when comparing masses by using a ballistic balance.

5–13 **O-M** What wind speed would be needed to overturn a saloon car?

5–14 **O-M** Estimate the power developed by the engine of a hovering helicopter.

Quantitative Problems

Impulse and Momentum

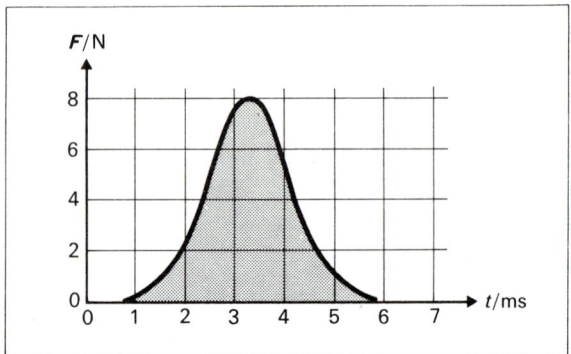

For Qu. **5–15**

†5–15 Refer to the diagram, which shows how the force acting on a body during a collision varies with time. Estimate from the graph the total impulse experienced by the body.

†5–16 A stationary billiard ball of mass 0.20 kg is hit by a cue which exerts an average force of 60 N over a time of 8.0 ms. Calculate the speed and k.e. of the ball after impact.

$$[2.4 \text{ m s}^{-1}, \ 0.58 \text{ J}]$$

5–17 Refer to the diagram opposite. The forces shown on the force-time graph are applied to a body of mass 10 kg.
   (*a*) Calculate the total impulse after 8.0 s.
   (*b*) Assuming that the body started at rest, draw (*i*) an acceleration-time graph, (*ii*) a velocity-time graph and (*iii*) a momentum-time graph.

$$[60 \text{ N s}]$$

5–18 A particle of mass 4.0 g is moving in a straight line at 1.0 km s^{-1}. It experiences a collision of duration 3.0 μs in which the direction of motion is deviated through 30°, but the speed is unchanged. Calculate the sizes of
   (*a*) the change of momentum, and
   (*b*) the force responsible.

$$[(a) \ 2.1 \text{ N s} \qquad (b) \ 0.69 \text{ MN}]$$

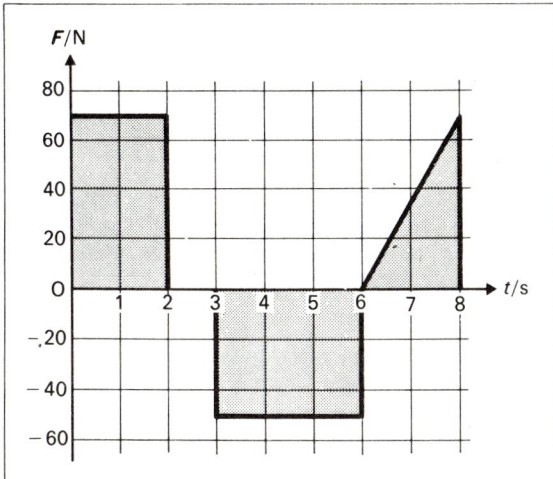

F/N

For Qu. **5–17**

5–19   A truck of mass *m* moving with velocity 3*v* collides with a truck of mass 2*m* moving with velocity *v* in the same direction, and the two move on together. Calculate

(*a*) the magnitude of the mutual impulse
(*b*) the loss of k.e. in the collision.

[(*a*) 4*mv*/3   (*b*) 4*mv*²/3]

Conservation of Momentum

5–20   A ball of mass 2.0 kg, travelling at 1.5 m s⁻¹, catches up and collides head-on with another ball of mass 3.0 kg travelling at 0.80 m s⁻¹ in the same direction. If the collision is perfectly elastic, calculate for each ball

(*a*) the velocity after collision, and
(*b*) the change in momentum.

[(*a*) + 0.66 m s⁻¹, + 1.4 m s⁻¹   (*b*) − 1.7 N s, + 1.7 N s]

5–21   *Radioactive decay.* An initially stationary ²³⁸U nucleus decays by changing to a ²³⁴Th nucleus and emitting an α-particle at a speed of 14 Mm s⁻¹ and a k.e. of 0.66 pJ. Calculate the recoil speed and the k.e. of the thorium nucleus.

[0.24 Mm s⁻¹, 11 fJ]

5–22   A nucleus, which is initially at rest, emits a positron of momentum 9.2 × 10⁻²³ N s and a neutrino of momentum 5.3 × 10⁻²³ N s. The angle between the directions of travel of the two particles is π/2 rad. Calculate the values of the following for the new nucleus, whose mass is 3.9 × 10⁻²⁵ kg:

(*a*) the magnitude and direction of its momentum,
(*b*) its velocity, and
(*c*) its kinetic energy.

[(*a*) 1.1 × 10⁻²² N s at 5π/6 rad to positron velocity
(*b*) 0.27 km s⁻¹   (*c*) 1.4 × 10⁻²⁰ J]

5–23   A nucleus of radium-226 decays by ejecting an α-particle of energy 0.77 pJ. What is the energy of recoil of the radon nucleus? Take the ratio of their masses to be 222:4.0.

[14 fJ]

5–24   Two pendulum bobs of equal mass hang side-by-side on the ends of equal suspensions of length *l* so that they just touch. One is pulled away from the other until its suspension is just horizontal, and is then released. When they collide, half the mechanical k.e. is converted to internal energy and sound energy. What happens to the two bobs after the collision?

5–25   A bullet of mass *m* and velocity *v* passes through the bob of a simple pendulum of mass *M*, and emerges with velocity *v*/2. If the length of the pendulum string is *l*, calculate the minimum value of *v* such that the bob will describe a complete circle.

[$2M\sqrt{5gl}/m$]

Force $F = d(mv)/dt$

†**5–26**   A machine gun fires bullets of mass 10 g at a speed of 1.2 km s⁻¹. If the person firing the gun can exert an average force of up to 80 N against the gun, calculate the maximum number of bullets that he can fire per minute. Sketch a force-time graph for the push of the gun on the person's shoulder and show what is meant by the *average* force.

[4.0 × 10²]

†**5–27**   A box moving at 0.40 m s⁻¹ on a horizontal frictionless track accumulates sand which falls into it vertically at the rate of 5.0 g s⁻¹.

(*a*) What horizontal force is required to keep the box moving with constant velocity?
(*b*) Describe quantitatively how sand could fall into the box so that no horizontal force is required to maintain motion.

[(*a*) 2.0 mN]

†**5–28**   The weight of a rocket is 80 kN at the instant it takes off, and the products of combustion are ejected at a velocity of 0.60 km s⁻¹ relative to the rocket. Calculate the minimum rate at which the rocket must be consuming fuel in order to take off vertically.

[1.3 × 10² kg s⁻¹]

5–29   Sand is deposited at a uniform rate of 30 kg s⁻¹ with negligible k.e. onto a belt moving horizontally at a constant speed of 2.0 m s⁻¹. Calculate

(*a*) the force required to maintain constant velocity,
(*b*) the power required to maintain constant velocity, and
(*c*) the rate of change of k.e. of the sand which is set into motion.

Why are the answers to (*b*) and (*c*) not the same?
(*It is interesting to note that this factor of* ½ *appears in other situations where energy is dissipated – see Qu.* **46–22** *for an example.*)

[(*a*) 60 N   (*b*) 0.12 kW   (*c*) 60 W]

5–30   *Pressure resulting from molecular bombardment.* A beam of molecules, each molecule having mass 4.8×10^{-26} kg and speed 0.50 km s^{-1}, strikes a wall perpendicular to its surface. For simplicity (but unrealistically) assume that all the molecules rebound straight backwards at the same speed. If the beam contains 2.0×10^{20} molecules m^{-3}, calculate the average pressure that the beam exerts on the wall.

[4.8 Pa]

5–31 A jet of water of cross-section A, density ρ and velocity v strikes a wall at right angles. Assuming that the water spreads out smoothly without backsplash, calculate the push of the water on the wall.

[$\rho A v^2$ parallel to the velocity]

5–32 Suppose that a bird of weight 2.5 N hovers by giving air of density 1.3 kg m^{-3} a uniform downward velocity v over an effective area of 5.0×10^{-2} m^2. Calculate the value of v. How do you think birds really hover?

[6.2 m s^{-1}]

5–33 A man on a sheet of smooth ice sets himself in motion by throwing successively his two boots, each of mass m, in the same horizontal direction with velocity v relative to himself. Calculate the man's final velocity if his mass without his boots is M.

$$\left[\frac{mv}{M}\left(\frac{2M + m}{M + m}\right)\right]$$

5–34   *Motion of a rocket.* A rocket and its fuel have, at any given time, a mass m. If the rocket and fuel have an initial mass m_0, and if v_e is the velocity of the exhaust gases relative to the rocket, show that the instantaneous velocity v of the rocket is given by $v = v_e \ln(m_0/m)$. If m_0 is 1.0×10^4 kg, the total mass of fuel is 2.0×10^3 kg and v_e is 4.0 km s^{-1}, calculate the maximum range of the rocket in the horizontal plane containing the position where all the fuel has been burned. State any necessary assumptions.

Assume g_0   [81 km]

Collisions

5–35 A golf club collides elastically with a golf ball and gives it an initial velocity of 60 m s^{-1}. If the ball is in contact with the club for 15 ms, and the mass of the ball is 2.0×10^{-2} kg, calculate the size of

(a) the impulse given to the ball,
(b) the impulse given to the club,
(c) the initial k.e. of the ball,
(d) the average force exerted on the ball by the club, and
(e) the work done on the ball. Note the relation between this answer and your answer to (c).

[(a) $+1.2$ N s   (d) 80 N   (e) 36 J]

5–36   *Neutron moderation.* Show that the fractional decrease in k.e. of a neutron of mass m_1 caused by a head-on elastic collision with a stationary atomic nucleus of mass m_2 is given by $4m_1m_2/(m_1 + m_2)^2$. Calculate this fraction for lead, carbon and hydrogen given that m_2/m_1 is 206, 12 and 1 respectively. Use your results to show that paraffin is a far better moderator than lead.

[0.020, 0.28 and 1.0]

5–37 A sphere is released from a height of 4.0 m so that it falls on to a plane horizontal surface and rises to 0.64 of its original height after the first bounce. Calculate the coefficient of restitution e for the surfaces.

[0.80]

5–38 Show that in an elastic collision between particles of equal mass, one of which is initially at rest, the recoiling particles always move off so that the angle between the directions of their velocities is $\pi/2$ rad. Can you reconcile your answer with the result of a head-on collision?

5–39 A smooth sphere of mass m strikes a second sphere of mass $2m$ which is at rest. After the collision their directions of motion are at right angles. Calculate the coefficient of restitution.

[0.50]

5–40 A large mass M collides elastically head-on with a small mass m which is initially at rest. What is the fractional decrease of velocity of the large mass?

[$2m/M$]

The Ballistic Balance

5–41   *Measurement of speed of a bullet.* A bullet of mass 20 g becomes embedded in a ballistic pendulum of mass 5.0 kg. Calculate the original speed of the bullet if the centre of mass of the pendulum rises a vertical distance of 0.20 m.

Assume g_0   [0.50 km s^{-1}]

5–42   *Comparison of masses.* Two unequal spheres are suspended by threads so that their centres of mass are at the same height. One sphere is displaced sideways and released so that it collides with the second sphere with a horizontal velocity 5.0 m s^{-1}. If its velocity is reduced to 3.5 m s^{-1} and the other sphere moves off with initial velocity 7.0 m s^{-1}, calculate the ratio of the masses of the two spheres. What is their coefficient of restitution?

[4.7 : 1, 0.70]

5–43 A bullet of mass 4.0 g is fired at a speed of 0.60 km s^{-1} into a ballistic pendulum of mass 1.0 kg and thickness 0.25 m. The bullet emerges from the pendulum with a speed of 0.10 km s^{-1}. Calculate

(a) the retarding force acting on the bullet in its passage through the block (which may be assumed constant), and
(b) the height to which the pendulum rises.

Assume g_0   [(a) 2.8 kN   (b) 0.20 m]

6 Rotational Dynamics

Questions for Discussion

6–1 Which points on a rigid body undergoing rotation about a fixed axis have the same angular displacement, speed and acceleration? What can you say about their linear values?

6–2 Derive a simple expression for the ratio of the angular speeds of a pair of coupled gears.

6–3 A moving body has a force acting on it that is always perpendicular to the direction of motion of the body and which increases uniformly in magnitude. Describe the path of the body and discuss any change in the speed of the body.

6–4 A yo-yo is at rest on a horizontal surface and is able to roll. Discuss the motion of the yo-yo when the string is pulled with a force at different inclinations to the horizontal.

6–5 Discuss the operation of a centrifuge from the viewpoint of (a) an inertial frame of reference, and (b) a rotating frame of reference.

6–6 What factors must be considered in accurate long-range gunnery? Discuss the difference in deviation between a shell fired towards the North from a place in the Northern Hemisphere and one fired from a place in the Southern Hemisphere.

6–7 When a mammal is running on level ground, a large amount of energy is used in accelerating the limbs. By considering moments of inertia, discuss the likely distribution of muscle in fast mammals.

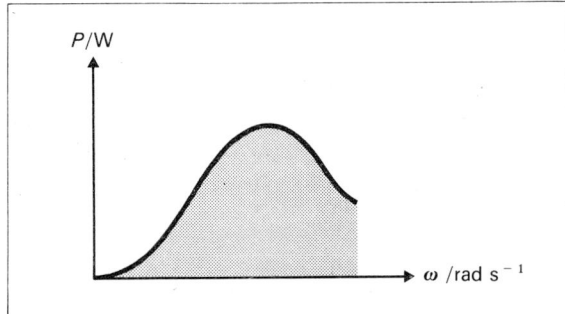

For Qu. 6–8

6–8 Refer to the diagram, which shows how the power of a motor car depends upon the angular velocity of the crankshaft. Use the graph to explain why maximum torque occurs at a lower engine speed than maximum power.

Discuss the ranges of road speed for which a motorist can obtain greater acceleration by engaging a lower gear.

6–9 Discuss the design of a flywheel and give some examples of practical applications.

6–10 A sphere, a cylinder and a ring, all with the same mass and radius, roll down a slope. Find the speed with which each reaches the bottom of the slope. (*Note that the speed of a homogeneous body rolling down a slope depends only on the shape of the body and not on its mass or dimensions.*)

6–11 A solid sphere rolls down two slopes of identical vertical height but different angles of inclination. Will it reach the bottom of the slope with the same speed in each case? Will it take the same time to roll down each slope? How would the speed and time compare with those of a sphere of smaller mass?

6–12 Two discs have the same mass and thickness but different densities. Which one, if either, will have the greater moment of inertia about an axis through its centre perpendicular to its plane?

6–13 Explain what is meant by the *radius of gyration* of a body, and give examples for some simple bodies. About what axis will a body have minimum moment of inertia? Can we consider the mass of a body to be concentrated at its centre of mass when we calculate its moment of inertia?

*6–14 Explain how it is possible to distinguish between a raw egg and a hard-boiled one by spinning each on a table.

*6–15 A dumb-bell is thrown through the air so that it has both translational and rotational motion. Describe these two motions carefully.

*6–16 Does a satellite orbiting round the Earth *in vacuo* conserve its angular momentum?

*6–17 When an Earth-satellite experiences resistive forces from the Earth's atmosphere, its speed increases (Qu. **20–31**). What happens to its orbital angular momentum? Try to make your answer quantitative.

*6–18 In the Earth-Moon system, what effect do tidal forces have on the spin angular momentum of the Earth? What is the consequence of this?

*6–19 If a cat is dropped upside down and with no rotation, how can it turn over so as to land on its feet?

*6–20 How can a skater and a diver increase their angular speeds?

*6–21   What is the effect on the Earth's rate of rotation of the accumulation of meteoritic matter from outer space?

*6–22   Explain why, if a billiard cue hits a billiard ball at a height (0.4 × radius) above the centre of mass, then the ball should roll without any sliding. What would be the ideal position for the cushioning on a billiard table?

*6–23   What is a *gyroscope*? What would you notice if a spinning gyroscope were (*a*) moved round the laboratory, and (*b*) left in the laboratory for some hours? Explain carefully how the properties of a gyroscope can be applied in the gyroscopic compass.

*6–24   A person is holding the horizontal shaft of a wheel which is rotating at high angular speed. What torques must he exert on the shaft if it is to be tilted upwards while remaining in the same vertical plane?

*6–25   Explain what is meant by *precession*, using the motion of a child's top to illustrate your discussion.

6–26   O-M Estimate the maximum acceleration of a point on the bit of an electric drill.

6–27   O-M Make estimates of the moments of inertia of the following bodies:

(*a*) the Earth about any diameter
(*b*) a cricket ball about any diameter
(*c*) a car flywheel about an axis perpendicular to its plane and passing through its centre of mass
(*d*) the suspended system of *Cavendish*'s apparatus for measuring *G*
(*e*) a small bar magnet about an axis perpendicular to the magnet and passing through its centre
(*f*) a hydrogen molecule about (*i*) a transverse axis, and (*ii*) a longitudinal axis.

6–28   O-M Estimate the rotational k.e. of the Earth assuming that it can be regarded as a sphere of uniform density. What mass of matter would be equivalent to this energy?

*6–29   O-M What fractions of (*a*) the total mass of the Solar system, and (*b*) its total angular momentum, are possessed by the Sun? Try to see whether the relative figures have any significance.

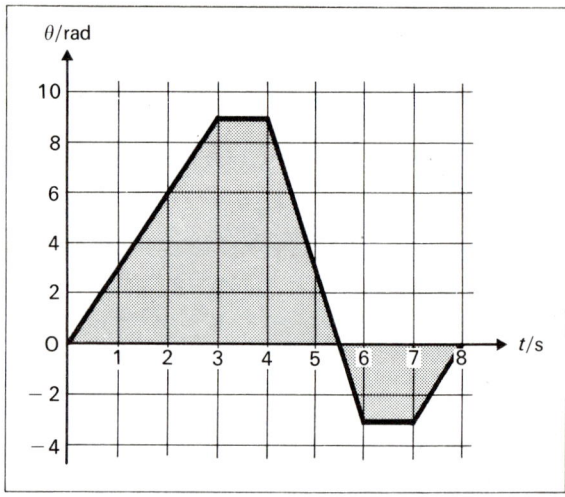

For Qu. **6–31**

6–31   Refer to the diagram. The graph shows how angular displacement θ varies with time t for a particular body which is capable of rotation. Draw a graph to show how its angular velocity ω varies with t.

6–32   When is the minute hand of a watch directly over the hour hand? How frequently are the hour, minute and sweep second hands all coincident?

[60 *H*/11 minutes after the *H*th hour]

6–33   The uniform acceleration of a wheel is 4.0 rad s^{-2}, and in a particular 5.0 s time interval its angular displacement is 0.35 krad. If the wheel was initially at rest, for how long had it been accelerating before the start of the 5.0 s interval?

[15 s]

6–34   A car has wheels of radius 0.30 m and is travelling at 36 m s^{-1} in a straight line.

(*a*) Calculate the angular speed of the wheels about the axle.
(*b*) Describe the path of a point on the rim.

The wheels now describe 40 revolutions while the car is being brought to rest. Calculate

(*c*) the wheels' angular acceleration
(*d*) the distance covered during braking.

[(*a*) 0.12 krad s^{-1} (*c*) − 29 rad s^{-2} (*d*) 75 m]

Quantitative Problems

Rotational Kinematics

†6–30   The angular speed of a car engine is increased uniformly from 0.12 krad s^{-1} to 0.40 krad s^{-1} in 14 s. Calculate the angular acceleration and the angular displacement during this time.

[20 rad s^{-2}, 3.6 krad]

Torque

†6–35   A tangential force of 8.0 N acts on a wheel which completes its first revolution from rest in 2.0 s. The force acts 0.30 m from the centre of the wheel. Calculate

(*a*) the angular acceleration, and
(*b*) the moment of inertia of the wheel.

[(*a*) 3.1 rad s^{-2} (*b*) 0.76 kg m^2]

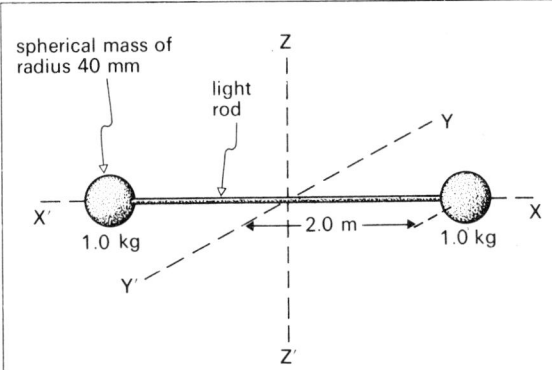

spherical mass of radius 40 mm

light rod

Z

Y

X'   1.0 kg   ←—2.0 m—→   1.0 kg   X

Y'

Z'

For Qu. **6-42**

6-36 A body of mass 2.0 kg and weight 20 N is attached to a cord which is wrapped round a uniform disc of radius 0.60 m and mass 15 kg, and this can rotate freely about a fixed horizontal axis passing through its centre. Calculate

(a) the angular acceleration of the disc when the weight is released from rest
(b) the angular acceleration if the force were applied by pulling the cord with a constant force of 20 N.

[(a) 3.5 rad s^{-2} (b) 4.4 rad s^{-2}]

6-37 A rotating wheel in the form of a uniform cylinder is slowed down by a frictional torque of 12 N m which acts on the axle. It has mass 80 kg, radius 0.50 m and is rotating initially at 0.20 krad s^{-1}. Calculate

(a) the time taken before the wheel stops, and
(b) its angular displacement during stopping.

[(a) 0.17 ks   (b) 17 krad]

6-38 A *Catherine-wheel* is in the form of a vertical uniform flat disc of mass 0.30 kg and diameter 0.12 m. It is freely pivoted through its centre about an axis perpendicular to the plane of the disc. Initially the combustion products are ejected tangentially at 2.0 m s^{-1}, and the disc is burned away at 10 g s^{-1}. Calculate the initial angular acceleration.

[2.2 rad s^{-2}]

*6-39 A cotton reel, radius r, is allowed to unwind under gravity, the upper end of the cotton being fixed. If the moment of inertia of the reel about its axis is Mk^2, calculate

(a) the linear acceleration of the reel, and
(b) the tension in the cotton.

[(a) $gr^2/(r^2 + k^2)$ (b) $Mgk^2/(r^2 + k^2)$]

*6-40 A uniform sphere rolls down a plane inclined at an angle α. Prove that the acceleration of the sphere is $(5/7)\, g \sin \alpha$.

Calculation of Moment of Inertia

†6-41 What is the radius of gyration of a flywheel in the form of a uniform cylinder of radius 0.43 m?

[0.30 m]

†6-42 *A model of a diatomic molecule.* Refer to the diagram.

(a) Treat the spheres as point masses, and calculate the m.i. of the dumb-bell about the axis YY'.
(b) What is the m.i. about the axis XX'?
(*That of a sphere about its own diameter is* $\frac{2}{5}Mr^2$.)
Bearing in mind the radius of the atomic nucleus ($\sim$ 1 fm) and the bond length of a diatomic molecule ($\sim$ 0.1 nm), discuss the relevance of your answers to the rotation of such a molecule.

[(a) 8.0 kg m^2 (b) 1.3 × 10^{-3} kg m^2]

†6-43 *Theorem of parallel axes.* Calculate the m.i. of a uniform metre rule of mass 0.10 kg about a transverse axis through (a) the 0.50 m mark, and hence (b) the 0.90 mark. Neglect the rule's width.

[(a) 8.3 × 10^{-3} kg m^2 (b) 24 × 10^{-3} kg m^2]

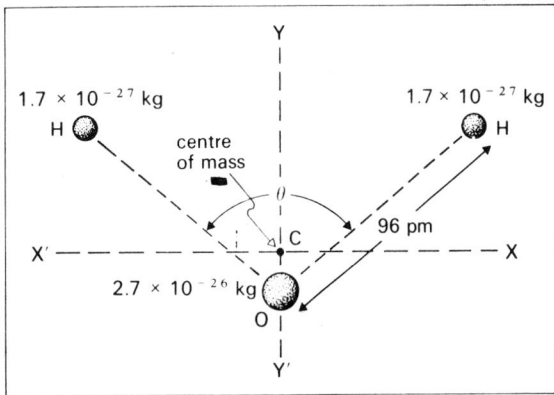

Y

1.7 × 10^{-27} kg   H   centre of mass θ 1.7 × 10^{-27} kg   H

X' —————   C   96 pm   ————— X

2.7 × 10^{-26} kg

O

Y'

For Qu. **6-44**

6-44 *Theorem of perpendicular axes.* Refer to the diagram, which is a schematic representation of a water molecule.

(a) *Estimate* the m.i. of an oxygen *atom* about an axis through its centre of mass.
(b) The m.i. of the water molecule about the axis YY' is 1.9 × 10^{-47} kg m^2. (*This is measured by infra-red spectroscopy.*) Calculate θ.
(c) The m.i. of the molecule about the axis XX' is 1.0 × 10^{-47} kg m^2. Calculate the m.i. about an axis through C perpendicular to the plane of the molecule.

[(b) 10(2)°   (c) 2.1 × 10^{-47} kg m^2]

Rotational k.e.

†6-45 If a shaft is rotating at 0.40 krad s^{-1} and is transmitting a power of 25 kW, what is the magnitude of the torque acting on the shaft?

[62 N m]

6-46 A pencil which is initially balanced on its point falls over, but its point does not move. If the length of the pencil is 0.18 m, calculate the speed with which the top of the pencil strikes the horizontal surface.

Assume g$_0$ [2.3 m s^{-1}]

6-47 An oxygen molecule consists of two atoms and has total mass 5.3×10^{-26} kg. Its average translational k.e. is 1.5 times its average rotational k.e., and its mean speed is 0.50 km s^{-1}. Its mean angular speed is 6.8 Trad s^{-1}. Calculate

(a) its m.i. about an axis through the c.m. and perpendicular to the line joining the atoms, and
(b) the linear separation of the atoms.

Can you suggest where the factor 1.5 originates?

[(a) 1.9×10^{-46} kg m^2　　(b) 0.12 nm]

6-48 The angular speed of a rotating uniform wheel is increased from 40 rad s^{-1} to 0.10 krad s^{-1} in 5.0 s. Calculate the work done by the torque acting on the body if the mass of the wheel is 10 kg and its radius is 0.45 m. What is the average power applied to the wheel during this time?

[4.2 kJ, 0.85 kW]

6-49 Calculate the average angular acceleration of a large flywheel which reaches an angular speed of 3.0 rad s^{-1} after 9.0 s. If the mass of the flywheel is 2.2×10^3 kg and its radius of gyration is 1.5 m, calculate the average power developed by the driving motor.

[0.33 rad s^{-2}, 2.5 kW]

6-50 A cylinder of mass 2.0 kg and radius 0.10 m is projected up a plane inclined at 30° to the horizontal so that it rolls without slipping and there is no rolling friction. The initial speed of the centre of mass of the cylinder was 6.0 m s^{-1}. Calculate

(a) the acceleration of the centre of mass of the cylinder
(b) the distance travelled up the plane before coming to rest momentarily.

Assume g_0　　　　　　　　[(a) $-$ 3.3 m s^{-2}　　(b) 5.5 m]

Angular Momentum

*6-51 In one simple model of the hydrogen atom the electron describes an orbit which is a circle of radius 53 pm with a rotational frequency of 6.6×10^{15} s^{-1}. Calculate the orbital angular momentum of hydrogen in this state. How does your answer compare with the *Planck constant*? (*Note that this model is oversimplified, and in practice the angular momentum of the electron is zero when the atom is in its ground state. This is because the electron does* not *follow a* circular *orbit.*)

Assume m_e　　　　　　　　　　[1.1×10^{-34} J s]

*6-52 A skater has moment of inertia 5.0 kg m^2 when she has both arms and one leg outstretched. She is turning at 4.0 rad s^{-1}. When she draws her arms and leg inwards her moment of inertia becomes 0.60 kg m^2. Calculate

(a) her final angular speed
(b) the change in her rotational k.e.

How do you account for (b)?

[(a) 33 rad s^{-1}　　(b) 0.29 kJ]

*6-53 An impulse of 40 N s is transmitted to the rim of a spinning wheel. If the wheel is a flat uniform disc of mass 24 kg and radius 0.60 m, calculate

(a) the angular impulse transmitted to the wheel, and
(b) the change in angular speed.

[(a) 24 N m s　　(b) 5.6 rad s^{-1}]

*6-54 An engine has rotating parts of mass 18 kg and radius of gyration 0.12 m. They reach an angular speed of 0.20 krad s^{-1} from rest in 4.0 s. Calculate

(a) the angular momentum and k.e. at this angular speed, and
(b) the torque and average power needed to reach it.

$$\begin{bmatrix} (a)\ 52\ \text{N m s}, & 5.2\ \text{kJ} \\ (b)\ 13\ \text{N m}, & 1.3\ \text{kW} \end{bmatrix}$$

*6-55 A bullet of mass 50 g moving at 1.2 km s^{-1} becomes embedded in the rim of a cylinder of radius 0.20 m and mass 20 kg. The cylinder is initially at rest and has a fixed horizontal axis of rotation. What can you say about the angular speed of the cylinder after the collision?

[$\omega < 30$ rad s^{-1}]

*6-56 A horizontal disc is rotating about a vertical axis through its centre at a uniform rotational frequency of 1.0 revolution per second. A piece of plasticine of mass 8.0 g is dropped onto the disc and sticks to it at a distance of 50 mm from the centre. If the speed changes to 0.80 revolution per second, what is the moment of inertia of the disc?

[8.0×10^{-5} kg m^2]

Compound Pendulum

*6-57 When a compound pendulum of mass 0.15 kg oscillates about parallel axes 0.22 m and 0.41 m from its centre of mass, its time period is 1.6 s in each case. Calculate

(a) the moment of inertia of the pendulum about a parallel axis through its centre of mass,
(b) a value for *g*,
(c) the minimum time period of the pendulum, and
(d) the length of the simple pendulum equivalent to this one (with a time period 1.6 s).

[(a) 1.4×10^{-2} kg m^2　(b) 9.7 m s^{-2}　(c) 1.5(7) s　(d) 0.63 m]

6-58 *Kater's reversible pendulum.* This is a compound pendulum which provides an accurate method of measuring *g*. Describe a standard form of this pendulum. In one experiment the period was 2.00 s when the distance from one knife edge to the centre of mass was 0.570 m, and was 1.98 s when the distance from the other knife edge to the centre of mass was 0.470 m. Calculate a value for *g* from these results.

[9.79 m s^{-2}]

6-59 *Graphical results for a compound pendulum.* The following results show how the time period *T* of a bar pendulum varies with distance *h* of the point of suspension from the centre of mass of the bar. Use them to plot a suitable graph from which you can determine

(a) a value for *g*, and

(b) the radius of gyration of the bar.

T/s	1.64	1.60	1.56	1.51	1.52	1.58	1.60	1.74
h/m	0.520	0.475	0.422	0.350	0.210	0.170	0.160	0.120

$[(a)$ 9.79 m s^{-2} (b) 0.275 m$]$

*6–60 *Centre of percussion.* If a cricket ball is driven perfectly by a cricket bat no jarring force is felt on the hands. This happens when the ball strikes the bat at the **centre of percussion**, the point at which the bat is held being the centre of oscillation. Show that the centre of percussion is at a distance from the centre of oscillation equal to the length of the simple equivalent pendulum. Make a reasonable estimate of this distance for a cricket bat.

7   Equilibrium

Questions for Discussion

7–1 If the resultant of all the forces acting on a body is zero, does this mean that the body must be in equilibrium?

7–2 A car driver turning a corner considers that he is in equilibrium; a pedestrian on the side of the road considers that the driver is not. Who is right and why?

7–3 Was a loaded washing line more likely to break if it was initially tightly stretched between two points or if it sagged quite a lot?

*7–4 *Equilibrium in a gravitational field.* Consider a gravitational equipotential surface on which there are several points where a particle can be in equilibrium. Discuss the shape of the surface at a point of (a) stable equilibrium, (b) unstable equilibrium and (c) neutral equilibrium. What is the shape of the surface near a point where a particle is in stable equilibrium with respect to one coordinate and in unstable equilibrium with respect to another coordinate?

7–5 The maximum reading on a spring balance is 100 N. How could you use this balance to measure forces far greater than this?

7–6 Give examples which illustrate that the centre of mass of a body need not necessarily lie within the body.

7–7 Show that the ratio of the distances of two particles from their centre of mass is the inverse ratio of their masses.

7–8 The centre of gravity of a body is usually considered to coincide with its centre of mass. Why is this justified? Describe the conditions under which it is not.

7–9 (a) The internuclear distance in the CO molecule is 113 pm. How far is the oxygen nucleus from the centre of mass?

(b) In the CO_2 molecule the c.m. coincides with the carbon nucleus. What can you say about the molecule?

(c) Refer to the diagram for Qu. **6–44** in which $\theta = 104°$. How far is the oxygen nucleus from the centre of mass of the molecule?

7–10 Discuss the conditions under which a suspended rigid body would be in (a) stable equilibrium (b) unstable equilibrium (c) neutral equilibrium.

7–11 Bodies with low centres of gravity are more stable than those with high centres of gravity. Thus it is much easier to stand a short pencil on its end than it is to balance a long thin stick. Discuss then why it is very much easier to balance the long stick on its point on the finger than it is to balance the short pencil on the finger. Describe why an inverted billiard cue should be even easier to balance than the long thin stick.

7–12 Comment on the following information about the muscles in the fore-legs of the horse and armadillo:

(a) the horse needs a low velocity ratio, and

(b) the armadillo needs a high mechanical advantage.

Quantitative Problems

†7–13 Two parallel forces of magnitudes 50 N and 80 N act in the same direction. Calculate the separation of the two forces if the distance from the resultant force to the smaller force is 0.24 m.

$[0.39$ m$]$

†7–14 A rigid uniform plank forms a bridge across a stream. The plank has weight 0.24 kN and is 6.0 m long. Derive expressions for the pushes of the ground on the ends of the plank when a man of weight 1.2 kN is a distance *x* from one end. Draw, on the same set of axes, two graphs to represent your results.

7–15   A light string is attached to a *light* ring threaded onto a horizontal bar. The coefficient of static friction between the ring and bar is 0.40. What is the least angle that the string can make with the bar when it is in a state of tension?

[68°]

7–16   In order to lift a metal wheel over a kerb of height 0.20 m, a force of 0.16 kN must be applied horizontally at the axle. The radius of the wheel is 0.50 m. Calculate

(a) the weight of the wheel
(b) the size of the minimum force applied at the axle that could lift the wheel.

[(a) 0.12 kN   (b) 96 N]

7–17   *Sliding and toppling.* A uniform cube rests on a horizontal surface with which it has a coefficient of friction of 0.4. A horizontal force is applied at the centre of, and perpendicular to, an upper edge. As the size of the force is steadily increased the cube will either slide or topple. Which?

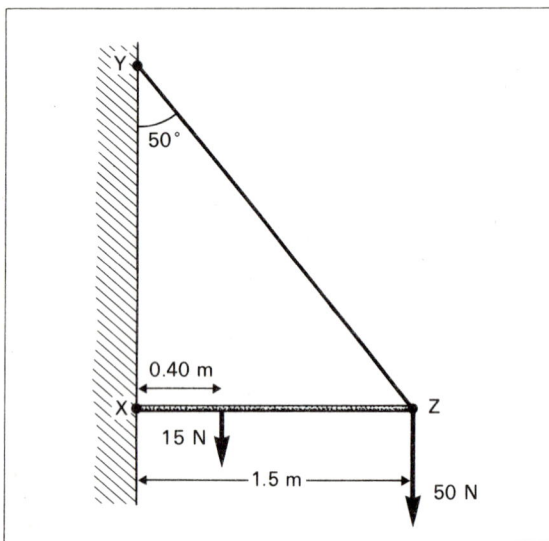

For Qu. **7–18**

7–18   Refer to the diagram. The non-uniform rod XZ has weight 15 N, length 1.5 m and is held in position by a length of string YZ. A weight of 50 N is hung at Z. Calculate the force exerted by the string, and the push of the hinge on the rod.

[84 N, 65 N at 9.8° to horizontal]

7–19   A uniform plank leans against a vertical wall. The coefficient of friction at each end of the plank is 0.30. Calculate the minimum angle at which the plank may be inclined to the horizontal without slipping.

[57°]

7–20   A uniform disc of radius R has a circular hole of radius r cut in it so that the centre of the hole is $R/2$ from the centre of the disc. Find the centre of mass.

[$Rr^2/2(R^2 - r^2)$ from centre]

*7–21   *Location of centre of mass.* Use the calculus definition to find the centre of mass of the following bodies:

(a) a solid cone
(b) a semi-circular lamina
(c) a hemisphere.

In each example indicate where you have used ideas of symmetry.

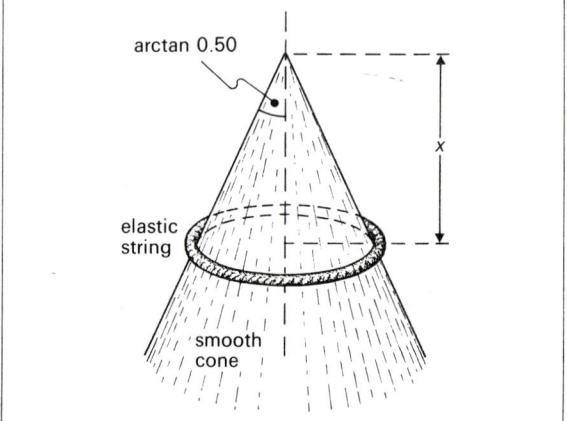

For Qu. **7–22**

*7–22   *The principle of virtual work.* Refer to the diagram. An elastic string of force constant 5.0 N m^{-1}, natural length 0.25 m and weight 2.0 N is slipped over the smooth cone.

(a) Write down, in terms of x, expressions for (i) the gravitational p.e. of the string (ii) its elastic p.e.
(b) Write down the changes in these quantities which occur if the string is lowered a further small distance δx. (*Watch the signs.*)
(c) According to the principle of virtual work the sum of these changes is zero at equilibrium. Find the value of x at which the string will rest in equilibrium.

Justify the first sentence of (c) without resorting to symbols.

[0.12 m]

8 Fluids at Rest

Questions for Discussion

8–1 Densities vary from $\sim 10^{-20}$ kg m^{-3} to $\sim 10^{+17}$ kg m^{-3}. Discuss substances to which these values might correspond, and draw up a table to show orders of magnitude of various densities between these extremes.

8–2 What are the causes of variation in atmospheric pressure?

8–3 Is air containing water vapour more dense or less dense than dry air at the same temperature and pressure?

8–4 Comment on the statement 'The pressure at a point in a fluid is the same *in all directions*'.

8–5 Explain how an aneroid barometer is used as an altimeter. Should any correction be made for temperature variations of the air?

8–6 Describe briefly the readings that you would take in order to measure the percentage composition of a lump of wax in which a block of metal is embedded. Outline the steps in your calculation from the readings.

8–7 If some metal girders slipped off a barge while it was in a lock, what would happen to the water level in the lock?

8–8 Why does a wooden stick never float vertically in water?

8–9 A straight rod has length l and small cross-sectional area a and is made of material density ρ. It is supported vertically above a liquid of density σ and is then lowered until it is partially submerged. Discuss the equilibrium conditions of the rod. (Neglect surface tension.)

8–10 A piece of cork floats half submerged in a non-volatile liquid at rest on the surface of the Earth. What would happen to the cork

   (*a*) on the surface of the Moon
   (*b*) in a satellite moving in a circular orbit
   (*c*) in a spaceship moving at constant velocity?

8–11 An enclosed box resting on a spring balance contains a live bird. Describe how the reading of the balance will vary depending on whether the bird is standing on the bottom of the box or flying in the air enclosed by the box. Discuss also the effects of taking off and landing.

8–12 What would be the difference in the apparent weight of a plastic bag when empty and when filled with air at atmospheric pressure? Would these two weights be different if they were measured in a vacuum?

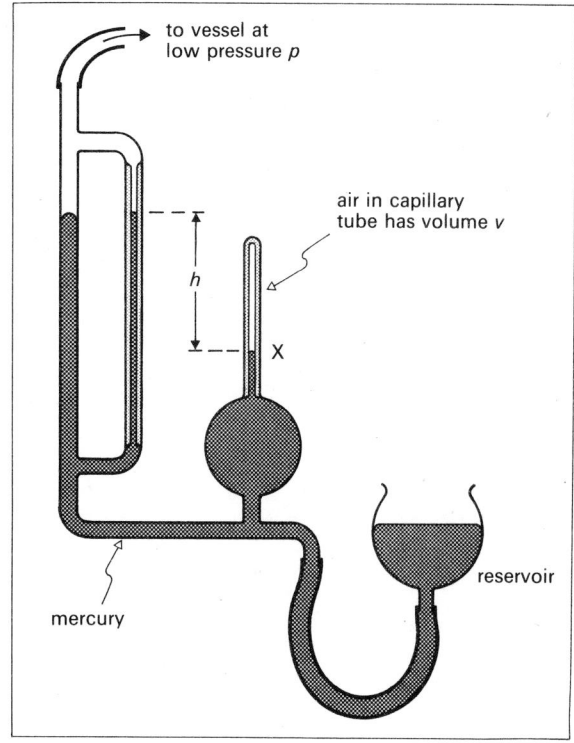

For Qu. **8–13**

8–13 *The McLeod gauge.* Refer to the diagram. The spherical vessel and its capillary tube have a total volume V. The gauge, which can measure pressures ~ 1 Pa, is operated as follows: (1) the reservoir is lowered until all the mercury has drained into it, and then (2) it is raised until the mercury reaches the fixed point X. The value of h is measured.

   (*a*) What is the purpose of the left-hand capillary tube?
   (*b*) How is p related to the known values of V and v, and the pressure exerted by the length h of mercury?

8–14 Describe briefly how a rotary pump operates. What other methods are there of obtaining low pressures?

8–15 If an inflated gas balloon starts to rise, it will reach a particular equilibrium height if no changes are made to the load or gas content of the balloon. Explain this and discuss what will happen to a submarine once it starts to sink in the sea.

   How can the vertical motion of a gas balloon and a submarine be controlled?

8–16   *Air pumps.* Sketch a possible design of exhaust pump and explain the action of the valves. Air at pressure p_0 is contained in a vessel of volume V and the volume of the barrel of the pump is v. Show that the pressure p_n at the end of the nth upstroke is given by

$$p_n = p_0 \left(\frac{V}{V + v} \right)^n.$$

How would you alter the pump to make it into a compressor? Derive an expression for p_n in the case of a compressor. What, in practice, will happen to the air temperature in each case, and what effect will this have on the pressure?

8–17   **O-M** What is the pressure at the bottom of the deepest ocean?

Quantitative Problems

†8–18   A wooden block floats in water with 80% of its volume below the surface. When the block is placed in oil, 95% of its volume is below the surface. Calculate the densities of the wood and the oil. (*Each step of your calculation must have a detailed explanation.*)

Assume ρ_{H_2O} $[(8.0 \text{ and } 8.4) \times 10^2 \text{ kg m}^{-3}]$

†8–19 10% of the total volume of an iceberg is above the surface of the sea. Calculate the density of the sea water given that the density of ice is 920 kg m^{-3}.

$$[1.02 \times 10^3 \text{ kg m}^{-3}]$$

8–20   *Relative density.* A liquid, which does not mix with water, is poured into a U-tube partly filled with water. The level of this liquid is 8.0 mm above the water level in the other limb,

which has risen 20 mm from its original position. Calculate the density of the liquid relative to that of water.

$$[0.83]$$

8–21   A container of liquid of density ρ is given an upward vertical acceleration $\boldsymbol{a}$. Apply *Newton's Second Law* to an element of liquid to find the pressure p at a depth h below the surface. What would the pressure become if the liquid were in a state of free-fall?

$$[h\rho (g_0 + a) + p_0, p_0]$$

8–22   A wooden block of volume $5.0 \times 10^{-4} \text{ m}^3$ floats in a liquid of density $8.0 \times 10^2 \text{ kg m}^{-3}$ so that it is 60% immersed. The block is placed in a liquid of density $1.2 \times 10^3 \text{ kg m}^{-3}$. Calculate the volume of metal of density $7.2 \times 10^3 \text{ kg m}^{-3}$ which should be attached to the block so that the combination just floats totally immersed in this liquid.

$$[6.0 \times 10^{-5} \text{ m}^3]$$

8–23   A balloon is filled with $1.5 \times 10^3 \text{ m}^3$ of hydrogen of density $9.0 \times 10^{-2} \text{ kg m}^{-3}$, and the weight of the envelope and attachments is 2.0 kN. The balloon is anchored to the ground by steel wire of density $7.6 \times 10^3 \text{ kg m}^{-3}$ and cross-sectional area 50 mm². How high will the balloon rise in air of density 1.3 kg m^{-3}? (*Assume, unrealistically, that the air density does not change.*)

Assume $\boldsymbol{g_0}$ $[4.3 \text{ km}]$

8–24   *Hydrometers.* Describe briefly how you would measure the relative density of a liquid using (*a*) a constant weight hydrometer (*b*) a constant volume hydrometer.

A constant weight hydrometer records a relative density 1.00 when placed in water in a measuring cylinder. The hydrometer displaces $2.40 \times 10^{-5} \text{ m}^3$ of water, and the distance between the 1.00 and 0.99 marks on the hydrometer is 6.00 mm. Calculate the cross-sectional area of the stem.

$$[40 \text{ mm}^2]$$

9   The Bernoulli Equation

Questions for Discussion

9–1   Under what conditions can the *Bernoulli equation* be applied exactly? To what extent are they observed in practice?

9–2   A tennis ball is thrown South in such a way that it has a clockwise spin when viewed from above. Draw a carefully

labelled flow pattern to show how the air moves relative to the ball, and decide which way the ball will be deflected.

9–3   Use the *Bernoulli principle* to explain the following. Include a carefully labelled flow pattern sketch for each example.

(*a*) A table tennis ball may be supported in a narrow jet of air.
(*b*) Gas and air together flow up the tube of a bunsen burner.

(c) A sliced or pulled golf ball drifts from its intended path.
(d) A golf ball attains maximum range when it is driven at an angle of less than 45° to the horizontal.
(e) Two nearby ships proceeding side by side might be in danger of colliding.
(f) The best position for the sail of a boat is when the plane of the sail bisects the angle between the keel of the boat and the wind direction.
(g) A boomerang glides back along a path similar to that which it took on the outward journey.
(h) A high wind destroys the roof of a house by lifting it *upwards.*

9–4   Why does the soft-top of a sports car become taut and tend to lift away from its supporting struts at high speed?

What effect does one notice when one is overtaken by another car, if the overtaking car passes rather close at a high relative speed?

9–5   Show how the *Bernoulli principle* predicts upward dynamic lift on an aerofoil. What is meant by the *angle of attack*? Discuss the possible dangers of ice accumulation on aeroplane wings.

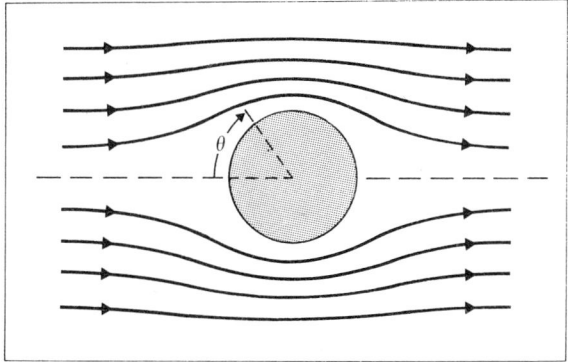

For Qu. **9–6**

9–6   The diagram shows the streamlines for frictionless fluid flow around a sphere. Sketch a graph to show the pressure distribution around the sphere by plotting the pressure variation Δp against the angle θ. Explain the shape of the graph.

*9–7   *Fields of flow.* A moving fluid can be described by a *field of flow.* Each point in the fluid has a flow velocity vector v associated with it, and this flow velocity can be related to a velocity potential. (Compare this with gravitational field strength and gravitational potential.) Use these ideas to sketch the following fields of flow:

(a) a uniform field
(b) a field from a point source
(c) a field flowing from a point source to a nearby sink.

Discuss any further analogies you can make between fluid fields and fields from other branches of physics.

*9–8   O–M A small leak in the top of a garden hose causes escaping water to cascade to a height of 1 m. What is the gauge pressure in the hose?

Quantitative Problems

(*In answering the questions that follow you should assume where necessary that the conditions required for the application of the Bernoulli equation are in fact observed.*)

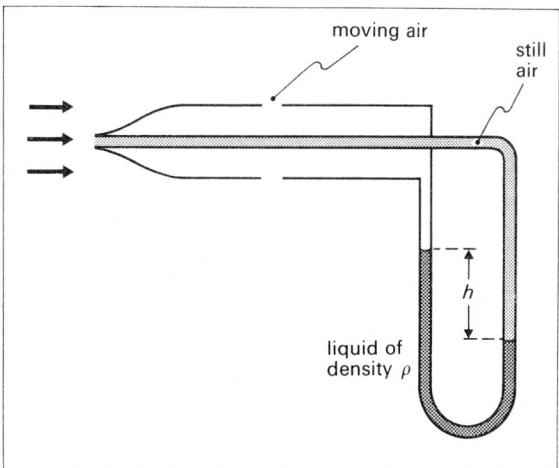

For Qu. **9–9**

*9–9   *The Pitot tube.* Refer to the diagram, which shows an instrument designed to measure the flow speed of a gas. Show that for a gas of (constant) density σ the flow speed is given by

$$v = \sqrt{2h\rho g_0/\sigma}\ .$$

Such a tube is being used as an air speed indicator for an aircraft, and an observer records $h = 0.12$ m using a liquid for which $\rho = 8.0 \times 10^2$ kg m^{-3}. The density of air is 1.3 kg m^{-3}. What is the speed of the aircraft?

Assume g_0 [38 m s^{-1}]

*9–10   A liquid of density 8.0×10^2 kg m^{-3} flows through a horizontal pipe which narrows at a constriction. The speed of the liquid at the constriction is 10 m s^{-1}, and the pressure drop from the original area to the constriction is 30 kPa. Calculate the area of the constriction relative to that of the original area.

[0.50]

*9–11   Water flows into a large tank at a rate of 8.0×10^{-4} m^3 s^{-1}, and flows out of a hole of area 6.0×10^{-4} m^2 which is in the bottom of the tank. At what depth does the water level in the tank stay constant?

Assume g_0 [91 mm]

*9–12   A tank of large cross-sectional area and with vertical sides contains water to a depth y_0. A hole is made in the tank a depth y below the surface, and this causes a jet of water to strike the ground a distance x from the tank.

(a) Show that $x = 2\sqrt{(y_0 - y)y}$.
(b) For what value of y is x a maximum?
(c) Calculate the corresponding values of y and x_{max} for a tank of depth 0.50 m.   [(c) 0.25 m, 0.50 m]

*9–13   A tank of base area 12 m^2 and vertical sides is filled with water to a depth of 2.0 m. A small hole of area 40 mm^2 is made in the base. How long will the tank take to empty? How long would it take if it were filled with alcohol? What assumptions do you make?   [0.19 Ms]

*9–14   *Aerodynamic lift on an aerofoil.* An aircraft wing has a weight 2.5 kN and an effective area 6.0 m^2. The air flow over the top surface is effectively horizontal, and of speed 60 m s^{-1}, while that under the bottom surface has a speed 50 m s^{-1}. The air density can be taken as constant at 1.2 kg m^{-3}. Calculate

(a) the aerodynamic lift on the wing
(b) the force exerted by the wing on the fuselage

[(a) 4.0 kN   (b) 1.5 kN upwards]

*9–15   *Thrust on a rocket.* A rocket chamber of cross-sectional area A contains a virtually incompressible gas of density ρ at pressure p. The gas escapes through a small hole of area a to the atmosphere, where the pressure is p_0. Calculate

(a) the speed of efflux of the gas
(b) the thrust exerted on the rocket.

(*Hint:* use $F = d(mv)/dt$ *in the form* $F = v \, (dm/dt)$.)

[(a) $\sqrt{2(p - p_0)/\rho}$ (b) $2a(p - p_0)$]

*9–16   *The Venturi meter.* In one type of *Venturi meter* for measuring the velocity of a fluid in a horizontal pipe, two pressure gauges are inserted in the pipe. One is inserted in the main body of the pipe and one at a contraction in the pipe. Show that the rate of volume flow of an incompressible fluid passing through any section of the pipe is given by

$$V/t = k \sqrt{p_1 - p_2}$$

where k is a constant depending on the size of the pipe and the density of the fluid, and p_1 and p_2 are the fluid pressures in the main pipe and the contraction respectively.

III Oscillations and Wave Motion

10 Simple Harmonic Oscillation

Questions for Discussion

10-1 $y = a \sin \omega t$ is the equation of s.h.m. Name quantities from four different branches of physics which could be represented by y in this context. In each case suggest an appropriate value for a and ω in the motion concerned.

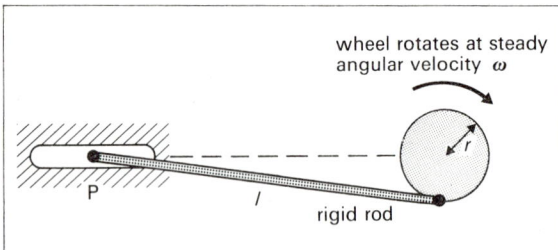

wheel rotates at steady angular velocity ω

P
l
rigid rod

For Qu. **10-2**

10-2 Refer to the diagram.
  (*a*) Show that, if $l \gg r$, P describes s.h.m.
  (*b*) What are the frequency and amplitude of the s.h.m.?

10-3 Can the two atoms of a diatomic molecule oscillate at different frequencies?

10-4 Is it possible for a motion to be **isochronous** (T independent of amplitude) but not true s.h.m.? Can an s.h.m. be non-isochronous?

10-5 Do *all* periodic motions approximate to s.h.m. at small amplitude?

10-6 Indicate, in a qualitative way, how the period of a periodic motion would vary with amplitude if the p.e. of the motion was described by (*a*) $E_p = kx$ (*b*) $E_p = kx^2$ (*c*) $E_p = kx^3$. (*You would find E_p–x sketch graphs helpful.*)

10-7 Does the period of oscillation of a pendulum increase or decrease when its angular amplitude is increased?

10-8 A pendulum clock is being operated in a lift. What happens to its time period when
  (*a*) the lift accelerates upwards
  (*b*) the lift descends at a steady speed
  (*c*) the lift cable snaps?

10-9 The bob of a simple pendulum consists of a massive hollow container with a small hole in the bottom. The container is filled initially with liquid. How does the frequency of oscillation vary with time?

10-10 When a mass oscillates vertically on a helical spring the time period is independent of g, but depends on m. For a simple pendulum the reverse is true. Discuss, with particular reference to the inertial and elastic characteristics involved.

10-11 A mass oscillating vertically on a spiral spring has a different mean position from the same mass oscillating on a horizontal frictionless surface. How are their time periods related?

10-12 A body of mass m hung on a particular spring gives a simple harmonic motion of period 1.0 s. What period would be obtained
  (*a*) using a body of mass 4m
  (*b*) on the Moon ($g_M \approx \frac{1}{6} g_E$)
  (*c*) using two springs identical to the first, connected in parallel
  (*d*) using a spring similar to the first, but twice as long
  (*e*) using two springs identical to the first, connected in series?

10-13 Refer to the diagram for Qu. **10-31**. What would be the nature of the oscillation if the two springs had different spring constants k_1 and k_2?

***10-14** Suppose you are provided with a pendulum bob and a supply of cotton. Devise a means of demonstrating a *Lissajous* figure with a frequency ratio 2:1.

Quantitative Problems

Kinematics of s.h.m.

†**10-15** A particle oscillates with an s.h.m. of amplitude 50 mm and a maximum speed of 0.25 m s^{-1}. Calculate the particle's
  (*a*) pulsatance,
  (*b*) period, and
  (*c*) maximum acceleration.
       [(*a*) 5.0 s^{-1} (*b*) 1.3 s   (*c*) 1.2 m s^{-2}]

†**10-16** The tip of a tuning fork prong has a maximum speed of 4.0 m s^{-1} while describing s.h.m. of frequency 512 Hz. What is its amplitude?
                                  [1.2 mm]

10–17 A torsional pendulum describes an angular s.h.m. of period 1.0 s and angular amplitude 2.0 rad. Calculate

(a) its maximum angular speed and acceleration

(b) the angular speed and acceleration when the angular displacement is 1.0 rad.

(*Hint: the kinematic equations of linear s.h.m. can be replaced by their angular analogues, but you are advised to use $\dot{\theta}$ rather than ω (for angular velocity) to avoid ambiguity.*)

$$\begin{bmatrix} (a) & 13 \text{ rad s}^{-1}, & 79 \text{ rad s}^{-2} \\ (b) & 11 \text{ rad s}^{-1}, & -39 \text{ rad s}^{-2} \end{bmatrix}$$

10–18 A particle describes s.h.m. in which the displacement is given by

$$x = (2.0 \text{ mm}) \sin (3\pi t/\text{s}).$$

Calculate, for this motion,

(a) the pulsatance (or angular frequency)

(b) the period

(c) the amplitude

(d) the maximum speed.

Find the following at the instant $t = (13/18)$ s:

(e) the displacement

(f) the velocity

(g) the acceleration.

[(e) 1.0 mm   (f) 16 mm·s^{-1} (g) $-$ 89 mm s^{-2}]

10–19 A point mass rests on a horizontal platform which can be made to describe vertical s.h.m. of amplitude 0.10 m and frequency 2.0 Hz. The mass makes contact with the platform as it rises from its lowest position. At what point, if any, will this contact be lost? (*Hint: above the mean position the platform experiences a downward acceleration. Will the pull of the Earth on the point mass enable it to achieve the same acceleration?*)

Assume g_0   [At the instant when $x = 62$ mm]

10–20 For the situation of the previous question and the same amplitude, what is the highest frequency at which the mass and platform do not separate?

[1.6 Hz]

Dynamics of s.h.m.

†**10–21** A mass 0.50 kg is attached to a spring which obeys *Hooke's Law* for both compression and extension. The spring has a force constant 50 N m^{-1}, and lies along a horizontal frictionless surface.

(a) What force acts on the mass if it is displaced x to the right?

(b) Write down the equation of motion for the mass.

(c) If it were set into oscillation, what would be the time period and frequency?

How would variations in g affect the time period? What effect do you think a temperature change would have on the time period?

[(b) $\ddot{x} = -(100 \text{ s}^{-2})\, x$ (c) 0.63 s, 1.6 Hz]

†**10–22** A helical steel spring 0.20 m long, and of spring constant 20 N m^{-1} is hung vertically, and a weight 0.98 N

suspended from the lower end. Calculate

(a) the attached mass

(b) the extension caused

(c) the time period of any subsequent oscillation.

What should be the maximum amplitude of the oscillation if it is to remain simple harmonic? Does g affect the time period?

Assume g_0   [(b) 49 mm   (c) 0.44 s]

†**10–23** The piston of a car engine has a mass 0.50 kg, and describes a motion which is approximately s.h.m., and which has an amplitude 50 mm. What is the greatest force that it experiences when the rev. counter shows 80 rotations s^{-1}?

[6.3 kN]

†**10–24** Near room temperature the particles of a solid undergo a periodic vibration of amplitude about 15 pm at a frequency of about 10 THz. The motion is nearly s.h.m.

(a) Calculate the maximum speed and acceleration of the particles.

(b) How many times g_0 is the maximum acceleration?

(c) How many times their weight is the maximum force exerted on each particle?

(d) What is the value of a typical force constant for these vibrations? (Take the mass of a particle to be 10^{-25} kg.)

Assume g_0

[(a) 0.94 km s^{-1}, 5.9×10^{16} m s^{-2} (d) 0.39 kN m^{-1}]

10–25 When a man of weight 0.75 kN walks to the end of a diving spring board, the end is depressed 0.30 m. If he were to jump lightly on the end, what would be the period of subsequent simple harmonic oscillation? (Take his mass to be 75 kg.)

[1.1 s]

*****10–26** A mass is hung on the end of a light spring in such a way that changes in its vertical deflection can be measured with great sensitivity. It is to be used as a device for detecting changes in g as small as 10 μm s^{-2}. The system has a natural period of 3.0 s. What is the minimum vertical deflection that must be detectable? (*When giving your answer, bear in mind the sort of distance changes that can be detected by using optical interferometry methods.*)

[2.3 μm]

*****10–27** *Reduced mass.* (a) Consider a molecule of HCl to consist of a hydrogen atom of mass 1.7×10^{-27} kg coupled to an infinitely massive chlorine atom by a spring of force constant 0.50 kN m^{-1}. What is the frequency at which the hydrogen atom will oscillate?

(b) In practice the chlorine atom also undergoes a vibration, and while the two atoms change their *relative* positions the centre of mass of the system remains in equilibrium. The system is equivalent to a single particle of mass μ, called the **reduced mass**, where

$$\frac{1}{\mu} = \frac{1}{m_1} + \frac{1}{m_2}.$$

Is the vibrational frequency of this system greater or less than that of the model used in (a)?

[(a) 86 THz]

10–28   *A model of a diatomic molecule.* Two bodies of masses 1.0 kg and 2.0 kg are joined by a spring of force constant 54 N m^{-1}, and rest on a frictionless horizontal surface. The spring is compressed, and the masses secured together by cotton. The cotton is then burned. Calculate

(a) the position of the centre of mass
(b) the effective force constant for the smaller mass
(c) the frequency of oscillation of the smaller mass, and hence of the system.

Check your answer by calculating the reduced mass (Qu. **10–27**), and using $f = (1/2\pi)\sqrt{k/\mu}$.

[(c) 1.4 Hz]

Energy in s.h.m.

†**10–29**   A simple pendulum bob of weight 0.25 N is drawn aside, and this raises its c.g. vertically by 12 mm. It is then released to describe s.h.m. What is the mean value of its k.e. during the motion?

[1.5 mJ]

10–30   A mass 0.20 kg undergoes an s.h.m. whose maximum speed is $\pi/2$ m s^{-1}. Calculate

(a) the average speed
(b) the average k.e.

[(a) 1.0 m s^{-1} (b) 0.12 J]

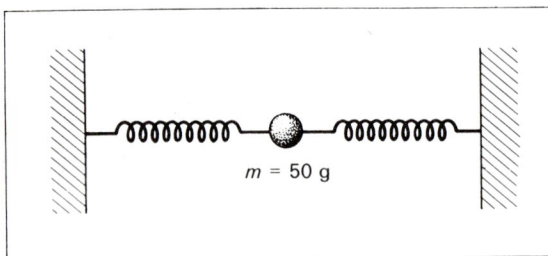

For Qu. **10–31**

10–31   Refer to the diagram, which shows a mass in a region free from a gravitational field. It is held by two identical springs, each of which has a force constant $k = 1.0$ N m^{-1}. When the mass is displaced by 0.20 m, its subsequent motion is found to be simple harmonic. Calculate

(a) the maximum value of the stored elastic p.e.
(b) the maximum speed of the mass
(c) the maximum acceleration of the mass
(d) its frequency of vibration
(e) its distance from the mean position when the energy of vibration is half kinetic, and half potential.

[(a) 40 mJ   (b) 1.3 m s^{-1} (c) 8.0 m s^{-2}
(d) 1.0 Hz   (e) 0.14 m]

***10–32**   *An energy approach to investigate s.h.m.* A point mass m lies in the base of a frictionless hemispherical bowl of radius R. It is given a horizontal displacement x and vertical displacement h, such that $R \gg h$, and it is then released to describe an oscillatory motion.

(a) Relate the variable h to the variable x and the constant R.
(b) Write down the instantaneous p.e. E_p in terms of x.
(c) Write the instantaneous k.e. K as $\frac{1}{2}m\dot{x}^2$, and differentiate (with respect to t) the equation
$$E_p + K = \text{constant}$$
to show that the motion is s.h.m.
(d) Calculate its period for $R = 0.50$ m.

Assume g_0

[(d) 1.4 s]

***10–33**   *Electrical oscillations.* When an inductor and a capacitor are connected into an oscillatory circuit, the energy W can be written
$$W = \frac{1}{2}LI^2 + \frac{1}{2}Q^2/C$$
where $I = dQ/dt$.

If we can ignore dissipative effects, then $W = \text{constant}$. Using the method of Qu. **10–32**, differentiate this equation to find the natural pulsatance (angular frequency) of the circuit.

[$\omega = 1/\sqrt{LC}$]

***10–34**   *The torsional pendulum.* The instantaneous k.e. of a rotating torsional pendulum is $\frac{1}{2}I\dot{\theta}^2$. The elastic p.e. of its twisted fibre is $\frac{1}{2}c\theta^2$. Use this information alone to show that its period T is given by $T = 2\pi\sqrt{I/c}$.

Superposition of two s.h.m.

***10–35**   A particle is subject to two simultaneous motions described by
$$x = (4.0 \text{ mm}) \cos 5(t/\text{s}), \text{ and}$$
$$y = (4.0 \text{ mm}) \sin 5(t/\text{s}).$$
Find

(a) its **trajectory equation** (the relationship between the variables x and y)
(b) the position of the particle at $t = 0$
(c) an equation which shows whether the speed varies with time, and hence the speed at $t = 0$ and $t = 1.0$ s
(d) the size of the particle's acceleration when $t = 2.0$ s.

[(a) $x^2 + y^2 = 16$ mm^2 (c) 20 mm s^{-1} (d) 0.10 m s^{-2}]

***10–36**   *Lissajous figures.* The electron beam of an oscilloscope traces out a path on the screen which can be described by
$$y = (12 \text{ mm}) \sin \omega_y t \text{ and } x = (5.0 \text{ mm}) \sin (\omega_x t + \delta).$$

(a) Suppose $\omega_x = \omega_y$. Describe the patterns observed when δ takes the values (i) 0 (ii) $\pi/2$ rad (iii) π rad. Derive their equations.
(b) Suppose $\delta = 0$. Sketch the patterns observed when (i) $\omega_x = \frac{1}{2}\omega_y$ (ii) $\omega_x = 3\omega_y$.

11 Damped and Forced Oscillation

Questions for Discussion

11-1 Sketch a graph of the energy of a damped oscillator plotted as a function of time. Use it to suggest a definition for the characteristic *lifetime* for the damped motion. For what kind of motion would this time period have (*i*) a maximum (*ii*) a minimum value?

11-2 If a violin string could vibrate undamped, how would it sound?

11-3 What is the function of a loudspeaker? Give *two* reasons why it is desirable for a loudspeaker to be heavily damped.

11-4 Why is the central tub of a washing machine often mounted on springs?

11-5 Why does the cab of a lorry often shake violently when it is stationary, but much less when it is moving?

11-6 It is found that on a long journey the tyres of a fully sprung motor car do not become as warm as those of a load which is largely unsprung (such as a field gun). Suggest reasons for this.

11-7 When an oscillating system is driven by a periodic force whose frequency varies from 0 through the resonant frequency f_0 to ∞ Hz, the amplitude of the oscillation is said to be **stiffness-controlled** one side of resonance, and **mass-controlled** the other. Suggest the origins of the two terms, and the frequency ranges to which they apply. (*Hint: try to explain the limits of the amplitude – frequency cur e at $f = 0$ and $f = \infty$ Hz.*)

11-8 Consider the forced vibration of a simple pendulum. At energy resonance what is the phase relation between

  (*a*) the applied force and bob velocity
  (*b*) the applied force and the displacement?

(*Hint: think in terms of the energy transfer.*)

11-9 *The quality factor.* The Q-factor of a forced oscillating system is one way of indicating the sharpness of resonance. Thus

$$Q = \frac{2\pi \text{ (energy of system)}}{\left(\begin{array}{c} \text{work done on the system} \\ \text{during one oscillation} \end{array} \right)}$$

An unforced system of high Q would be expected to keep vibrating for a long time, since the energy taken from the system during each oscillation is small. Note that Q is dimensionless.

Which system, in the following pairs, would you expect to have the higher Q-factor?

  (*a*) Loaded and unloaded paper cones of the *Barton* pendulums.
  (*b*) A mass oscillating on a helical spring, and a loaded paper cone oscillating as a pendulum.
  (*c*) An American car and an English car.
  (*d*) An oscillatory electric circuit with $R = 20\ \Omega$ and an otherwise identical one with $R = 40\ \Omega$.

Comment on the Q-factor of the *Tacoma Bridge* which was destroyed by a resonant vibration.

11-10 What is the unit of the Q-factor? Explain the significance of your answer. What is the Q-factor of an undamped oscillator?

*11-11 O-M Estimate the Q-factors of (*a*) a violin string (*b*) an excited atom.

Quantitative Problems

11-12 The resistive force experienced by a moving object is proportional to its speed, being given by
$$F = -k\boldsymbol{v},$$
where $k = 50$ N s m^{-1}. What power is needed to keep the object moving at the steady speed of 16 m s^{-1}?

[13 kW]

11-13 A simple pendulum is set vibrating with an angular amplitude 50 mrad. What will the angular amplitude have become when half the total energy has been dissipated?

[35 mrad]

11-14 *Timing a forced oscillation.* A point mass 0.20 kg performs s.h.m. of amplitude 0.10 m on a spring of force constant 80 N m^{-1}. A fixed *impulse* of $+ 0.40$ N s is to be applied to the mass. What change of energy does it cause if it is applied at

  (*a*) $x = 0$, $\dot{x}$ positive
  (*b*) $x = -0.10$ m?

Discuss qualitatively the effect of applying the impulse at points between these two. (*Hint: calculate the velocity change from the momentum change.*)

[(*a*) 1.2 J   (*b*) 0.40 J]

*11-15 *Logarithmic decrement.* Suppose that the amplitudes of successive swings of a damped motion on opposite sides of the mean position are $a_1, a_2, a_3 \ldots$ etc., and are related by
$$\frac{a_1}{a_2} = \frac{a_2}{a_3} = \frac{a_3}{a_4} \ldots = e^{\lambda},$$

where λ is called the **logarithmic decrement**. (*Such a decay is exponential.*)

(a) By expanding e^{λ}, show that if λ is small then
$$a_1 \approx a_2 (1 + \lambda).$$
(b) In a ballistic galvanometer experiment the uncorrected first throw is θ_1. Write the equation relating θ_1 and θ_1', the value θ would have taken in the absence of damping.

*11–16 *Correcting a galvanometer throw.* The coil of the galvanometer in Qu. 11–15 gives an observed first throw of 50.0 mm on a screen, and the eleventh deflection on the same side (after 20 complete swings) is 10.0 mm.

(a) Show that if n is the number of swings, then $\theta_{n+1}/\theta_1 = e^{-n\lambda}$, and find the value of λ.
(b) Calculate θ_1', the corrected first throw.
[(a) $8.0(5) \times 10^{-2}$ (b) 52.0 mm]

*11–17 Calculate the Q-factor (see Qu. 11–9) of an electrically maintained tuning fork from the following information: electrical energy is supplied at the rate 50 mW to maintain a fixed amplitude at frequency $f = 512$ Hz, and when the tuning fork passes through its equilibrium shape it has k.e. 0.10 J.
[6.4×10^3]

12   Wave Motion

Questions for Discussion

12–1 Refer to the diagram. Suppose mass A is given a rapid longitudinal impulsive push to the right.

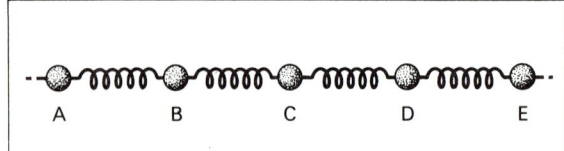

For Qu **12–1**

(a) What is E's *instantaneous* response?
(b) What is E's eventual response?
(c) Does mass A interact with mass E?
(d) Does this interaction obey *Newton's Third Law* at each instant of time?
(e) How does A transfer momentum to E?
(f) Is it possible for any momentum transfer to occur without the help of an intervening mechanical medium?

(*This question points to momentum conservation as a more fundamental idea than Newton's Third Law, to the concept of a field, and the possibility of interaction at a distance through wave motion.*)

12–2 The speed c of a mechanical wave motion is given, in general terms, by an equation of the form
$$c = (\text{constant}) \left(\frac{\text{an elasticity factor}}{\text{an inertial factor}} \right)^{\frac{1}{2}}.$$

Discuss the elasticity and inertial factors for the following wave types:

(a) slinky spring waves
(b) earthquake waves
(c) sound waves
(d) ocean waves (large λ)
(e) ripple tank waves (small λ).

12–3 A large rubber sheet is stretched on a trampoline frame, and one end is given a sudden transverse displacement which causes a wave pulse to cross the sheet.
(a) What are the elasticity and inertial factors which control the speed of the pulse? (*Hint: these are surface waves.*)
(b) Use the method of dimensions to find how the pulse speed is related to these two factors.

12–4 When we analyse the movement of an elastic wave motion through a material medium we usually ignore the molecular structure of matter. Under what conditions can this be justified?

12–5 Which properties of a travelling wave are altered, and which remain unchanged, when it crosses the boundary between two media with different physical properties?

**12–6 *Earthquake waves.* Earthquake waves that pass through the body of the Earth can be longitudinal or transverse, and have speeds
$$c_l = \sqrt{(K + \tfrac{4}{3}G)/\rho} \text{ and } c_t = \sqrt{G/\rho},$$
where G and K are shear and bulk moduli.

(a) When there is a distant earthquake, one kind of wave always activates a detector before the other. Which?
(b) The transverse waves are not propagated through the core. What information about the core's physical properties can be deduced from this?

12–7 Discuss qualitatively the different effects that concave, plane and convex reflecting surfaces have on the curvature of an incident wavefront, and indicate how your discussion leads on to the adoption of a consistent sign convention for the radii of curvature of the respective spherical surfaces involved.

12–8 Give three examples of **anisotropic media** through which different kinds of waves can be propagated.
In an isotropic medium a point source produces spherical wavefronts. What shape would the wavefronts be in these media?

*12–9 *Group and phase velocity.* If you observe a group of three or four waves crossing a water surface, you will see that waves are being created at the rear of the **group**, and others are being absorbed at the front – the crests move faster than the group. How would you define a speed for these waves?

12–10 How would you investigate experimentally whether a stretched string is a dispersive medium for small amplitude transverse waves? What result would you expect to find, and why?

12–11 In shallow water of depth h waves of all wavelengths (as distinct from ripples) propagate at a speed $c = \sqrt{gh}$. By writing c in terms of the inertial factor ρ, suggest the physical origin of the wave elasticity factor. How would you demonstrate that the shallow water is acting as a non-dispersive medium?

12–12 *Ripple tank analogies.* Describe, with appropriate wavefront sketches, ripple tank arrangements which would illustrate wave behaviour analogous to that in the following phenomena:

(a) a star appears to be 'higher' in the sky than it really is
(b) to a fish in a pond, a hovering insect appears further above the surface than it really is
(c) the dispersion of a beam of white light passing through a prism
(d) the total reflection of sound waves passing over a flat water surface
(e) the passage of sound waves through a child's balloon filled with hydrogen (ignore diffraction)
(f) a converging glass lens being used as a magnifying glass.

12–13 *Inverse square law.* Under what conditions can one use an inverse square law for (a) the intensity, and (b) the amplitude of the disturbance of a wave motion? Discuss real physical situations in which the law can be applied, stating whether the relationship is exact or only approximate.

*12–14 Can an infinite sinusoidal wavetrain be used for propagating *signals* (i.e. information)? – discuss.

*12–15 *A one-dimensional wave equation.* The equation of a one-dimensional sinusoidal wave travelling along a string in the positive x-direction is given by
$$y(x, t) = a \sin(\omega t - kx).$$

(a) Explain in detail the meanings of the symbols y, a, ω and k, and give a consistent unit for each quantity that they represent.
(b) Which of the quantities in (a) are determined entirely by the source?
(c) If λ is the distance between points having the same phase, how are λ and k related?
(d) Use the method of dimensions to suggest how the wave speed c might be related to ω and k.
(e) What would be the equation of a wave travelling in the negative x-direction?

*12–16 *The wave equation.* (a) Verify that the equation
$$y = y_0 \sin(\omega t - kx)$$
is a valid solution of the equation
$$d^2y/dt^2 = c^2(d^2y/dx^2)$$
(a simplified form of the **wave equation** for one dimension). c is the wave speed.
(b) It can be shown that under certain conditions the pressure of an ideal gas obeys the equation
$$d^2p/dt^2 = \left(\frac{\gamma p_0}{\rho_0}\right) d^2p/dx^2,$$
where p_0 and ρ_0 are the undisturbed values of the gas density and pressure. What conclusions can you draw?

12–17 O-M What is the minimum speed of surface water waves?

Quantitative Problems

(*For other quantitative questions on waves, refer to* Section VII (*Light waves) and* Section VIII (*Sound waves*).)

Reflection and Refraction

†12–18 *Use of a stroboscopic lamp.* Linear wavefronts of wavelength 80 mm are generated in a ripple tank by a source of frequency 2.5 Hz. They travel normally across a boundary in a ripple tank which separates a deep water region A from a region B made shallow by using a thick plate of glass, and their speed is now reduced to 0.12 m s^{-1}.

(a) Calculate (i) the wave speed in A (ii) the wavelength in B (iii) the relative refractive index of the boundary.
(b) Describe the pattern in the two regions when the tank is illuminated by a lamp flashing at frequencies of 1.25, 2.0, 2.5 and 5.0 Hz respectively.

[(a) (iii) 1.7]

12–19 *One-dimensional reflection.* A transverse pulse of amplitude 50 mm moves along string A, and is incident on a join with a second string B. The linear density of A is greater than that of B, and this causes the *reflected* pulse to have an amplitude 30 mm.

(a) What are the phases of the reflected and transmitted waves relative to that of the incident wave?

(b) How is the energy carried by a wave pulse related to its amplitude?

(c) Use the law of conservation of energy to find the fraction of the incident energy carried by the transmitted pulse.

(d) Which wave has the greater speed?

What sort of information would you need to know to calculate the amplitude of the transmitted pulse?

[(c) 0.64]

†12–20   *Total internal reflection.* (a) Waves travelling in air at 330 m s^{-1} are first totally reflected back into the air from a water surface when the angle of incidence reaches 13.0°. At what speed do the waves travel in water?

(b) What is the relative refractive index for sound waves passing from air into water?

(c) Indicate, with reasons, whether you would expect more or less partial reflection for sound waves incident at 45° on a water-air interface than for light waves incident at 45° on an air-water interface.

[(a) 1.47 km s^{-1} (b) 0.22(5)]

Energy and the Wave Equation

12–21   *Inverse square law.* A spherical source of diameter 1.0 m emits spherical waves with a power 10 W through an isotropic medium which dissipates no wave energy.

(a) The source is not a point: discuss whether an i.s. law is applicable.

(b) What is the wave intensity at a point distant 9.5 m from edge of the source?

[(b) 8.0 mW m^{-2}]

*12–22   A progressive wave travelling along a taut string is described by the equation

$$y = (25 \text{ mm}) \; \sin 2\pi \, (t/20 \text{ ms} + x/4.0 \text{ m}).$$

(a) In which direction is the wave travelling? – justify your answer.

(b) What are the (i) amplitude (ii) wavelength (iii) frequency and (iv) wave speed?

*12–23   Write down an equation to describe the following wave: it travels at 50 m s^{-1} along a string in the negative z-direction, with a transverse disturbance of frequency 20 Hz and amplitude 5.0 mm which is confined to the x–z plane.

*12–24   Refer to the diagram. The left end L of the string is made to undergo s.h.m. of amplitude 20 mm at a frequency 4.0 Hz, and this causes a wave to propagate along the string at 8.0 m s^{-1}. Write down the equations which give the variation with time of the displacements of the string particles at P, Q and R. (*Assume that of L to be given by y = a sin ωt.*)

*12–25   A long taut horizontal string is held at one end. It has a linear mass density of 0.12 kg m^{-1}, and is subjected to a tension of 48 N. The hand holding it describes s.h.m. with an amplitude of 40 mm and a frequency of 2.0 Hz.

(a) Calculate the wave speed and wavelength.

(b) Write down, in terms of x and t, the variations with time of the following quantities for an element of rope at horizontal displacement x: (i) the vertical displacement y, (ii) the vertical velocity, and (iii) the vertical acceleration.

(c) What is the size of the maximum transverse acceleration of the element?

(d) What is the largest vertical force experienced by the element if its length is 10 mm? Would it be possible for this force to have the same order of magnitude as the tension force in the string?

[(a) 20 m s^{-1}, 10 m   (c) 6.3 m s^{-2} (d) 7.6 mN]

*12–26   *Calculation of wave power.* An external agent shakes the end of a long horizontal string of linear mass density 80 g m^{-1} subject to a tension force of 12 N. His movement has a frequency 2.0 Hz and amplitude 50 mm.

(a) He does work on the element of string which he holds, and yet it gains no energy. Explain.

(b) What is the energy of a length λ (the wavelength) of string?

(c) How long does it take for this energy to be taken from that length of string?

(d) Calculate the power of the wave motion.

Ignore all dissipative effects.   [(d) 0.19 W]

*12–27   A wave is propagated along a long string so that each kilogram of string has energy 5.0 J.

(a) What is the new specific energy of the string if the amplitude is doubled?

(b) What would it be if, using the original amplitude, the frequency were doubled?

(c) The original frequency was 20 Hz. What was the original amplitude?

[(a) 20 J kg^{-1} (b) 20 J kg^{-1} (c) 25 mm]

*12–28   An oscillator of frequency 50 Hz generates travelling waves of amplitude 2.0 mm in a string of linear mass density 50 g m^{-1} when it is subject to a tension force of 15 N. What is the power of the oscillator?

[0.17 W]

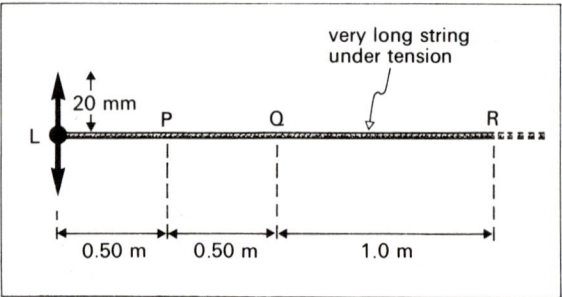

For Qu. **12–24**

13　Huygens's Construction

13–1　To what extent is the idea put forward by *Huygens* a *principle*? Does it have any physical meaning, or is it entirely geometrical in nature?

13–2　What sort of body could convert a plane light wavefront into a letter S? Describe the subsequent behaviour of such a wavefront.

13–3　A large number of loudspeakers are mounted round the inner surface of a large spherical bowl and are energized by the same source (i.e. they emit coherent radiation). Use the *Huygens construction* to draw the secondary wavelets emitted by the surface, and hence determine the shape of the wavefront after some time interval Δt. To what point is the wave energy concentrated?

13–4　A boat is moving through still water at the same speed as that with which the waves that it generates travel. What is the form of the wavefront produced?

13–5　Plane waves are incident normally on a circular disc. Use the *Huygens construction* to show that there will always be wave energy concentrated on points along the axis of the disc on the far side. (*In optics this phenomenon results in a bright spot at the centre of the geometrical shadow of a circular object.*)

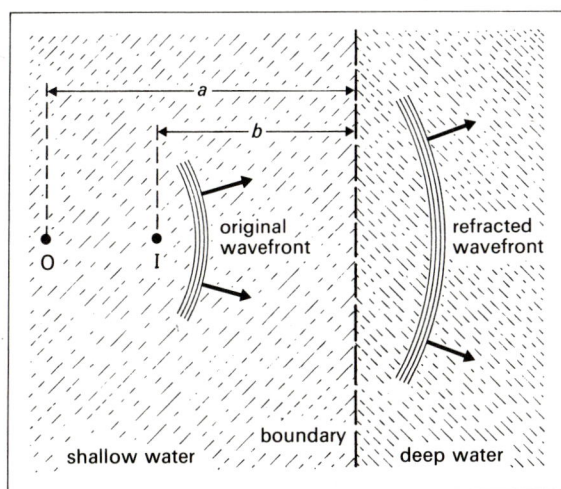

For Qu. **13–6** *(plan view of ripple tank)*

13–6　Refer to the diagram. Waves generated at O cross the boundary and appear, after refraction, to have originated at I. Show how the *Huygens construction* enables us to draw the refracted wavefront, and use your diagram to find the relationship between the ratio of the wave speeds (n), a and b.

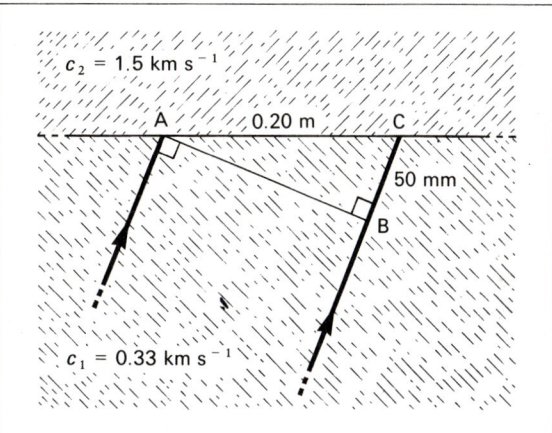

For Qu. **13–7**

†13–7　Refer to the diagram.
  (a) How long does the point B on the wavefront AB take to reach C?
  (b) What then is the radius of the secondary wavelet that originated at A?
  (c) Describe the behaviour of the wave at the boundary.

*13–8　*The spherical reflecting surface.* The **curvature** of a spherical surface equals the reciprocal of its radius of curvature. Thus a spherical wavefront of radius u has a curvature $1/u$. Show, using the *Huygens construction* and *Malus's theorem*, that a concave mirror of radius of curvature r changes the curvature of an incident wavefront by $2/r$, and hence evaluate the curvature $1/v$ of the wavefront that it reflects. (*Treat all distances as positive, and assume that the mirror aperture ≪ u, v or r so that the sagitta relationship can be used.*)

*13–9　*The lens.* Show, by the method implied in Qu. **13–8** that when spherical wavefronts are refracted by a thin spherical converging lens of small relative aperture whose focal length is f, they experience a change of curvature $1/f$. (*Experiment bears out our prediction that the change of curvature is the same for all such wavefronts. To what angular relationship in geometrical optics does this correspond?*)

*13–10 Use the *Huygens construction* to draw the wavelets reflected from

(a) a 'smooth' surface, and
(b) a 'rough' surface,

and use these diagrams to suggest, quantitatively, how rough the surface can be before the process is better described as *scattering* rather than *reflection*. (*Evaluate at which stage the waves become incoherent because their phase relationships have become random.*)

14   The Principle of Superposition

Questions for Discussion

14–1 What theoretical justification exists for the principle of superposition for wave disturbances?

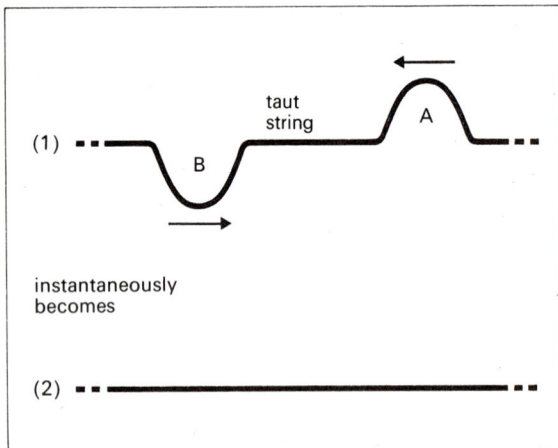

For Qu. **14–2**

14–2 Refer to the diagram.

(a) When the two wave pulses shown are superposed there is an instant when all points on the string show zero transverse displacement. What has become of the pulses' energy?
(b) How are the original pulses apparently regenerated from the straight string? – (as they must be if each is ultimately unaffected by the other).
(c) Discuss how this example illustrates the law of conservation of linear momentum (both longitudinal and transverse).

14–3 Refer to the diagram, which shows the configuration of a stretched string the instant before two travelling pulses A and B are superposed.

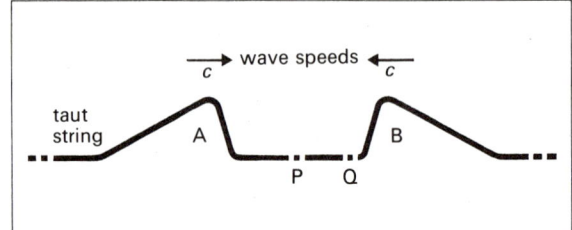

For Qu. **14–3**

(a) Draw displacement-position curves (like photographs) at five successive instants of time to show the stages in the superposition of the pulses. (*You will find it helpful to draw dotted outlines of the two waves being superposed.*) What is the shape of pulse A after superposition?
(b) Draw displacement-time curves to show the approximate behaviour of the points P and Q while the pulses are passing.

14–4 A particular point on a string is subjected to a periodic vibration described by

(1) $y_1 = a \sin \omega t$, then
(2) $y_2 = a \sin 2\omega t$, and finally
(3) $y = y_1 + y_2$.

Draw accurate sketch graphs to display y_1 and y_2 as functions of time, and then apply the principle of superposition to show the effect on the point of the two disturbances together.

14–5 Sketch displacement-position and kinetic energy-position curves for a standing wave at the instants

(a) $t = 0$, with displacement $= 0$ everywhere, and
(b) $t = T/4$.

14–6 Can a progressive wave be travelling along a stretched string if some of its particles permanently show zero displacement?

14–7 How would you measure the speed at which transverse waves propagate along a given metallic string in a musical instrument?

14–8 A wave incident on a boundary has amplitude a_1, and gives rise to a reflected wave of amplitude a_2. The quantity *standing wave ratio* is defined to be $(a_1 + a_2)/(a_1 - a_2)$.

(a) Calculate the s.w.r. for (i) complete reflection, and (ii) zero reflection.

(b) Draw the envelope of a standing wave pattern for which $(a_1 + a_2)/(a_1 - a_2) = 2$.

14–9 For a two-source ripple tank experiment, will the anti-nodal lines have *zero* disturbance? – discuss.

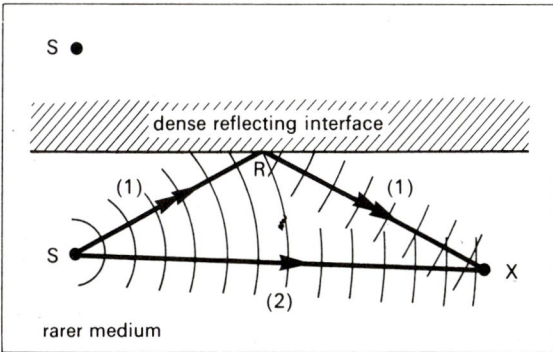

For Qu. **14–10**

14–10 Refer to the diagram. Waves are travelling from S to X along both paths (1) and (2). What is the condition for their superposition to give zero disturbance if the waves are

(a) water waves

(b) sound waves

(c) electromagnetic waves?

14–11 (a) What is the shape of a nodal line in a two-source ripple tank interference experiment? Which part of the line most nearly approximates to a straight line? Under what conditions could we call it straight?

(b) If one moves along a line parallel to the twin sources, are the nodal lines equally spaced? Qualify your answer.

14–12 A pair of sources separated by 2λ show a steady phase difference $\pi/2$ rad. What is the smallest angle made by an antinodal line with the axis of symmetry at points relatively far from the sources? What do you think *relatively* implies here?

14–13 A point X in a ripple tank receives waves from a pair of identical sources A and B. Suppose that $AX - BX = p$. Write down the conditions for (a) constructive superposition, (b) de-structive superposition and (c) the crest from A to arrive with zero disturbance from B, when the sources show the following phase differences: (i) zero (ii) π rad (iii) A leads B by $\pi/2$ rad. (Nine answers.)

14–14 When we superpose two disturbances of frequencies f_1 and f_2, the amplitude of the resultant has a frequency $\frac{1}{2}(f_1 \sim f_2)$. Why is this *not* the beat frequency?

*14–15 Draw on the same sheet of graph paper, and reason-ably accurately, the graphs which represent

$$y_1 = a \sin \omega t, \quad y_3 = \frac{a}{3} \sin 3\omega t \quad \text{and} \quad y_5 = \frac{a}{5} \sin 5\omega t$$

for the time interval $t = 0$ to $t = 2\pi/\omega$. Superpose these waves, and use your result to suggest how it might be possible to synthesize a square waveform.

*14–16 A piston vibrates through a circular aperture in a baffle. Discuss, illustrating your answer by the relative sizes of the piston and the emitted wavelength, how the intensity of the sound wave emitted varies as a function of direction. (*Hint: think of the diffraction of plane waves through a circular aperture.*)

Quantitative Problems

(For other quantitative questions on waves, especially on inter-ference and diffraction, refer to Section VII (Light waves) and Section VIII (Sound waves).)

Stationary Waves

†14–17 A vibrator is driven from the a.c. mains at 50 Hz, and sends transverse waves at a speed 60 m s^{-1} along a stretched string. What length of string would show a standing wave pattern with three displacement antinodes? Draw sketches of the string configuration at times $t = 0$, 5.0 ms and 10 ms, taking $t = 0$ to correspond to a maximum displacement.

†14–18 A microwave transmitter is directed normally at a plane metallic reflector. A small detector moving along the normal to the reflector travels 0.12 m from the 1st to the 9th successive minimum of intensity. Calculate the frequency of the microwave oscillator.

Assume c   [10 GHz]

14–19 A helical spring is stretched round the curved surface of a smooth cylinder whose axis is horizontal, and an electric vibrator of variable frequency can be used to send waves along the spring at a speed of 4.0 m s^{-1}. The stretched length of the spring is 1.0 m.

(a) How long would a crest take to return to its starting point?

(b) What frequency of the vibrator would enable a crest to be exactly superposed on its successor? (i.e. to result in constructive superposition).

(c) What other frequencies would give constructive super-position?

*(These frequencies correspond to the **allowed modes of oscillation**, and the situation discussed in this question corresponds closely to a simple model of electron behaviour in the wave mechanics atom.)*

*14–20 A sinusoidal wave travelling along a string has the equation

$$y_1 = (30 \text{ mm}) \sin 10\pi \, [4(t/\text{s}) - (x/\text{m})].$$

(a) Calculate the wavelength of the wave.
(b) Write down the equation of a wave which would give a standing wave when superposed on the original. In which direction is this wave travelling?
(c) What is the amplitude of the standing wave?
(d) What is the separation of adjacent string particles which satisfy the condition $y = 0$ *permanently*? Compare this answer with that to part (a).

[(d) 0.10 m]

*14–21 The vibration of a string satisfies the equation

$$y = (40 \text{ mm}) \times \cos(\pi x/\text{m}) \times \sin(100 \, \pi t/\text{s}).$$

(a) The location-dependent term does not involve t – what can be deduced from this?
(b) What is the node-node separation?
(c) Draw sketches of a 1 m length of the string for the instants $t = 0$ s, $t = 5.0$ ms and $t = 10$ ms.
(d) Write down the equations of the two waves whose superposition generated the given vibration, and say what you can about the waves.

[(b) 1.0 m]

*14–22 The vibration of a string wave satisfies the equation

$$y = (60 \text{ mm}) \times \cos(2 \, \pi x/\text{m}) \times \sin(50 \, \pi t/\text{s}).$$

Find the transverse displacement y and velocity $\dot{y}$ of the string particles at (a) $x = 0$ and (b) $x = 0.25$ m for the time instants (i) $t = 0$ and (ii) $t = 10$ ms (four answers).

Interference

14–23 Refer to the diagram.

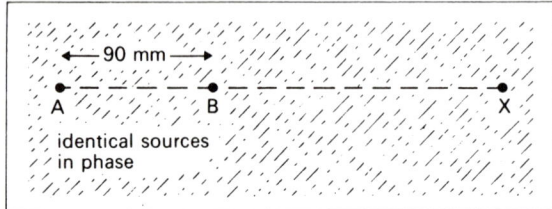

For Qu. **14–23** (plan view of ripple tank)

(a) Calculate *three* different wavelengths for which the point X would permanently show (i) minimum displacement (ii) maximum displacement.
(b) The water waves travel at 0.36 m s⁻¹. What is the *lowest* source frequency which produces destructive superposition?

[(b) 2.0 Hz]

14–24 A pair of microwave transmitters of frequency 10 GHz are placed 0.30 m apart. What, at some distance from the sources, will be the angular separation of the two nodal lines nearest the centre of the transmitted wave pattern?

Assume c   [0.10 rad]

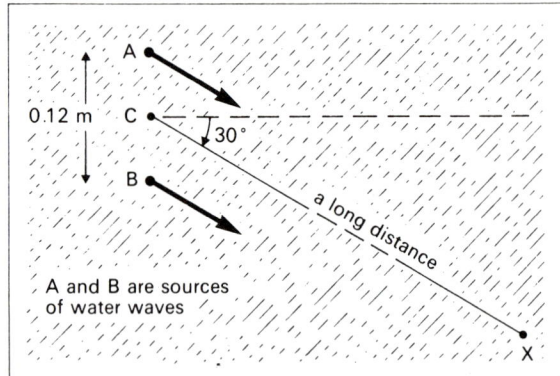

For Qu. **14–25** (plan view of ripple tank)

14–25 Refer to the diagram. The sources A and B have a frequency 20 Hz, and generate waves in phase together. It is found that X lies on the third antinodal line (the *order of interference* is 3). Calculate the speed of the waves.

[0.40 m s⁻¹]

Beats

14–26 *Calculation of beat frequency.* Two masses of 0.80 kg and 1.0 kg are hung from identical helical springs of force constant 4.0 π^2 N m⁻¹. Both are released simultaneously from a position of maximum stretch to describe s.h.m. Calculate

(a) the frequencies of the two systems
(b) the shortest time in which one system describes exactly one vibration more than the other (the **beat period**)
(c) the frequency at which the two systems show maximum stretch at the same instant (the **beat frequency**).

[(b) 8.5 s   (c) 0.12 Hz]

14–27 Two sinusoidal waves of amplitude a and frequencies 10.0 Hz and 11.0 Hz are passing along a string in the same direction at the same speed.

(a) Sketch the resultant disturbance y of a point on the string as a function of time over a period of two seconds. (Take the amplitude and y to be zero at $t = 0$.)
(b) What is the amplitude of the resultant when $t = \frac{1}{6}$ s, $\frac{1}{4}$ s and $\frac{1}{2}$ s?

[(b) a, 1.4a and 2a]

15 Doppler Effect

Questions for Discussion

15–1 *Moving observer.* A periodic source of water waves in a large tank generates waves which have a frequency f_s and which travel through the water at speed c. The waves are intercepted by a model boat which approaches the source at speed v_0. Calculate

(a) the wavelength of the waves as measured by an observer moving with the boat,

(b) the relative velocity of the wave crests and the boat, and

(c) the apparent frequency f_0 measured by the observer.

15–2 *Moving source.* Suppose that in the previous question the observer is at rest in the water, and that the source approaches him at speed v_s. Calculate

(a) the relative velocity of the wave crests and the source,

(b) the wavelength measured by the observer, and

(c) the apparent frequency f_0 measured by the observer.

15–3 A cowbell sounds on one side of a valley, and the sound is carried *downwind* to a stationary observer on the other side. Will the observer hear the same wavelength and frequency as that given out by the bell?

15–4 Suggest a method whereby one can demonstrate the *Doppler effect* for electromagnetic radiation in the laboratory using charged particles. Bear in mind that high source speeds are necessary.

15–5 One member of a binary star system emits visible light. Sketch a graph of the variation with time of the *Doppler* frequency shift observable at the Earth's surface.

15–6 O-M *Doppler broadening.* (See also Qu. **15–17**.) At what temperature would the D-lines from atomic sodium show a broadening of 4.0 pm?

*15–7 O-M What is the largest *Doppler* shift that could be produced in a wavelength of 550 nm by an astronomical body such as a quasar? To which part of the spectrum would the line be shifted? Can non-relativistic expressions be used in this instance?

Quantitative Problems

Mechanical Waves

15–8 A source of periodic sound waves approaches an observer at rest relative to the air at a speed of 100 m s^{-1}. The observer measures an apparent wavelength of 1.00 m, and the speed of the waves relative to him is 350 m s^{-1}.

(a) Calculate the apparent frequency (the number of wave crests that pass the observer in each second).

(b) How many wave crests separate source and observer when they are 500 m apart?

(c) How many 1.00 s later?

(d) Calculate the true frequency of the source by noting the rate of decrease in the number of wave crests in the space that separates source and observer.

[(c) 400 crests   (d) 250 Hz]

15–9 *The importance of a material medium.* (a) An observer is at rest relative to the air, and listens to the note emitted by a source of frequency 500 Hz while the source approaches him at 40 m s^{-1}. The sound travels through still air at 340 m s^{-1}. What effective wavelength and frequency would be measured by the observer?

(b) Suppose now that the source is stationary relative to the air, and that it is approached by the observer at 40 m s^{-1}. What wavelength and frequency would he measure?

Do your answers to (a) and (b) differ? (*If your answer is yes, that would indicate that the speeds of source and observer relative to the medium (not just to each other) have significance for sound waves.*)

[(a) 0.600 m, 567 Hz]

15–10 *Moving medium.* A fixed hooter sounds a note of frequency 200 Hz. A motorist is driving due West towards the hooter at a speed of 20.0 m s^{-1}, and there is a wind blowing from the North West at 28.3 m s^{-1}. Calculate

(a) the wavelength of the waves as they appear to the motorist,

(b) the resolved part of the relative velocity of the wave compressions and the motorist along a direction due West, and

(c) the apparent frequency f_0 measured by the observer.

The speed of sound in still air is 330 m s^{-1}.

[(c) 211 Hz]

15–11 A train moves through a station blowing its hooter, and to an observer standing on the platform the frequency varies over the range 280 Hz to 240 Hz. Draw a sketch graph showing the form of the variation of frequency with time, and calculate the speed of the train. The speed of sound is 330 m s^{-1}.

How would your sketch graph be modified if the observer were standing some distance away from the railway track?

[25 m s^{-1}]

15–12 *The moving reflector.* Sound waves of frequency 1.00 kHz travel at 330 m s^{-1} through the still air towards a reflector which is approaching the stationary source at 20 m s^{-1}.

(a) With what frequency do the wave compressions strike the reflector?

(b) This reflector now acts as a moving source. What is the

frequency of the echo heard by an observer stationed at the original source?

(*This method can be used to produce beats with light waves.*)

[(b) 1.13 kHz]

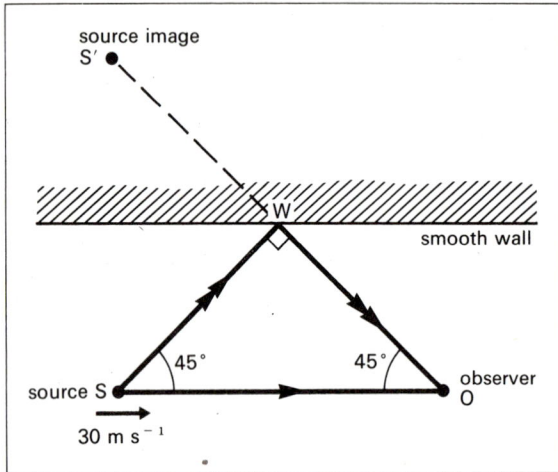

For Qu. **15–13**

15–13 Refer to the diagram. The moving car S approaches O emitting sound waves of frequency 500 Hz. Their speed through still air is 330 m s^{-1}. Calculate the beat frequency between waves travelling to O along the paths SO (direct) and SWO (after reflection). Indicate qualitatively how the observed frequency of beating varies with time.

[1(6) Hz]

Electromagnetic Waves

*15–14 *The use of radar waves to measure speed.* Microwaves of wavelength 100 mm are transmitted from a source so as to strike an aeroplane which approaches the source, and the reflected wave is found to give a beat frequency of 6.00 kHz when superposed on the transmitted wave. Use simple (first order *Doppler* effect) methods to calculate the approach speed of the aeroplane. (*For a hint refer to* Qu. **15–12**.)

Assume c   [300 m s^{-1}]

15–15 Because of the Sun's rotation, light waves received from opposite ends of a diameter show equal but opposite *Doppler* shifts. If the relative speed of source and observer is 2.0 km s^{-1}, what wavelength shift would be expected in the hydrogen F-line ($\lambda = 486.1$ nm)? (*In practice observations of such shifts allow the speeds of rotating bodies to be measured.*)

Assume c   [3.2 pm]

15–16 *The red shift.* A hydrogen source which is stationary relative to the observer emits a red light (the hydrogen C-line) of wavelength 656.3 nm. Observations on a star give the wavelength 660.0 nm when the same line is analysed. Calculate a value for the resolved part of the star's relative velocity along the line of sight, using a simple (non-relativistic) expression.

Assume c   [1.7 Mm s^{-1}]

*15–17 *Broadening of spectral lines.* (a) Use the expression $\frac{1}{2}mv^2 = \frac{3}{2}kT$ to find a value for the root-mean-square speed of hydrogen atoms at 500 K.

(b) Calculate the broadening to be expected in the hydrogen C-line ($\lambda = 656.3$ nm) at this temperature.

Assume c, m_p, k [(a) 3.52 km s^{-1} (b) 7.7 pm]

Shock Waves

15–18 A bullet travels at 1.02 km s^{-1} through air in which the speed of sound is 340 m s^{-1}. Draw the *Mach cone* which results, and calculate the semi-angle of the cone.

[19°]

15–19 In shallow water of depth d waves of all wavelengths travel at a speed $\sqrt{gd}$. With what speed must a boat travel in water of depth 2 m if the semi-vertical angle of its wake is to be $\pi/6$ rad? Take $g = 10$ m s^{-2}. (*Do not distinguish between group and phase velocities.*)

[9 m s^{-1}]

15–20 *Čerenkov radiation.* A charged particle moves through a medium in which light travels at 0.22 Gm s^{-1}, and emits radiation such that the *Mach* cone has a semi-vertical angle of 52°. What is the speed of the particle?

[0.28 Gm s^{-1}]

IV Structure and Mechanical Properties of Matter

16 Structure of Matter

Questions for Discussion

16–1 For most simple purposes we visualize the volume occupied by an atom to consist mainly of empty space. Solids are made up of atoms and yet often behave as impenetrable bodies. Discuss this apparent contradiction.

16–2 *Estimating the mass of an air molecule.* Make the following assumptions: the average speed of a gas molecule must be greater than the speed of sound (say 0.5 km s^{-1}), the mass of a smoke particle is $\sim 10^{-14}$ kg, and when it has the same k.e. as an air molecule its mean Brownian movement speed is about 1 mm s^{-1}.

(*a*) Calculate an approximate value for the mass of an air molecule.
(*b*) How does this compare with 30 m_u ($\sim 5 \times 10^{-26}$ kg)?
(*c*) Can you suggest a more precise macroscopic method of estimating the average speed of the gas molecule?

16–3 What is meant by the *ionization potential* of an ion, atom or molecule?

16–4 What simple demonstration indicates that atoms must repel one another at small separation?

16–5 Many plastic objects can be stretched to several times their original length without breaking. Suggest a possible molecular structure for plastic materials that could explain such behaviour.

16–6 Sketch graphs to illustrate the variation in interaction p.e. as a function of the separation of their centres for

(*a*) two ions of unlike charge,
(*b*) two ions of like charge, and
(*c*) two neutral atoms.

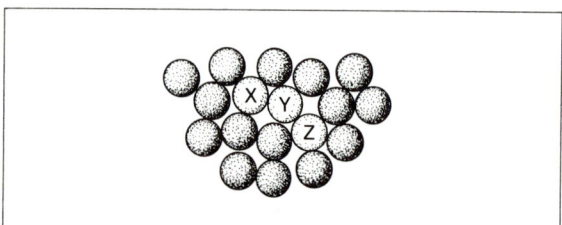

For Qu. **16–7**

16–7 The diagram shows a group of atoms in a liquid. (*a*) X and Y are in the equilibrium position and almost touching (nearest neighbours), and their interaction p.e. is $-\varepsilon$. (*b*) X and Z are next nearest neighbours and their p.e. is $-\varepsilon/30$.
What can you deduce from this information?

16–8 Can the molecules of solids or liquids be said to *collide*? If so, how often?

16–9 In what ways is the molecular structure of a liquid similar to that of (*a*) a solid, or (*b*) a gas?

16–10 Why are two *oppositely* charged ions held apart in a crystalline structure?

16–11 Why do atoms in crystal and molecular structures have a spacing which is approximately uniform?

16–12 Give examples of materials in which the bonds are mainly (*a*) ionic, (*b*) covalent, (*c*) metallic. To what extent do the properties of these materials depend on the nature of the bonds?

16–13 *Coordination number.* The coordination number n is the number of nearest neighbours which make effective contact with a given atom or molecule. What is the maximum value of n for a close-packed solid? The difference of coordination number between solid and liquid is quite small, whereas between liquid and gas it is large. What can be deduced from this regarding specific latent heats of transformation?

16–14 Discuss reasons why real materials have lower breaking strains than those predicted by the theory of ideal crystals.

16–15 Explain why the specific latent heat of fusion of a solid is usually much smaller than the specific latent heat of vaporization of a liquid.

16–16 Discuss the measurements you would make in order to obtain an approximate value for molecular diameters by the following methods:

(*a*) oleic acid drop on water
(*b*) cleavage of mica
(*c*) thickness of soap film.

16–17 O-M How many iron nuclei are there in a human body of mass 80 kg?

16–18 O-M Discuss the relative magnitudes of the main types of intermolecular forces.

16–19 O-M Spherical particles of polyvinyl chloride may be used to illustrate *Brownian motion*. It can be shown that the translational k.e. of such a particle is given by $\frac{1}{2}m\overline{v^2} = \frac{3}{2}kT$, where m is the mass of the particle, $\overline{v^2}$ the mean square speed, k the Boltzmann constant and T the temperature.
Estimate the r.m.s. speed of a polyvinyl chloride particle used in the *Brownian motion* experiment.

Quantitative Problems

†16–20   0.06 mm³ of stearic acid formed a circular film on water of radius 0.1 m. What *valid* conclusions can be drawn?

†16–21   To dissociate 1.0 mol of molecular hydrogen we must provide 0.45 MJ of energy. Express the dissociation energy in aJ per molecule.

Assume N_A   [0.75 aJ per molecule]

†16–22   When 1.0 mg of polonium decays completely, 3.0×10^{18} particles are emitted. If one particle per atom is emitted, what is the mass of a polonium atom? What is the volume occupied by one polonium atom if its density is 1.0×10^4 kg m⁻³?

†16–23   The relative atomic mass of copper is 63.5, and its density is 8.9×10^3 kg m⁻³. Calculate the ionic separation in solid copper.

Assume N_A   [0.23 nm]

†16–24   The constant b in the *van der Waals* equation can be written $b = \frac{2}{3}\pi N_A r_0^3$, where r_0 is the '*diameter*' of a molecule. Calculate r_0 for helium, for which $b = 2.3 \times 10^{-5}$ m³ mol⁻¹.

Assume N_A   [0.26 nm]

16–25   *Estimation of ε from surface tension.* Surface tension is closely related to surface energy per unit area, and the two can be considered equal at low temperatures. The surface tension can be written $\gamma = \frac{1}{4} n\varepsilon N/A$, where n is the coordination number (Qu. **16–13**) and N the number of molecules in surface area A. Estimate the **molecular interaction** p.e. ε for carbon tetrachloride given that $\gamma = 26$ mN m⁻¹, $N/A = 3.5 \times 10^{18}$ m⁻², and that for a liquid n is ~ 10.

[3×10^{-21} J per molecule]

16–26   *Estimation of ε from specific latent heats.* It is a useful approximation to say that the binding energy of a substance has the same order of magnitude as its latent heat of vaporization. At low temperatures, the binding energy L is equal to $(\frac{1}{2}nN\varepsilon)$, where n is the coordination number (the number of nearest neighbours), N is the number of molecules and ε is the size of the molecular interaction energy.

Estimate the value of ε for carbon tetrachloride given that its specific latent heat of vaporization is 0.210 MJ kg⁻¹, its molar mass is 0.153 kg mol⁻¹, and n is ~ 10.

Assume N_A   [1.1×10^{-20} J per molecule]

16–27   The specific latent heat of sublimation, which is closely related to the heat required to separate each of the molecules of 1 kg of a solid substance from all its neighbours, gives a useful measure of the strength of intermolecular forces.

The binding energy between one water molecule and one neighbour in the solid phase is 4.0×10^{-20} J, and the molar mass of water is 18×10^{-3} kg mol⁻¹. Calculate the specific latent heat of sublimation of ice assuming that each ice molecule has 4 nearest neighbours. What other assumption do you make?

Assume N_A   [2.7 MJ kg⁻¹]

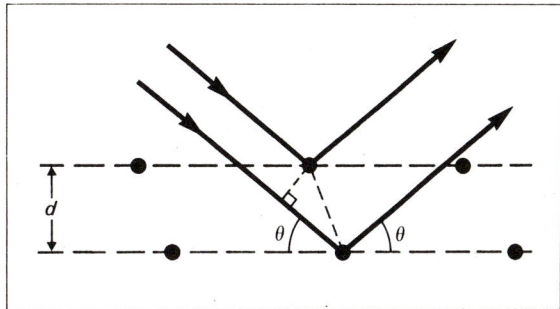

For Qu. **16–28**

*16–28   *Estimation of atomic diameter from X-ray diffraction.* Refer to the diagram, which shows the diffraction of an X-ray beam from a set of parallel planes of atoms. The *Bragg equation* giving the direction of a diffracted beam is $2d \sin \theta = m\lambda$, where λ is the X-ray wavelength. (See Qu. **63–6**.) For a particular reflection from an argon crystal, $m = 1$, $\theta_1 = 16.52°$, and $\lambda = 0.154$ nm. Calculate the spacing d of the planes of atoms. Given that d is related to the 'diameter' a of an argon atom by $a = 2d/\sqrt{2}$, evaluate a.

[$d = 0.27$ nm, $a = 0.38$ nm]

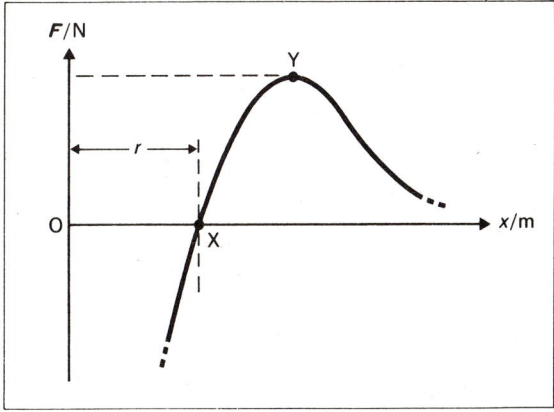

For Qu. **16–29**

*16–29   Refer to the diagram. The graph shows a force-displacement curve for calculating the theoretical brittle strength of a crystal. At Y the force is great enough to pull the crystal apart. If x is the particle separation, the potential energy E_p of an ion in an ionic crystal is given by

$$E_p = - \text{constant} \left[1/x - r_0^8/9x^9 \right].$$

Find, in terms of r_0, the value of x at which the crystal will disintegrate. What will have been the percentage volume increase when this happens?

[$1.23 \, r_0$, 85%]

*16–30   The interaction energy E_p of a pair of molecules is given by

$$E_p = -A/x^6 + B/x^{12}$$

where A and B are constants, and x their separation. Calculate

(a) their equilibrium separation distance r_0, and
(b) the value of E_p at this separation.

Draw a rough sketch of E_p as a function of x, and use your sketch to explain why it is that a solid expands on heating.

$$[(a)\ (2B/A)^{\frac{1}{6}} \qquad (b)\ -(A^2/4B)]$$

*16–31   The force exerted on a molecule of small mass M by a more massive neighbouring molecule is given by $F = A/x^m - B/x^n$, where A, B, m and n are constants. Given that equilibrium occurs when $F = 0$ at a separation $x = x_0$, show that for small displacements about x_0 the less massive molecule undergoes s.h.m., and derive an expression for the frequency.

$$\left[f = \frac{1}{2\pi} \left[\frac{A}{M} \left(\frac{n-m}{x_0^{m+1}} \right) \right]^{\frac{1}{2}} \right]$$

*16–32   The force between the two atoms of a hydrogen molecule is given approximately by $F = A/x^3 - B/x^{10}$.

If $x_0 = 74$ pm, and the dissociation energy per molecule is 0.70 aJ, then calculate the vibrational frequency of the two atoms about their common centre of mass. To what part of the electromagnetic spectrum does this frequency correspond? (Note that if you use the result of Qu. 16–31, the effective value of M is $\frac{1}{2}m_p$, since both atoms vibrate. See reduced mass – Qu. 10–27.)

$$[2.6 \times 10^{14}\ \text{Hz}]$$

17   Elasticity

Questions for Discussion

17–1   What is the difference between conventional stress and true stress? Discuss how a conventional stress-strain graph would differ from a true stress-strain graph for a particular sample under test.

17–2   Sketch a graph to show the variation of intermolecular force with intermolecular distance. To which region of the graph does Hooke's Law apply?

17–3   Explain why any body of constant mass subjected to a resultant force which obeys Hooke's Law will execute simple harmonic motion.

17–4   (a) Use the ideas of intermolecular forces to explain why, when a wire is put into a given state of tension, the extension is inversely proportional to the area of cross-section.
(b) Why, similarly, is the extension proportional to the length of the wire?

17–5   Describe qualitatively how you would expect the Young modulus to vary with temperature. Suggest the simplest possible equation showing this relationship, and hence define a temperature coefficient of the Young modulus.
In what way are rubber-like solids different from practically all other solids? Predict what would happen to the volume of a balloon if the surrounding temperature altered.

17–6   How does the Young modulus for polyethylene compare with that for diamond? Explain your answer in terms of the structure of each.

17–7   What would you find in a book of tables if you looked up (a) the bulk modulus of an ideal gas, or (b) the shear modulus of water?

17–8   It is required to punch a hole through a sheet of metal. How does the force required depend on (a) the diameter of the hole, and (b) the thickness of the plate?

17–9   What is meant by saying that a rod has two elastic limits? Sketch a graph to show how the change in length varies with the force applied. Indicate the region in which Hooke's Law is obeyed.

17–10   Refer to the diagram opposite, which is a mechanical hysteresis loop showing the behaviour of a solid when it is stretched and compressed beyond its elastic limits. Describe what is happening throughout the process, starting at point P and finishing at Q having passed through T.
(a) What is the name given to PR?
(b) What is the name given to PS?
(c) What does the area of the loop represent?
(d) How can a mechanical hysteresis loop be used to describe the frictional energy dissipation from rolling bodies?

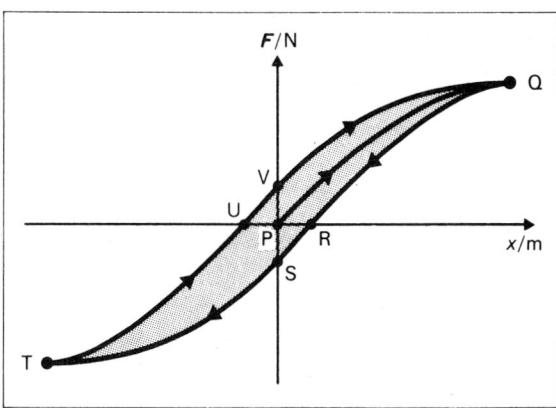

For Qu. **17–10**

17–11 Brick, stone, cast iron and glass are all classified as *brittle* materials. Why is it that many buildings are made of these materials?

17–12 Explain why girders have the familiar **H**-shaped cross-section, and why bicycle frames are made from *hollow* tubes.

17–13 Why do brittle materials fail long before the theoretical breaking stress is reached?

17–14 Explain what is meant by *fatigue* fracture. Discuss some situations in which this type of fracture is likely to occur.

17–15 The *Poisson ratio* μ is 0.29 for steel and 0.48 for rubber. (The value for rubber falls for very large loads.) What can be deduced from these figures? What could you say about a hypothetical new substance which had μ greater than 0.50?

17–16 Explain what is meant by the two *principal* bulk moduli of a gas which can be considered as nearly ideal, and why *Hooke's Law* is not obeyed except for very small changes.

17–17 The bulk modulus of a gas which can be considered as nearly ideal is greater under adiabatic conditions than under isothermal. Other things being equal, under which conditions would a fixed pressure change produce the larger volume change?

17–18 What is the nature of the distortion of the coil of a helical spring when the spring is stretched or compressed? By which modulus of elasticity are the properties of the spring determined?

17–19 The force constant k of a closely-coiled helical spring is given by

$$k = Gr^4/4nR^3,$$

where r is the wire radius, n the number of turns, and R the radius of these turns. By what factor would k change if (separately)

    (*a*) the length of the spring were doubled
    (*b*) the wire radius were doubled
    (*c*) the turn radius were halved
    (*d*) a glass helix ($G = 25$ GN m^{-2}) were substituted for a steel one ($G = 81$ GN m^{-2})?

Quantitative Problems

The Young Modulus

†**17–20** Calculate the stress in a human hair of 50 μm diameter which is supporting a weight of 0.20 N. How does this compare with the breaking stress of a steel wire?

[0.10 GN m^{-2}]

†**17–21** Calculate the stress required to increase the length of a wire by 0.1% assuming that the *Young* modulus for the wire is 0.12 TN m^{-2}.

If the cross-sectional area of the wire is 2.0 mm^2, calculate the tension required to produce this extension.

[0.12 GN m^{-2}, 0.24 kN]

†**17–22** A rod of original length 1.2 m and area of cross-section 1.5×10^{-4} m^2 is extended by 3.0 mm when the stretching tension is 6.0 N. Calculate the energy density of the stretched rod (the elastic p.e. per unit volume).

[50 J m^{-3}]

†**17–23** Calculate the tension required to extend a vertical steel wire of cross-sectional area 0.40 mm^2 by the same amount as would result from a temperature rise of 20 K. For steel $\alpha = 1.2 \times 10^{-5}$ K^{-1}, $E = 0.20$ TN m^{-2}.

[19.2 N]

17–24 A body of weight 15 N produces an extension of 0.40 mm when attached to the end of a vertical rigidly supported uniform wire. Calculate

    (*a*) the elastic energy stored in the wire, and
    (*b*) the change in gravitational p.e. of the load.

Comment on your two answers.

[(*a*) 3.0 mJ   (*b*) −6.0 mJ]

17–25 Refer to the diagram on the next page, which shows a rod, with a small sphere at each end, rotating about a vertical axis through its centre. Calculate the constant angular speed at which the rod should be rotated in order to be stretched by 1.0 mm. Each sphere has mass 4.0 kg and the rod, whose mass can be neglected, has cross-sectional area 3.0 mm² and *Young* modulus 0.20 TN m⁻². Ignore the bending of the rod.

[35 rad s⁻¹]

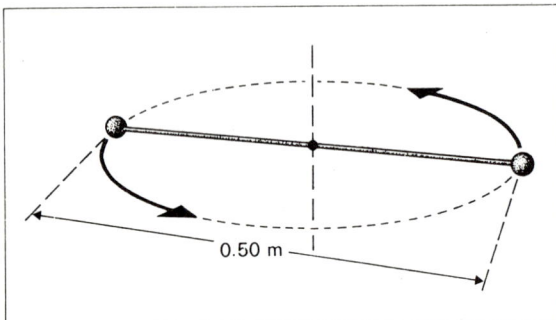

For Qu. **17–25**

17–26   A light steel wire 1.5 m long and of cross-sectional area 2.5 mm² is supported horizontally between two points 1.5 m apart. A load attached to its mid-point causes a vertical displacement of 60 mm. Calculate

(a) the extension of the wire,
(b) the tension in the wire, and
(c) the size of the load (in newtons).

The *Young modulus* for steel is 0.20 TN m⁻².

[(a) 4.8 mm   (b) 1.6 kN   (c) 0.26 kN]

17–27   The strain in a rubber ring on a rim of a wheel of radius 0.40 m is 3.0 × 10⁻³ when the wheel is stationary. The normal push of the rim on the ring just becomes zero when the wheel is rotating at angular speed ω. Calculate the value of ω if the *Young modulus* for the rubber is 0.50 GN m⁻², and its density is 9.4 × 10² kg m⁻³.

[0.10 krad s⁻¹]

*17–28   A wire of unstretched length l_0 and area of cross-section *A* is made of material of density *ρ* and has *Young modulus E*. The wire is freely suspended at one end. Calculate

(a) the average strain, and
(b) the elastic p.e. stored in the wire as it stretches under its own weight.

[(a) $\rho g l_0/2E$ (b) $\rho^2 g^2 A l_0^3/6E$]

Bulk Modulus

17–29   At standard pressure the isothermal bulk modulus of any ideal gas is 0.10 MPa. What volume change would 1.0 m³ of air undergo if the pressure were to be reduced by 2.0 kPa? Is your answer exact or approximate?

[+ 2.0 × 10⁻² m³]

17–30   The **compressibility** of sea water (the reciprocal of its bulk modulus) is 4.3 × 10⁻¹⁰ m² N⁻¹.

(a) What pressure increase would cause its density to change by 0.10%?
(b) At what depth (approximately) would this occur in the ocean?

[(a) 2.3 MPa]

17–31   An aluminium sphere of bulk modulus 70 GN m⁻² has a volume 4.0 × 10⁻⁶ m³ at atmospheric pressure (0.10 MPa). Calculate the volume change if the sphere is placed

(a) in a vacuum
(b) in a liquid in which the pressure is 10 MPa (about 100 times atmospheric pressure).

[(a) + 5.7 × 10⁻¹² m³   (b) − 5.7 × 10⁻¹⁰ m³]

17–32   A sphere of radius 20 mm undergoes a volume change of + 5.0 × 10⁻¹¹ m³. What was its change of radius? (*Hint :* treat the volume change as that of an area $4\pi r^2$ moving through a distance δr.)

[+ 10 nm]

Shear Modulus

17–33   A rubber cube of side 80 mm is made of material of shear modulus 75 kN m⁻². It is placed on a rough flat horizontal surface, and a force applied to the upper surface parallel to a side has a horizontal resolved part of 12 N. By how much is the top of a horizontal end-face displaced relative to the bottom? Draw a free-body diagram for the cube.

[2.0 mm]

*17–34   A hollow metal tube is 2.0 m long, has internal and external radii 40 mm and 43 mm and is rigidly clamped at one end. Calculate the torque required at the other end to produce an angle of shear of π/6 rad if the shear modulus of the metal is 25 GN m⁻².

[8.8 kN m]

*17–35   A galvanometer suspension of length 0.10 m is to have a constant c (defined by $T = c\theta$) of 3.1 μN m rad⁻¹. The only metal available has a shear modulus 40 GN m⁻². What would be the radius of the cylindrical wire?

[47 μm]

*17–36   A wire of shear modulus 40 GN m⁻² is twisted through an angle π/2 rad. If the radius of the wire is 1.0 mm and the length is 0.80 m, calculate the work done on the wire.

[97 mJ]

General

†17–37   Two springs of force constants 5.0 N m⁻¹ and 10 N m⁻¹ are joined end to end. What is their effective force constant? (*Hint: consider the detailed analysis of similar problems like capacitors in series and resistors in parallel. Look for a common factor.*)

[3.3 N m⁻¹]

17–38   *Tensile strength.* The tensile strength or breaking stress of hard-drawn pianoforte wire of density 8.0×10^3 kg m^{-3} is 2.0 GN m^{-2}.

(*a*) What is the greatest tension to which such wire of cross-sectional area 0.20 mm^2 can be subjected?

(*b*) What is the greatest length of this wire that can be hung vertically? Do you think that the weight of a wire is relevant in the design of (say) a ski-lift?

Assume g_0 　　　　　　[(*a*) 0.40 kN　　(*b*) 26 km]

17–39   *Calculation of an atomic force constant.* A bar has *Young* modulus E, length l_0 and area of cross-section A, and its equilibrium molecular separation is r_0. When a tension F is applied to the bar it causes a macroscopic extension Δl.

(*a*) Calculate

(*i*) the number of chains of atoms in any cross-section

(*ii*) the number of atoms in a chain of length l_0

(*iii*) the microscopic extension per atom (say δx)

(*iv*) the tensile force per atom (say f)

(*b*) By writing $f = k\delta x$, show that $k = Er_0$.

(*c*) Calculate the value of k for a typical metal, using $r_0 \sim 0.1$ nm, and $E \sim 0.1$ TN m^{-2}.

[(*c*) 10 N m^{-1}]

17–40   The frequency of vibration of the particles in a solid is given by $f = \left(\dfrac{1}{2\pi}\right)\sqrt{\dfrac{2k}{m}}$, where k is the force constant for the interaction of a particle with a single neighbour and m is the mass of a particular particle. Calculate the frequency of vibration of the ions in a copper lattice at room temperature if their force constant is 23 N m^{-1}. M_r for copper is 63.5.

Assume m_p 　　　　　　　　　[3.3 THz]

18   Surface Tension

Questions for Discussion

18–1   Experiment shows that the free surface energy of water in bulk is the same as that of a film of water 2 nm thick. What conclusion can we draw about the molecular interaction that causes surface tension?

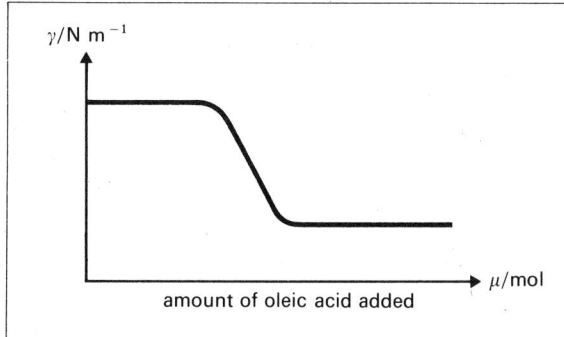

*For Qu. **18–2***

18–2   Refer to the diagram. The graph shows how the surface tension of water varies with the amount of oleic acid added to it. Suggest an explanation for its shape.

18–3   Why are soap films so much more stable than films of water? (*Hint: the main condition for stability seems to be that if the film is stretched, its surface tension at the stretched region should increase.*)

18–4   When the separation of two molecules is slightly increased the attractive force that each exerts on the other is increased. Yet when the surface area of a thick soap film is increased under isothermal conditions, the surface tension does not change. Why not?

18–5   The total free surface energy W of an area A of a liquid of free surface energy σ is given by

$$W = \sigma A.$$

Under what conditions is it true that the extra energy ΔW required to create a further area ΔA is given by

$$\Delta W = \sigma \Delta A?$$

Are the conditions that you specify observed by (say) a rubber membrane?

*18–6   At temperature T a liquid surface of area A and surface tension γ has a *total* surface energy W given by

$$W/A = \gamma - T(d\gamma/dT).$$

(*a*) How is it that W/A sometimes exceeds γ?

(*b*) Is this always so?

18–7 Why does aniline form spherical droplets in salt solution? Describe and explain what would happen to the aniline if a heater were placed underneath a beaker containing aniline and salt solution.

18–8 If water drops are formed on a wax surface, some will be nearly spherical and some will be much flatter. Why is this?

18–9 Describe in detail what happens when a soap bubble bursts.

18–10 What is the source of energy responsible for capillary rise?

18–11 How does the curvature of a liquid surface control the equilibrium vapour pressure above the surface? Explain your answer in terms of excess pressures.

18–12 A liquid is contained between two parallel vertical plates which are separated by a small distance d. The angle of contact between the liquid and the material of the plates is zero. Derive an expression for the liquid rise.

If the plates remained vertical, but were inclined so that they met at a small angle α, show that the line of contact of the liquid on the plates would be a rectangular hyperbola.

18–13 Make use of free-body diagrams to show why a greased needle is more likely to float in water than is a perfectly clean needle. Discuss the equilibrium conditions for each needle.

18–14 A small model boat can be propelled by placing a small block of camphor at the stern of the boat in contact with the water surface. Explain the mechanism.

18–15 Suggest why it is that soap bubbles can exist for long periods of time, especially if the surrounding air is very damp.

18–16 In what ways does surface tension play an important part in the working of cloud and bubble chambers? (*Hint: consider the excess pressure of a charged drop compared with that of an uncharged drop.*)

18–17 *Drop-weight measurement of* γ. In some situations this is a quick and useful method of measuring γ. Liquid drops grow at and fall away from the end of a tube. The surface tension can be calculated from $\gamma = G/3.8r$ where G is the weight of a drop and r the external radius of the tube.

How would you perform the measurement? Explain why this method is particularly useful for *comparison* of surface tensions.

18–18 The diagram shows part of a large liquid drop – drops shaped like this are called **sessile drops**. The surface tension of a liquid may sometimes be measured by the sessile-drop method. It can be shown that

$$\gamma = \tfrac{1}{2}\rho gh^2, \text{ and } (1 - \cos\theta) = \rho gH^2/2\gamma.$$

Indicate how you would measure γ by this method. Mention the precautions that you would take and errors that you might expect.

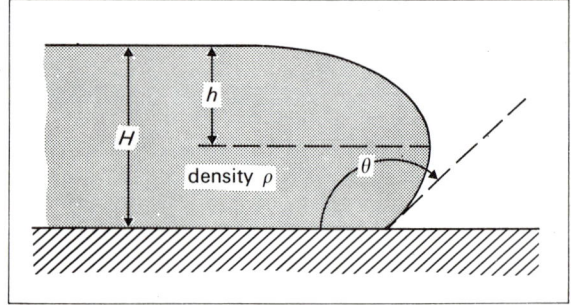

For Qu. **18–18**

18–19 A spherical body of large density is gradually lowered into a liquid, whose angle of contact with the material of the body is zero. Describe and explain, using sketch diagrams, how the surface tension forces on the sphere vary as it is lowered.

18–20 In which of the following situations is it preferable to have a large angle of contact, and in which to have a small angle of contact?

(*a*) Water on tent fabric.
(*b*) Solder on metal.
(*c*) Liquid paint on surfaces to be painted.
(*d*) Water on dirty clothes when being washed.
(*e*) Rain on a car.

*18–21 When a liquid of density ρ and surface tension γ rises to a height h in a capillary tube, the p.e. of the liquid can be written $E_p = \pi r^2 h^2 \rho g - 2\pi r h \gamma$, where r is the radius of the tube. Find an expression for h by considering the condition for E_p to be a minimum.

18–22 O-M Certain water-beetles are able to walk on water. Draw a clearly labelled diagram, showing the forces acting on the water-beetle, to explain this.

Make an estimate of the maximum weight that an insect could have to be supported in this way.

Quantitative Problems

Pressure

†18–23 A hemispherical soap bubble is formed at the end of a tube whose internal diameter is 12 mm. If the excess pressure in the bubble is 20 Pa, calculate the surface tension of the soap solution.

[30 mN m^{-1}]

18–24 Two spherical soap bubbles of the same material have radii 40 mm and 60 mm. They coalesce so that they share a common surface.

(a) Draw a diagram to illustrate the arrangement.
(b) What, in terms of γ, is the pressure difference across the common surface?
(c) What is the radius of curvature of the common surface? Assume the radii of the bubbles to remain unchanged.

[0.12 m]

18–25 A soap bubble is made of 8.0 mg of soap solution, and is filled with hydrogen of density 0.090 kg m^{-3}. It just floats in air of density 1.29 kg m^{-3}. What is the excess pressure inside the bubble? The surface tension of soap solution is 25 mN m^{-1}.

[8.6 Pa]

18–26 In the *Jaeger* method for measuring γ of a liquid, the lower end of a capillary tube of radius 0.20 mm is 25 mm below the surface of the liquid whose surface tension is required, and whose density is 8.0×10^2 kg m^{-3}. The excess pressure in a hemispherical bubble formed at the end of the tube is measured as 40 mm on a water manometer. Calculate γ for the liquid.

Assume g_0, ρ_{H_2O} [20 mN m^{-1}]

18–27 *Jamin's tubes.* If a narrow tube contains a number of liquid drops, it may be impossible to remove the liquid by blowing because of the large pressure difference which this establishes between the ends of the tube. Explain why.

Calculate the maximum pressure difference for a tube of radius 1.0 mm which contains 20 drops of liquid of surface tension 0.47 N m^{-1}. How does this compare with atmospheric pressure?

[38 kPa]

18–28 A thin film of water of thickness 80 μm is sandwiched between two glass plates, and forms a circular patch of radius 0.12 m. Take the surface tension of water to be 72 mN m^{-1}, and the angle of contact to be zero, and calculate the normal force needed to separate the plates.

[81 N]

Capillarity

†18–29 Two clean glass capillary tubes are partly submerged in the same liquid. The liquid rises to 80 mm and 50 mm respectively. What can be deduced from this information?

18–30 How high does water rise in a capillary tube of internal radius 0.30 mm if the surface tension of the water is 72 mN m^{-1}? Assume that the angle of contact is zero.

Describe what happens when the tube is lowered slowly into the water so that only 12 mm of the tube remain above water.

Assume g_0, ρ_{H_2O} [49 mm]

18–31 The internal diameters of the limbs of a clean glass U-tube are 1.0 mm and 10 mm respectively. Water, of surface tension 72 mN m^{-1} and zero angle of contact with the glass, is poured in. Calculate the difference in the water levels in the two limbs.

Assume g_0, ρ_{H_2O} [26 mm]

18–32 Draw a diagram to show the equilibrium position of a length of mercury poured into a U-tube formed from two lengths of capillary tubing of internal radii 0.10 mm and 0.40 mm. Find the difference between the two mercury levels if γ for mercury is 0.47 N m^{-1} and its angle of contact with the material of the tubing is 135°.

Assume g_0, ρ_{Hg} [37 mm]

Forces and Energies

†18–33 How much energy is expended in blowing a soap bubble of radius 10 mm from a soap solution for which $\gamma = 25$ mN m^{-1}? What becomes of this energy? Indicate any assumptions made in arriving at your answer.

18–34 A spherical drop of surface area 10 mm^2 and free surface energy 17 mJ m^{-2} is split, under isothermal conditions, into eight identical droplets. Calculate

(a) the surface areas of the individual droplets, and
(b) the free surface energy supplied when the drop was split up.

[(a) 2.5 mm^2 (b) 0.17 μJ]

18–35 *The liquid-drop nuclear model.* A nucleus of radius 6.0 fm undergoes fission into two identical nuclei (of equal volume). We imagine that we can treat the material of the nucleus as a liquid of free surface energy 1.0×10^{17} J m^{-2}. Calculate

(a) the total surface area of the new nuclei, and
(b) the free *surface* energy that must be added.

(*We take no account in our calculation of the binding energy of nucleons, or the energy resulting from Coulomb forces.*)

[(b) 12 pJ]

18–36 A circular ring of mean radius 50 mm is suspended from the arm of a balance so that it is horizontal and in the surface of a liquid of surface tension 30 mN m^{-1}. If the angle of contact between ring and liquid is 20°, calculate the *additional* force necessary to pull the ring clear of the liquid. Why is your answer necessarily approximate?

[18 mN]

*18–37 Calculate the smallest radius that a stable drop of water can have if it is free to evaporate but is thermally isolated. The specific latent heat of vaporization of water is 2.5 MJ kg^{-1} at room temperature and its free surface energy is 73 mJ m^{-2}. (*Thus a small enough drop has sufficient surface energy to enable it to evaporate spontaneously.*) Comment on your answer.

Assume ρ_{H_2O} [88 pm]

19 Viscosity

Questions for Discussion

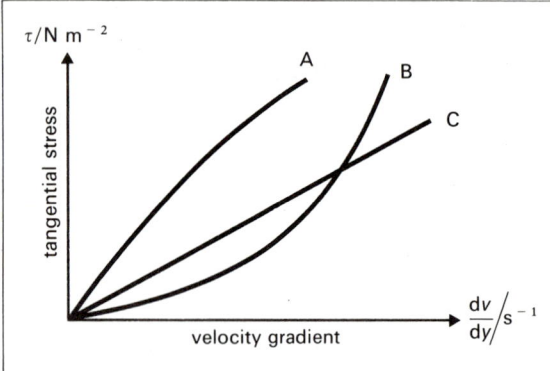

For Qu. **19–1**

19–1 Refer to the diagram, which illustrates the variation of tangential stress with velocity gradient for different types of liquid. Interpret the graphs and suggest which line would best describe the behaviour of (*a*) water (*b*) wet sand which appears to become dry when walked upon (*c*) quicksand (*d*) paint (*e*) a simple organic liquid (*f*) ink and (*g*) glue.

19–2 Why should lubricating liquids have small temperature coefficients of viscosity?

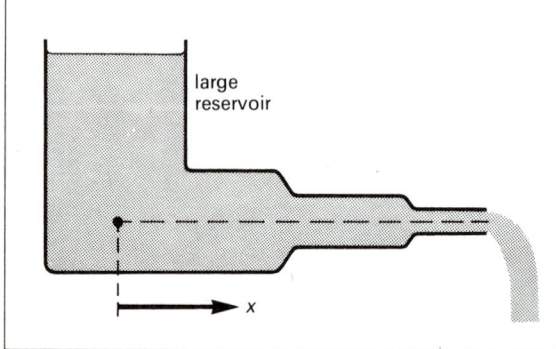

For Qu. **19–3**

19–3 Refer to the diagram. Sketch a graph which shows how the pressure along the dotted line varies with *x* when the liquid is (*a*) ideal, or non-viscous (*b*) viscous, such as water. Indicate on both your graphs regions in which there is a pressure gradient, and say what causes it.

19–4 Describe and account for the difference between laminar and turbulent flow as illustrated by the smoke rising from a cigarette in a still room. (*Hint: consider the size of the Reynolds number.*)

19–5 In the dimensional analysis of the flow of a viscous fluid through cylindrical pipes we have the following possible variables: liquid viscosity η and density ρ, pipe length l and radius r, and the pressure difference Δp between the ends of the pipe. Yet when we analyse

(*a*) the maximum speed of flow, we ignore ρ
(*b*) the rate of volume flow, we ignore ρ
(*c*) the critical speed, we ignore Δp and l.

Justify these procedures.

19–6 Why could *Millikan*, in his oil-drop experiment, not use *Stokes's Law* in the form $\boldsymbol{F} = 6\,\pi\eta r\boldsymbol{v}$?

19–7 Explain what is meant by a *boundary layer*, and describe how the nature of the boundary layer controls the drag on a body.

19–8 Viscosity is often referred to as a *transport phenomenon*. What physical property is being transported? Name three other transport phenomena and their corresponding properties.

19–9 Theory indicates that, except at very low and very high pressures, the coefficient of viscosity of a gas is independent of pressure. Yet when a feather is allowed to fall through a glass tube in which the pressure is progressively reduced, it falls at a greater speed for smaller pressures. Explain.

*19–10 *The viscosity of an ideal gas.* The kinetic theory of the ideal gas gives us the equation $\eta = \frac{1}{3}\rho\bar{c}\lambda$, in which the symbols have their usual meanings.

(*a*) Use this equation to predict how η is controlled by (*i*) pressure (*ii*) thermodynamic temperature.
(*b*) Why are the predictions of (*i*) likely to fail at *both* high and low pressures?
(*c*) Use order-of-magnitude values for η, ρ and $\bar{c}$ to calculate a value for λ, and compare your answer with 0.1 μm, that for air molecules at s.t.p.

*19–11 *The Ostwald viscometer.* Refer to the diagram. The viscosities of different liquids can be compared by measuring the time taken for the top surface of the liquid under test to fall from X to Y. Show that

(*a*) for any one liquid the average rate of volume flow between the points is proportional to ρ/η, and hence
(*b*) for two liquids used successively $\eta_1/\eta_2 = \rho_1 t_1/\rho_2 t_2$.

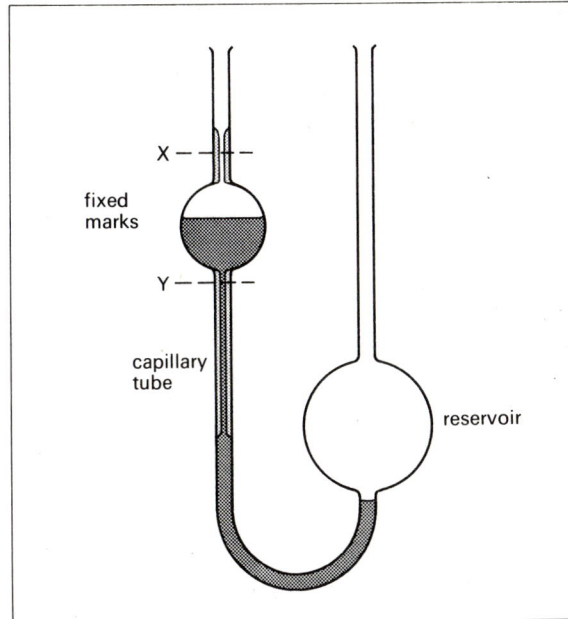

For Qu. **19–11**

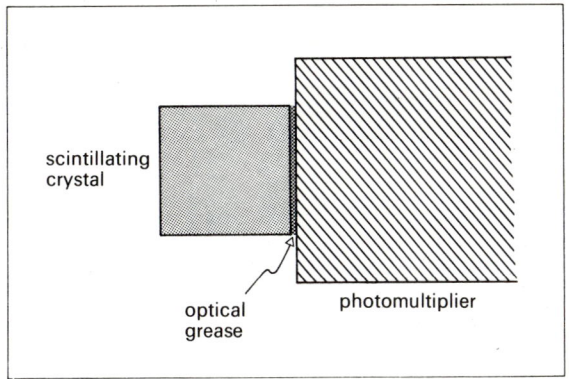

For Qu. **19–15**

19–12   O-M What is the terminal velocity of a spherical stone of radius 50 mm falling through water? (*Stokes's Law may not be applicable.*)

19–13   O-M Estimate the terminal velocity of a raindrop. State your assumptions explicitly.

Quantitative Problems

The Coefficient of Viscosity

*19–14   Water, whose coefficient of viscosity is 1.0 mPa s flows over a horizontal plate 5.0 m long and 3.0 m wide. The velocity of the water increases from zero at the plate surface to 0.80 m s^{-1} at a distance of 12 mm from the plate. Calculate the tangential force on the plate, making clear your assumptions.

[1.0 N]

*19–15   Refer to the diagram. The scintillating crystal has a weight of 1.5 N, and the area in contact with the photomultiplier face is 8.0 × 10^{-4} m^2. It is found that, after two weeks, the crystal has moved down 5.0 mm. Calculate the coefficient of viscosity of the grease if the layer is 2.0 μm thick. How does your answer compare with the coefficient of viscosity of golden syrup?

[0.91 MPa s]

*19–16   *The Reynolds number and the onset of turbulence.*
When liquid of density ρ and viscosity η moves with a speed of bulk flow v through a tube of diameter $(2r)$, the *Reynolds number* is given by $(Re) = 2vr\rho/\eta$.

(*a*) Calculate (Re) for water flowing at an average speed of 0.22 m s^{-1} through a tube of radius 5.0 mm. Its viscosity is 1.0 mPa s and density 1.0 × 10^3 kg m^{-3}. (*At this speed the flow is becoming turbulent.*)

(*b*) Use your answer from (*a*) to calculate the critical speed for air of viscosity 18 μPa s and density 1.3 kg m^{-3} flowing through the same pipe.

[(*a*) 2.2 × 10^3 (*b*) 3.0 m s^{-1}]

*19–17   (*a*) An aeroplane rudder moves through the air at 0.20 km s^{-1}. How far along the rudder will the *Reynolds number* reach 5.0 × 10^5? The density of air is 1.3 kg m^{-3} and the coefficient of viscosity is 20 μPa s.

(*b*) Calculate the corresponding distance for a submarine fin moving through water of coefficient of viscosity 1.0 mPa s at a speed of 20 m s^{-1}.

Assume ρ_{H_2O}

[(*a*) 38 mm   (*b*) 25 mm]

Dimensional Methods and the Poiseuille Equation

*19–18   Calculate the pressure required to push 6.0 × 10^2 mm^3 of grease per second into a grease nipple of a car if the diameter of the base of the nipple is 0.40 mm and the length of the nipple is 3.0 mm. The viscosity of the grease is 80 Pa s. How does your answer compare with atmospheric pressure? How would it compare (roughly) for a diameter of 4 mm?

[0.23 GPa]

*19–19   Water flows along a uniform horizontal tube which is 1.5 m long and has circular cross-section of radius 1.0 mm. If there is a pressure difference of 5.3 kPa maintained between its two ends, and the viscosity of water is 1.0 mPa s, calculate

(*a*) the rate of bulk flow, and

(*b*) the average speed of the water.

[(*a*) 1.4 × 10^{-6} m^3 s^{-1} (*b*) 0.45 m s^{-1}]

*19–20   A pressure difference Δp across the ends of a capillary tube causes a liquid to flow through the tube at a steady streamlined rate. The diameter of half the length of the tube is halved. What pressure difference across the ends of the tube will now be required to maintain the same rate of flow as before?

[$8.5\,\Delta p$]

*19–21   A circular disc of radius r rotates with angular velocity ω in a liquid of viscosity η. If it rotates coaxially at a distance d above a similar fixed disc, use the method of dimensions to find how the viscous torque on the disc depends upon η, ω and r. For your analysis you should assume that the torque is inversely proportional to d.

*19–22   A liquid of viscosity η and density ρ flows through a tube of circular bore and diameter d. Experiment shows that the k.e. content per second, P, of the emerging fluid is proportional to $(\Delta p/l)^3$, where $(\Delta p/l)$ is the pressure gradient. Use dimensional analysis to find a functional relation for P in terms of η, ρ, d and $(\Delta p/l)$.

*19–23   Refer to the diagram. The liquid of density ρ and coefficient of viscosity η leaves the container of area A by a capillary tube of radius r. Use the *Poiseuille equation* for the rate of flow of liquid along a capillary tube to show that the height h of the liquid after a time t is given by

$$h = h_0\, \exp\left(\frac{-\pi g\rho r^4}{8A\eta l}\,t\right)$$ where h_0 is the initial height of liquid.

Terminal Velocity

*19–24   A suspension is formed by shaking up spherical particles of radius 0.50 μm and density 4.0×10^3 kg m^{-3} in water. If the depth of water is 50 mm and the particles are distributed uniformly throughout the water, calculate the percentage of particles still in suspension one hour after the mixture has been left to stand. The viscosity of water is 1.0 mPa s.

Assume g_0, ρ_{H_2O}   [88%]

*19–25   An oil drop of density 9.0×10^2 kg m^{-3} reaches a terminal speed of 0.20 m s^{-1} when falling through a gas. If the gas viscosity is 15 μPa s, what is the radius of the drop?

Assume g_0   [39 μm]

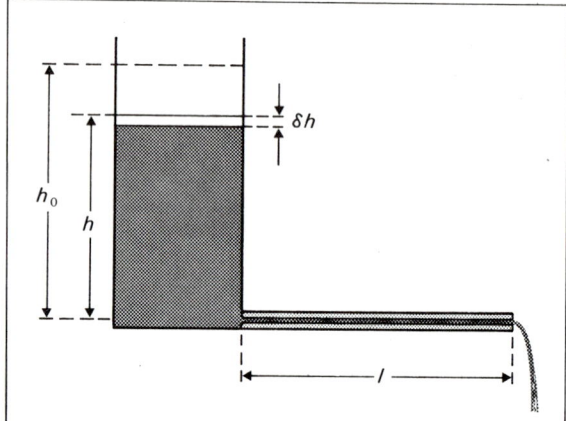

For Qu. **19–23**

20 Gravitation

Questions for Discussion

The Gravitational Field

20–1 How, in principle, would you set about calculating the mutual gravitational attraction of two cubic solids placed relatively close together?

20–2 A particle of mass m is placed midway between two fixed spheres each of mass M. Describe the equilibrium of m with respect to displacements along a line

(a) joining the two fixed bodies, and

(b) at right angles to the previous line, and through its mid-point.

20–3 Why, in problems on gravitation, is it often easier to consider potential energy or potential rather than field strength?

20–4 Sketch the gravitational field lines and equipotential surfaces corresponding to two equal spherical masses separated by a given distance. Attempt a similar sketch for the Earth/Moon system indicating where the resultant gravitational field strength is zero.

20–5 Sketch the variation of the gravitational field strength g as a function of distance r from the centre of a spherical mass

(a) of uniform density

(b) whose density increases considerably towards its centre (as does that of the Earth).

20–6 A solid homogeneous sphere of mass M is placed at the centre of a thin spherical shell of radius r, density ρ and thickness Δr. Calculate the net force exerted by

(a) the sphere on the shell

(b) the shell on the sphere.

Are your answers consistent with *Newton's Third Law*?

20–7 Sketch a graph to show the gravitational p.e. E_p of a body of mass m as a function of its distance r from the centre of a spherical shell of mass M and radius R. Take $E_p = 0$ at $r = \infty$ m. How would your graph have been different if the gravitational force had been repulsive?

20–8 What is the gravitational force exerted by a spherical shell on a particle of mass m inside it? What would be the gravitational force on the particle if the shell were itself in some other gravitational field of strength g? How does the gravitational potential vary inside the isolated shell?

*20–9 Attempt a quantitative statement, for gravitation, of *Gauss's Law*. (*Think carefully about a quantity analogous to ε_0, and remember that gravitational forces are attractive.*)

*20–10 How, in principle, and without using *Gauss's Law*, would you set about proving that a sphere attracts point masses outside itself as though all its mass were concentrated at its centre?

20–11 Suppose the value of G were suddenly to be altered by an order of magnitude. What effects would we observe?

20–12 Would it be possible to perform the *Cavendish* experiment for the measurement of G on another planet? Would the value differ from that obtained on the Earth?

20–13 What is the effect on the mutual gravitational force exerted by a pair of bodies of (a) the nature of the intervening medium (b) their orientation (c) their temperatures? Suggest how your answers can be verified by experiment.

Terrestrial Gravitation

20–14 Explain carefully what *you* understand by *weightlessness*. In what way is the term confusing? Give some examples of situations in which a person might experience weightlessness. If a person is weightless, is he also *momentumless*?

20–15 Why is the gravitational interaction more important than the electrostatic interaction in keeping the Earth in orbit round the Sun?

20–16 It is possible for a river to flow uphill – how?

20–17 Explain why a harbour on the coast usually experiences two tides a day. What is the cause of spring and neap tides?

20–18 If we drop a stone, it falls to the ground. We do not constrain air molecules – why do they not all do the same?

20–19 The Sun exerts a very large attractive gravitational force on the Earth.

(a) Why does the average Earth-Sun separation not decrease significantly with time?

(b) What effect does this force have on our measurement of weight by a spring balance?

20–20 The gravitational field of the Sun at the Earth is far greater than that of the Moon at the Earth. How is it then that tidal action on the Earth is caused primarily by the Moon?

*20–21 If a *Foucault pendulum* is oscillating freely, the plane of oscillation slowly rotates relative to a line on the ground. Explain why this happens and find an expression for the period T of rotation at a latitude of θ. Discuss the values of T at the poles and the equator.

20–22   Explain carefully why the gravitational field strength of the Earth increases on going down a mine.

20–23   Distinguish between *gravitational* mass and *inertial* mass. By considering the equation of motion for a particle in free-fall near the Earth's surface, derive an expression for the ratio of its two masses.

What behaviour would we observe in bodies of different mass if the two masses were not proportional?

*20–24   *Motion relative to the Earth.* The gravitational acceleration of a body in a state of free-fall measured by an observer rotating with the Earth differs from the acceleration as measured by a non-rotating observer for two reasons.

(1) The body is moving in a circular path and therefore has a centripetal acceleration. The effective (observed) gravitational acceleration deviates slightly from the radial direction and its magnitude will vary with latitude.

(2) The Earth is rotating underneath the falling body and, to the rotating observer, the body appears to have another acceleration. This is called the **Coriolis effect**.

(*a*) Explain why, in (1), a freely falling body is deviated to the South in the Northern hemisphere and to the North in the Southern hemisphere.

(*b*) Explain why, in (2), the deviation of a freely falling body is to the East in both the Northern and Southern hemispheres.

(*c*) Describe the effect, on the falling body, of (1) and (2) combined.

(*d*) Explain why a body moving in a horizontal plane is deviated by the *Coriolis effect* from its straight line path to the right in the Northern hemisphere and to the left in the Southern hemisphere.

(*e*) Suggest how the *Coriolis effect* will combine with a low pressure centre in the atmosphere to produce a whirling motion of the wind. How will the sense of rotation depend upon the hemisphere?

Satellites

20–25   If an astronaut leaves his orbiting capsule and is disconnected from it, will he be left behind? Is he capable of turning round or moving in a straight line?

20–26   Describe the path of a spaceship travelling from Earth to Mars.

20–27   Will a low-flying Earth satellite escape if its kinetic energy is doubled?

20–28   A body is taken to a height above the Earth's surface at which satellites can describe circular orbits at a speed 6.0 km s^{-1}. Describe qualitatively its subsequent motion if it is then given an initial horizontal velocity of (*a*) zero (*b*) 5.0 km s^{-1} (*c*) 6.0 km s^{-1} (*d*) 8.0 km s^{-1} (*e*) 6.0 $\sqrt{2}$ km s^{-1} (*f*) 12 km s^{-1}.

20–29   Consider a satellite which is in a circular orbit round a planet. What change in trajectory would be caused by firing its rocket for a short time interval so as to impart an impulse which is (*a*) forward (*b*) reverse (*c*) downward (*d*) upward?

*20–30   Does the total mechanical energy of a planet-satellite system vary with time if the satellite describes a highly eccentric elliptical orbit? What is the significance of such an energy having a negative value? Does the angular momentum of the satellite vary with time? Reconcile your answer with the fact that the satellite's angular speed is varying.

*20–31   A satellite of mass *m* describes a circular orbit of radius *r* about a planet of mass *M*. Write down, in terms of *G* and these variables only, expressions for (*a*) its k.e., (*b*) its gravitational p.e., and hence (*c*) its total energy.

Use your answer to (*c*) to explain how it is, that when such a satellite experiences a resistive force from the air molecules of the atmosphere, both its k.e. and its speed are *increased*.

Orders of Magnitude

20–32   O-M   What is the average gravitational attraction and gravitational p.e. between a neighbouring pair of

(*a*) air molecules at s.t.p.
(*b*) liquid-phase water molecules?

20–33   O-M   At what height from the Earth's surface could a satellite describe a circular orbit, and appear to be stationary when viewed from the Earth?

20–34   O-M   Use the values of *G*, g_0 and the Earth's radius to find a value for the Earth's mass, and hence estimate its mean density.

*20–35   O-M   Calculate the distance from the Moon at which the gravitational field strength due to the Earth-Moon system is zero. Use this to calculate the minimum launching speed of a rocket from the Moon's surface if the rocket is to reach the Earth.

*20–36   O-M   An Earth satellite of mass 2.0 kg is circling the Earth at a height of 40 km at the time of re-entry. Estimate the increase in internal energy of the satellite during re-entry, and its temperature rise.

20–37   O-M   Calculate the position of the centre of mass of the Sun and the Earth. What effect does the Earth have on the motion of the Sun?

20–38   O-M   The Moon subtends an angle $\pi/360$ rad at the Earth. Use this information, together with values for *G*, the Earth's mass and the period of the lunar month, to estimate

(*a*) the Earth-Moon distance, and hence
(*b*) the Moon's radius.

*20–39   O-M   What is the escape speed for a particle trying to leave (*a*) the Earth (*b*) the Moon (*c*) the Sun?

20-40 O-M *The Boys torsion balance.* *Boys* used a quartz fibre of length 0.43 m, radius 6.2 μm and shear modulus 30 GN m^{-2}. The transverse beam had a length 23 mm, and the suspended system a time period of 96 s. Estimate the following quantities:

(a) the torsional constant c of the fibre ($T = c\theta$)
(b) the moment of inertia of the suspended system
(c) the mass of the small suspended spheres (treat them as point masses).

Quantitative Problems

Newton's Law

†20-41 Two spherical masses of 1.0 kg and 3.0 kg are at rest on a table. The distance between their centres is 2.0 m. Calculate the gravitational force that each exerts on the other. If the table were frictionless, what would be the acceleration of each mass? Why do they exert zero electrical force?

Assume G   [50 pN]

†20-42 Calculate the maximum gravitational attraction between two lead spheres each of 0.20 m diameter. The density of lead is 11×10^3 kg m^{-3}.

What value of the coefficient of friction would give a maximum frictional force of this size for one of the spheres resting on a table top?

Assume G, g$_0$ [3.5 μN, 7.7×10^{-9}]

20-43 Calculate the size of the mutual gravitational attractive forces of the Sun and the Earth. The radius of the Earth's orbit is 0.15 Tm, and their masses are 2.0×10^{30} kg and 6.0×10^{24} kg respectively.

What equal charges would the Earth and the Sun have to carry to exert mutual electrostatic forces of the same size?

Assume G, 1/4πε$_0$ [3.6×10^{22} N, 3.0×10^{17} C]

20-44 Calculate the periodic time of a hypothetical simple pendulum which is held at a distance of 1.6 Mm from a fixed point on the surface of the Earth, if its periodic time at the Earth's surface is 2.0 s. The radius of the Earth is 6.4 Mm. Ignore the Earth's rotation.

[2.5 s]

20-45 *The Poynting measurement of G.* A small spherical body of mass 22 kg was balanced on a bullion balance, and a large spherical mass of 1.5×10^2 kg was swung on a special pivot to bring it vertically beneath the small sphere. Their centres were then 0.23 m apart. It was found that a downward force of 4.0 μN applied independently to the same pan caused the same angle of tilt as the gravitational force. What value does this experiment give for G? (*In practice many corrections are necessary to obtain an accurate result.*)

[64 pN m^2 kg^{-2}]

The Gravitational Field

†20-46 A mass 5.0 kg is taken from a point X, at which the gravitational potential is 10 J kg^{-1}, to Y where it is 25 J kg^{-1} Calculate

(a) the potential difference between X and Y,
(b) the gain of gravitational p.e., and
(c) the work done.

[(c) 75 J]

20-47 Make the (false) assumption that the density of the Earth is uniform. By how much would the gravitational field strength change if one went 1.0 m

(a) away from the Earth
(b) into the Earth?

The Earth's radius is 6.4 Mm.

Assume g$_0$ [(a) $- 3.1$ μN kg^{-1} (b) $- 1.5$ μN kg^{-1}]

*20-48 *Tunnels through the Earth.* Imagine a hypothetical straight frictionless tunnel along a diameter of the Earth, and a particle of mass m in the tunnel a distance r from the Earth's centre.

(a) Assume (falsely) that the Earth has uniform density, and calculate
  (i) the size of the resultant gravitational force on the particle
  (ii) the time period of the s.h.m. that would result if it were released.

(b) What would the period become if the particle were released
  (i) along a diameter, but from the Earth's surface
  (ii) along a chord which subtends an angle θ at the Earth's centre?

(c) How do your answers compare to the period of a satellite in circular orbit very close to the Earth's surface?

Assume g$_0$, R$_E$ = 6.4 Mm   [(a) (i) + $mg_0(r/R_E)$ (ii) 5.1 ks]

Orbital Motion

20-49 In one simple model of the hydrogen atom the tangential speed of the electron in its circular orbit is 2.2 Mm s^{-1}

(a) Calculate the gravitational force of the proton on the electron if the radius of the orbit is 53 pm.
(b) How does this compare with the centripetal force on the electron revolving in its orbit?

Comment on the difference between your two answers.

Assume G, m$_e$, m$_p$ [(a) 3.6×10^{-47} N   (b) 4.3×10^{-40}]

†20-50 The two components of a binary star move in concentric circles with radii in the ratio k. Calculate the ratio of their masses.

[1/k]

20-51 A satellite is in orbit just above the surface of a spherical planet whose mean density is ρ. If the periodic time for each orbit is T, find an expression for ρT^2 and comment on its value.

[3π/G]

20–52 (a) Calculate the minimum period of rotation that the Earth would need to have about its axis if a body at the equator were to experience zero normal contact force with the ground. The radius of the Earth is 6.4 Mm. *Assume g_0*

(b) Tidal friction causes the length of each day (at present) to be 50 ns longer than the previous day. Assume that the rate of slowing down has not changed, and calculate the time that has elapsed since the rotation of the Earth had this time period. Give your answer in years.

How does your answer compare with the age of the Earth? Comment on the assumption made in (b).

[(a) 5.1 ks   (b) 4.(5) × 10^9 years]

Gravitational p.e. and Escape Speed

20–53 The gravitational p.e. of two point masses a distance r apart is $- Gm_1m_2/r$. For a hydrogen molecule, we have two atoms each of mass 1.7×10^{-27} kg separated by a distance of 74 pm. What is their mutual gravitational p.e.? In the light of your answer, comment on the fact that the experimental value for the dissociation energy of a molecule of hydrogen is 0.72 aJ.

Assume G   [-2.6×10^{-54} J]

*20–54 Would a planet of radius 0.30 Mm and mean density 4.0×10^3 kg m^{-3} be able to keep nitrogen in its atmosphere if the mean temperature were 7.0×10^2 K? The molar mass of nitrogen is 2.8×10^{-2} kg mol^{-1}.

Assume R, G

*20–55 (a) A meteorite originates from rest a distance r from the centre of the Earth. With what speed v would it strike the Earth's surface? (Neglect friction.)

(b) Calculate a numerical value for v for a meteorite originating with negligible k.e. a great distance from the Earth. The Earth's mass is 6.0×10^{24} kg and its radius 6.4 Mm. How does this compare with the escape speed?

Assume G   [(b) 11 km s^{-1}]

*20–56 The radii of Earth and Mars are 6.4 Mm and 3.4 Mm respectively. The mass of the Earth is 9.5 times the mass of Mars. Calculate

(a) the ratio of the mean density of Mars to that of Earth,

(b) the gravitational field strength on the surface of Mars, and

(c) the ratio of the escape speed from Mars to the escape speed from Earth.

Assume G, g_0   [(a) 0.70   (b) 3.7 N kg^{-1} (c) 0.45]

*20–57 *Gravitational p.e. of a solid sphere.* A sphere of mass M, radius R and uniform density ρ condenses from matter which was originally infinitely dispersed. Calculate

(a) (in terms of r) the gravitational potential at the surface of the condensed sphere when it has radius r,

(b) the decrease of gravitational p.e. when a further shell of thickness δr is added, and

(c) the total energy released by the formation of the sphere. What do you think would happen to this energy?

[(c) $3GM^2/5R$]

V Thermal Properties of Matter

21 Temperature

Questions for Discussion

21–1 Consider two systems called A and B. Without actually putting A into thermal contact with B, what experimental test could you carry out to discover whether they would exchange heat if they were brought into contact?

21–2 What is the optimum number of macroscopic parameters (or measurable variable physical quantities) for a system intended for use as a thermometer?

21–3 How would you test experimentally whether a particular physical property changes uniformly with changes in temperature? (*Think carefully about the theoretical implications.*)

21–4 Could the unit *joule per particle* be used for measuring temperature?

21–5 Is the thermodynamic temperature scale more fundamental than the ideal-gas temperature scale?

21–6 How is the *kelvin* defined? Is it true to say that the interval between the steam point and ice point of water is *exactly* 100 kelvins? Discuss.

21–7 Do the results obtained from a constant volume gas thermometer depend upon the gas used? If they do, what factors would you consider when selecting a gas?

21–8 Describe what is meant by the *International Practical Temperature Scale*, and discuss how the whole temperature range is covered by different thermometers.

21–9 Comment on the statement: '500 °C is five times the boiling point of water'.

21–10 Why is the *Celsius* temperature interval θ *not* defined by the relationship
$$\theta = T - T_{ice},$$
where T is the thermodynamic temperature?

21–11 It is thought that the size of a particular thermometric property X varies with the *Celsius* thermodynamic temperature interval θ according to the parabolic relationship
$$X = X_0 (1 + \alpha\theta + \beta\theta^2).$$
What procedure is needed to calibrate a thermometer based on the thermal variations of X?

21–12 How would you measure the temperatures of the following? (*a*) Liquid oxygen (*b*) the photosphere of the Sun (*c*) liquid in a method of mixtures experiment (*d*) the air in a room (*e*) liquid in a constant-flow calorimeter (*f*) a body whose temperature is fluctuating rapidly.

21–13 *A magnetic thermometer.* For a paramagnetic salt which obeys *Curie's Law*, the magnetic susceptibility χ_m is related to the thermodynamic temperature T by
$$\chi_m = constant/T.$$
(*a*) In what temperature range would you expect χ_m to be most useful as a thermometric property?
(*b*) In practice the law is not obeyed exactly. Is there any validity in using a **magnetic temperature** T^*, defined by
$$T^* = constant/\chi_m?$$

21–14 'The temperature of a body increases with the mean k.e. of its molecules. Therefore a bullet moving past an observer is hotter than an identical bullet which is at rest relative to the same observer.' Comment.

21–15 What physical property determines the temperature (if there is one) of the following? (*a*) A volume of empty space, (*b*) the same volume occupied by one molecule, and (*c*) the volume occupied by a mole of ideal-gas molecules.

*21–16 Say in your own words why it is not possible to attain a temperature of 0 K. How close have experimenters come to this temperature? What can you say about the k.e. of molecules at absolute zero?

Quantitative Problems

Temperature Scales

21–17 *Corrections to mercury thermometer.* A mercury-in-glass thermometer reads 1.20 °C at the ice point (273 K) and 98.3 °C at the steam point (373 K) under standard pressure.

(*a*) Draw a graph to show the variation of the necessary correction with temperature. Mention any assumption that you need to make in drawing the graph.
(*b*) Use the graph to find the correct temperatures when the thermometer reads − 12.0 °C, 28.0 °C and 64.0 °C.
(*c*) At what temperature is the necessary correction equal to zero?

[(*b*) − 13.6 °C, 27.6 °C, 64.7 °C   (*c*) 41.(5) °C]

21–18 The numerical values r and s of two thermometric properties R and S are observed at a number of fixed points. These numbers are related for all temperatures by the equation $s = x + ry$, where x and y are constants. Show that the *Celsius* scales based on the properties R and S are identical, even if R and S do not vary linearly with thermodynamic temperature.

*21–19 Constant volume gas thermometer scale. The pressure and volume of the gas in the bulb of a constant volume gas thermometer are related by the equation $pV = aT + bpT$, where T is the ideal-gas temperature and $b = 4.0 \times 10^{-7}$ m^3 K^{-1}. If the volume of the bulb is 9.0×10^{-2} m^3, and the thermometer is calibrated at the ice and steam points (273 K and 373 K), what temperature will it record when the ideal-gas scale reads 350 K?

[349 K]

*21–20 Resistance thermometer scale. The resistance of the coil of a resistance thermometer at a temperature θ on the ideal-gas *Celsius* scale is given by $R_\theta = R_0 (1 + a\theta + b\theta^2)$, where R_0 is the resistance at $0\,°C$, and $b = -1.54 \times 10^{-4} a$ K^{-1}.

(a) Write down expressions for R_θ at the ice point (273 K) and the steam point (373 K), and
(b) calculate the temperature on the resistance thermometer scale when $\theta = 40.0\,°C$.

[40.4 °C]

Thermometers

†21–21 Constant volume gas thermometer. The low pressure of a dilute gas kept at constant volume is measured at the normal b.p. of sulphur, and at the triple point of water. The ratio of the pressures is 2.627. Calculate the normal b.p. of sulphur.

Assume T_{tr}

†21–22 Mercury thermometer. The length of the mercury column in a glass tube of uniform bore is 30 mm at the ice point (273 K) and 280 mm at the steam point (373 K). Calculate the temperature recorded by this thermometer when the length of the column is 180 mm.

[333 K]

21–23 The thermocouple thermometer. The e.m.f. of a thermocouple is given approximately by $\mathscr{E} = \alpha\Delta T + \beta(\Delta T)^2$, where ΔT is the thermodynamic temperature difference between the junctions, $\alpha = 6.93\ \mu$V K^{-1}, and $\beta = -2.10$ nV K^{-2}. The cold junction is kept at 273 K. Calculate

(a) the value of $\mathscr{E}$ when the hot junction is at 373 K
(b) the probable temperature of the hot junction when $\mathscr{E} = 3.25$ mV.

On what scale is the temperature value in (b) given?

[(a) 0.67(2) mV   (b) 839 K]

†21–24 When the cold junction of a thermocouple is at 273 K, and the hot at 373 K, its e.m.f. is balanced by 960 mm of a potentiometer wire. When the hot junction is placed in naphthalene at its m.p., balance is then achieved across 768 mm of wire. Calculate the m.p. of naphthalene on the scale of this thermocouple thermometer.

[353 K]

22   Heat, Work and Energy

Questions for Discussion

22–1 The original concept of a *mechanical equivalent of heat* is now recognized as being no longer necessary. What, if any, is the present-day significance of the pioneering experiments of men like *Rumford, Joule* and, more recently, *Callendar and Barnes*?

22–2 Describe briefly the energy changes which take place in the following situations:

(a) a bullet is fired from a gun and then hits a target
(b) an internal combustion engine
(c) a meteorite reaches the Earth's surface from outer space
(d) a moving truck is stopped by applying its brakes
(e) an immersion heater raises the temperature of a liquid
(f) a coin is dropped from a great height onto a hard surface.

22–3 Give two reasons for a sports car having wire wheels.

22–4 A room can be warmed by opening the door of an oven but it cannot be cooled by opening the door of a self-contained refrigerator. Discuss and explain.

22–5 What is the essential distinction between *heat* and *work*? What fundamental property separates them both from the concept of internal energy? In what way are all three quantities similar?

22–6 Attempt an exact definition of the *internal energy* of a system. What distinction is there between your definition and the *total energy* of the system?

22–7 If we leave out rest-mass energy, then in physics we recognize only two fundamental energy classifications – *kinetic* and *potential*. How do heat, work, internal energy and chemical energy fit (if at all) into this scheme?

22–8 Is it possible to increase the internal energy of a system bound by perfectly rigid adiabatic walls?

22–9 O-M Estimate a minimum satisfactory melting point for the heat-shield of a satellite whose re-entry speed is 8 km s^{-1}.

22–10 O-M If the Earth stopped spinning in such a way that all its rotational k.e. were converted into its own internal energy, what would be the mean temperature change of the Earth?

Quantitative Problems

22–11 *Rotating drum.* The tangential frictional force exerted by a band brake on a rotating metal drum of circumference 0.25 m is found to be 20 N. If the mass of the drum is 0.40 kg and its specific heat capacity is 0.35 kJ kg^{-1} K^{-1}, calculate the number of complete revolutions of the drum required to increase its temperature by 5.0 K.

[1.4×10^2]

22–12 A bullet of mass 30 g travelling at 0.20 km s^{-1} becomes embedded in a fixed target.

(*a*) Describe the energy changes which take place.
(*b*) What is the increase in internal energy of the target and the bullet?

(*c*) Calculate the temperature increase of the bullet if it absorbs 75% of this internal energy. The specific heat capacity of the metal is 0.13 kJ kg^{-1} K^{-1}.

[(*b*) 0.60 kJ   (*c*) 1.2×10^2 K]

22–13 Water flows at 4.0 m^3 s^{-1} over a waterfall of height 50 m.

(*a*) Discuss the energy changes involved.
(*b*) Calculate the temperature difference between the top and the bottom of the fall if 70% of the p.e. is converted into internal energy.
(*c*) What maximum theoretical power is there available?

Assume ρ_{H_2O}, c_{H_2O}, g_0 [(*b*) 82 mK   (*c*) 2.0 MW]

22–14 A car of weight 15 kN is moving uniformly at 12 m s^{-1} down a 1 in 6 hill with the engine switched off. If the brake drums have mass 30 kg and specific heat capacity 0.40 kJ kg^{-1} K^{-1}, calculate their rate of increase of temperature. What assumptions have you made?

[2.5 K s^{-1}]

22–15 A drill, using a current of 2.0 A when connected to a 0.24 kV mains supply, makes a hole in a piece of iron of mass 0.80 kg. Calculate the temperature rise in 20 s if 60% of the electrical energy is converted to the iron's internal energy. Iron has a specific heat capacity of 0.46 kJ kg^{-1} K^{-1}. Discuss any assumptions you have made.

[16 K]

23   Expansion of Solids and Liquids

Questions for Discussion

23–1 Give two examples, one for a solid and one for a liquid, of substances which *contract* on being heated. Attempt to explain each in terms of the behaviour of microscopic particles.

23–2 What could cause a body to be distorted by a change of temperature?

23–3 Suggest a method for the *accurate* measurement of the *cubic* expansivity of a small quartz crystal.

23–4 If we heat a solid, and plot a graph of its density against temperature, we find a significant change in slope as we approach the melting point. Sketch such a graph, and suggest reasons for its shape.

23–5 *The harmonic oscillator.* Refer to the diagram opposite, in which the p.e. E_p of a particle in a hypothetical solid is plotted as a function of its distance r from a second particle.

(*a*) What does the shape of the graph tell you about the nature of the interatomic forces?
(*b*) How does the interatomic separation depend upon the amplitude of vibration? (*The simple pendulum is a helpful analogy.*)
(*c*) Predict the linear expansivity α of a solid whose particles are bound by purely harmonic forces.
(*d*) For such a solid what would be the relation between α and the specific heat capacity?

23–6 *The anharmonic oscillator.* Refer to Qu. **23–5**. In practice two atoms can be pulled apart more easily than they can be pushed together.

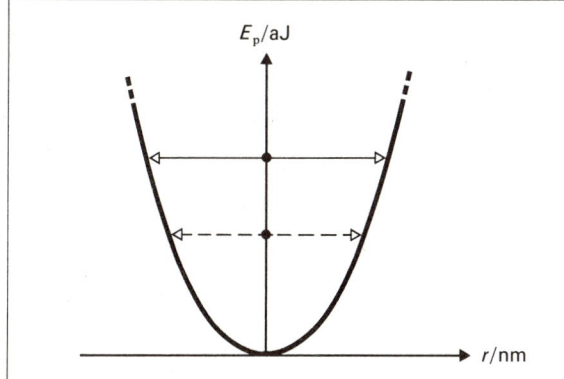

E_p/aJ

r/nm

For Qu. **23–5**

(a) Sketch the non-parabolic E_p–r curve which describes the anharmonic motion of real atoms.
(b) By considering the old and new limits of oscillation, indicate how a particle's mean position will alter when its temperature is increased, and hence show how α for this type of solid differs from that for Qu. **23–5**. (*Hint: refer to Qu.* **4–35**.)
(c) According to the **Grüneisen relation**, linear expansivity is proportional to specific heat capacity at constant pressure for any substance at all temperatures. Is this consistent with the curve that you have drawn in (b)?

23–7 If you were asked to undertake an experimental test of the *Grüneisen relation* (Qu. **23–6**), what temperature range would you choose, and why?

Quantitative Problems

†**23–8** *Microscopic expansion.* A particular metal has a *cubic* expansivity 6×10^{-5} K^{-1}, and its ions are about 0.1 nm apart.

(a) What is the change in their mean separation when the temperature increases by 100 K?
(b) How does this compare with the amplitude of their vibratory motion?

[(a) 0.2 pm]

†**23–9** The diameter of a homogeneous metal disc increases by 0.15% when its temperature is raised by 80 K.

(a) Calculate the percentage increase in its surface area, thickness and volume.
(b) How have the mass and density of the disc altered?
(c) Calculate the linear, superficial and cubic expansivities of the metal of the disc.
(d) What would be the fractional change in the moment of inertia of the disc?

[(b) $\Delta\rho/\rho = -4.5 \times 10^{-3}$ (c) $\gamma = 5.6 \times 10^{-5}$ K^{-1} (d) 0.30%]

†**23–10** *Balancing column method.* Prove that the change Δh in height h of a liquid barometer at constant pressure when the temperature changes by ΔT is given by $\Delta h = \gamma h \Delta T$, where γ is the cubic expansivity. In the balancing column method for measuring γ, two vertical glass tubes filled with liquid maintained at different temperatures are connected at their lower ends by a horizontal capillary tube. Calculate γ if the length of the column at 273 K is 1200 mm, and that of the column at 290 K is 1220 mm. What is the advantage of using a capillary tube?

[9.8×10^{-4} K^{-1}]

23–11 (a) Show that, for a given substance, the cubic expansivity (γ) is approximately 3 × the linear expansivity (α).
(b) What is the largest value that α may have if this is to be true to an accuracy of 1%?

[(b) 6×10^{-2} K^{-1}]

23–12 *The pendulum clock.* A simple pendulum whose suspension has a linear expansivity α has a period T at length l_1. When a temperature increase $\Delta\theta$ causes its length to become l_2, the period becomes $(T + \Delta T)$.

(a) Show that (i) $(T + \Delta T)/T = \sqrt{1 + \alpha\Delta\theta}$, and hence
(ii) $\Delta T/T = \frac{1}{2}\alpha\Delta\theta$.
(b) A brass pendulum for which $\alpha = 1.9 \times 10^{-5}$ K^{-1} keeps correct time at 288.0 K. By how much will it be wrong at the end of a day when the temperature is 293.0 K, and which way?

[4.1 s]

*****23–13** Water has a bulk modulus 2.0 GN m^{-2} and an apparent cubic expansivity of 1.8×10^{-4} K^{-1} relative to the container into which it is poured. When the container has been sealed it contains water only, and is then heated by 5.0 K. By how much would the pressure inside the container increase? (*Ignore the fact that the container would expand under the increased stress.*)

[1.8 MPa]

24 Specific Heat Capacity

Questions for Discussion

24–1 Try to compile an exhaustive list of all the different phenomena which can contribute to the internal energy of (a) a metal lattice (b) a crystal made of polyatomic molecules.

24–2 Repeat Qu. **24–1** for the liquid phases of (a) mercury (b) water.

24–3 Is it possible for a system of finite mass to have a heat capacity of zero? If so give an example.

24–4 Discuss the methods of cooling of a hot body placed in a room temperature environment. Distinguish between rate of loss of heat and rate of fall of temperature, and describe the factors that control each.

*24–5** A given mass of gas has two principal heat capacities, measured at constant pressure and volume. Volume and pressure are, respectively, examples of **extensive** and **intensive** quantities. Can you suggest suitable variables for a *surface* (such as a soap film) with reference to which we could define analogous principal heat capacities?

24–6 Heat capacity and specific heat capacity are examples of *intensive* and *extensive* quantities (see Qu. **24–5**). Which is which?

24–7 When would you expect (a) copper, and (b) diamond to have a specific heat capacity which is (i) high, and (ii) low?

*24–8** Look up the **Debye temperature** in a reference book, and show how it can be used to illustrate that apparent exceptions to the *Dulong and Petit rule*, such as diamond, do in fact conform to a general relationship.

Quantitative Problems

General

†**24–9** *Error caused by thermometer.* An accurate thermometer of heat capacity 20 J K^{-1} reads 18.0 °C. It is then placed in 0.25 kg of water and both reach the same final temperature of 50.0 °C as recorded by the thermometer. Assuming that there are no other heat exhanges, calculate the temperature of the water before the thermometer was placed in it.

Assume c_{H_2O} [50.6 °C]

24–10 *Cooling correction.* A metal container of heat capacity 20 J K^{-1} holds 0.15 kg of liquid. An immersion heater is placed in the liquid and a current of 3.0 A with a p.d. of 12 V flows for 4.0 minutes. The temperature/time values are tabulated below:

temperature θ/°C	15.0	20.0	24.7	29.3	33.9	32.8	31.7	30.6
time t/minutes	0.0	1.0	2.0	3.0	4.0	5.0	6.0	7.0

Use these figures to calculate
(a) an approximate cooling correction, and
(b) the specific heat capacity of the liquid.

[(b) 2.6 kJ kg^{-1} K^{-1}]

†**24–11** (a) The temperature of 0.45 kg of water in a vessel of heat capacity 80 J K^{-1} is increased from 288 K to 352 K in 0.48 ks. Neglect heat losses, and calculate the power of the heater.
(b) When this heater is placed in 0.50 kg of paraffin in a similar vessel at the same initial temperature, the temperature reaches 341 K in 0.24 ks. Calculate the specific heat capacity of the paraffin.

Assume c_{H_2O} [(a) 0.26 kW   (b) 2.2 kJ kg^{-1} K^{-1}]

†**24–12** A 10 W immersion heater is placed in 0.25 kg of a liquid in a calorimeter of heat capacity 50 J K^{-1} and is switched on. After a time the temperature of the liquid reaches a constant value. The heater is now removed and the initial rate of fall of temperature is measured as 15 mK s^{-1}. What is the specific heat capacity of the liquid?

[2.5 kJ kg^{-1} K^{-1}]

24–13 *The Nernst calorimeter.* This method is used to measure the specific heat capacity of a solid that is a good conductor. Heat exhanges are greatly reduced by using a vacuum round the solid, and the only correction is for the radiation loss. At the low temperatures used, the latter may be very small. In a determination using a block of mass 0.30 kg the temperature rise was 1.2 K when a current of 3.0 A with a p.d. of 2.0 V flowed for 22 s. When a current of 2.7 A with a p.d. of 1.8 V flowed for the same time, the rise was 0.97 K. Calculate, for this temperature,
(a) the specific heat capacity of the block, and
(b) the (small) power loss.

[(a) 0.36 kJ kg^{-1} K^{-1} (b) c. 6 × 10^{-2} W]

24–14 *Vacuum flask method.* A 15 W heating coil is immersed in 0.20 kg of water and switched on for 0.56 ks, during which time the temperature rises by 10 K. When the water is replaced by the same volume of another liquid of mass 0.15 kg, the power required for the same temperature rise in the same time is 8.3 W. Calculate the specific heat capacity of this liquid. What is the significance of using the same volume of liquid? Are heat losses *eliminated*?

Assume c_{H_2O} [3.1 kJ kg^{-1} K^{-1}]

24–15 *Continuous-flow method.* Water flows at a steady rate of 6.0 g s^{-1} through a *Callendar* and *Barnes* calorimeter when the p.d. across the coil is 11 V and the current is 5.0 A. The difference between the inflow and outflow temperatures is 2.0 K. When the flow changes to 2.0 g s^{-1}, the current is adjusted to 3.1 A to produce the same temperature rise. Calculate

(*a*) the new p.d. across the heating coil, and hence the new power input, and
(*b*) the specific heat capacity of the water.

[(*a*) 21 W   (*b*) 4.2 kJ kg^{-1} K^{-1}]

24–16 The steady rate of flow of the liquid is 20 g s^{-1} and the electric heating element dissipates 0.15 kW. Under these conditions, the outflow temperature of the liquid is 3.0 K higher than the inflow temperature. When the rate of flow is halved, the power required to maintain the same temperature difference is 81 W. Calculate

(*a*) the specific heat capacity of the liquid, and
(*b*) the power loss to the surroundings.

[(*a*) 2.3 kJ kg^{-1} K^{-1} (*b*) 12 W]

†24–17 *Simple use of Newton's Law.* A body cools under forced convection from 350 K to 330 K in 20.0 minutes inside an enclosure at a temperature 280 K.

(*a*) Calculate (*i*) the average excess temperature, and (*ii*) the average rate of cooling.
(*b*) What would be the approximate rate of cooling for a temperature excess of 40 K?
(*c*) Roughly how long (in minutes) would the body take to cool from 330 K to 310 K?

The *correct* answer for (*c*) is 30.5 minutes. What percentage error is introduced by your approximate calculation?

[(*c*) 30.0 minutes]

24–18 *Method of cooling.* Two identical containers each of heat capacity 12 J K^{-1} cool from 340 K. One holds 8.0 × 10^{-5} m^3 of water and takes 0.15 ks to cool from 325 K to 320 K, and the other holds an equal volume of an unknown liquid which takes 50 s to cool over the same range. If the density of the liquid is 8.0 × 10^2 kg m^{-3}, what is its average specific heat capacity over the range 325 K − 320 K?

Assume c_{H_2O}, ρ_{H_2O} [1.6 kJ kg^{-1} K^{-1}]

*24–19 · *Newton's Law of cooling.* A 1.5 kW immersion heater is required to keep 1.2 × 10^3 kg of liquid at a constant temperature of 340 K under conditions of forced convection. If the temperature of the surroundings is 290 K, calculate the time taken for the liquid to cool to 320 K after the heater is switched

off, given that the specific heat capacity of the liquid is 3.2 kJ kg^{-1} K^{-1}.

[66 ks]

*24–20 *The Debye equation.* At very low temperatures the specific heat capacities of solids vary greatly with temperature and obey the relationship $c_V \approx kT^3$. For lead $k = 18$ mJ kg^{-1} K^{-4}. Calculate *approximate* values for

(*a*) the mean specific heat capacity of lead between 5 K and 10 K, and
(*b*) the heat required to increase the temperature of 0.30 kg of lead over this temperature range.

(*It should be noted that the expression given above does not take account of the free electrons in the metal, and so the answers are approximate only.*)

[(*a*) 8.4 J kg^{-1} K^{-1} (*b*) 13 J]

Molar Heat Capacity

24–21 *The Dulong and Petit Law.* The table below lists the specific heat capacities at constant pressure and the relative atomic masses A_r of five elements. Calculate their molar heat capacities near room temperature, and comment on your answers.

substance	diamond	Be	Al	Fe	Pb
A_r	12	9.0	27	56	207
c_p/kJ kg^{-1} K^{-1}	0.51	1.8	0.88	0.44	0.13
measured at T/K	295	323	273	273	273

Use these answers to show that the heat required to raise the temperature of a given number of ions in many metallic lattices through a given range is independent (to a first approximation) of the ionic mass.

24–22 The specific heat capacity of aluminium at room temperature is 0.90 kJ kg^{-1} K^{-1}, and its molar mass is 27 × 10^{-3} kg mol^{-1}. Calculate the molar heat capacity of aluminium. Use this value to predict the specific heat capacity of tungsten, given that its molar mass is 0.18 kg mol^{-1}.

24–23 ($C_{p,m} - C_{V,m}$) ***for a liquid.*** Water has a relative molecular mass of 18. Over the range 273–283 K its average cubic expansivity is 2.0 × 10^{-4} K^{-1}, its mean density is 1000 kg m^{-3}, and its mean specific heat capacity (at constant pressure) is 4.2 kJ kg^{-1} K^{-1}. Calculate

(*a*) the molar volume and molar heat capacity of water,
(*b*) the change in volume of 1.0 mol of water when heated from 273 to 283 K,
(*c*) the work done by this water against an external pressure of 0.10 MPa, and
(*d*) the difference between the principal molar heat capacities.

(*For most liquids, including water at higher temperatures, the expansivity is considerably larger, and the value of $C_{p,m} - C_{V,m}$ becomes significant.*)

[(*d*) 0.36 mJ mol^{-1} K^{-1}]

*24–24 *Equipartition of energy.* Consider a single ion in a metallic lattice to which we may apply *Newtonian mechanics.* Calculate

(a) the number of degrees of freedom it has
(b) its mean vibrational energy at (i) 400 K (ii) 500 K
(c) its *molecular* heat capacity (i.e. the heat capacity of this typical ion).

Use your answer to (c) to calculate the *molar* heat capacity of the metallic lattice, and compare your value with that of the *Dulong and Petit Law* (about 25 J mol^{-1} K^{-1}).
(*The free electrons make a negligible contribution to the heat capacity.*)

Assume k, N_A [(c) 4.1×10^{-23} J K^{-1} molecule^{-1}]

25   Ideal Gases: Kinetic Theory

Questions for Discussion

25–1 A mole of oxygen molecules and a mole of hydrogen molecules *always* have one common property. What is it?

25–2 Two containers of equal volume are joined by a capillary tube. One contains air at temperature T_1, and the other contains air at T_2 ($T_2 > T_1$). What will happen if the temperature of both is increased through the same range ΔT?

25–3 What is the fundamental distinction between a real gas and an ideal gas? Is there (in principle) a unique ideal gas, or can we postulate the existence of several with different chemical properties?

25–4 What do you understand by the 'diameter' of a gas molecule?

25–5 What is meant by a *flux* of molecules?

25–6 Explain how it is that a gas exerts a force on its container wall which is essentially constant, even though its molecules strike the walls randomly.

25–7 What do you understand by an *elastic* collision between two gas molecules? Suggest what might happen to the translational k.e. of two gas molecules if they were to collide inelastically. Would this sort of process, if it occurred throughout the whole volume of the gas, result in a change of temperature?

25–8 What, if any, is the difference between *thermodynamic* temperature T and *ideal-gas* temperature? Other things being equal, how is T related to (a) the r.m.s. speed of gas molecules (b) their mass?

25–9 The velocity of the molecules of a gas sample can be measured relative to (a) a laboratory frame of reference, or (b) the centre of mass of the sample. What can you say about

the sample if the average velocity is not zero (a) in the first frame (b) in the second?
Can the average speed of the molecules ever be zero?

25–10 At room temperature the mean speed of a gas molecule is typically $\sim 10^2$ m s^{-1}. Explain why a gas may take many minutes to diffuse across a closed room.

25–11 Indicate similarities and differences between the behaviour of electrons in a metal and the molecules of a gas. What phenomena are analogous to (a) gaseous diffusion, and (b) the drift of electrons, when an electric field is applied to the metal?

25–12 The temperature gradient in a gas controls the rate of conduction of heat. Suggest the factor that controls the rate of diffusion of one gas through another.

*25–13 Consider a fixed amount of gas. If all other variables are held constant, what would be the effect on the mean free path λ of its molecules of doubling (independently)

(a) the mean molecular speed
(b) the gas density
(c) the diameter of its molecules?

Use your answer to (a) to predict how λ depends upon temperature.

*25–14 *Force on a moving surface.* A flat plate moves normal to its plane at velocity v through a low pressure gas whose molecules have an average speed c_x measured parallel to v.

(a) How do the pressures at the leading and trailing faces depend upon c_x and v?
(b) How does the pressure difference between them depend on c_x and v?
(c) Show that the plate experiences a drag force $F \propto c_x v$.

25–15 O-M Make reasoned estimates of the following:

(a) The number of molecules in 1 mm^3 of a typical liquid.
(b) The number density of air molecules in a good laboratory vacuum (pressure $\sim$ 10 μPa).
(c) The mean separation of gas molecules (i) at s.t.p. (ii) in interstellar space (density $\sim 10^{-20}$ kg m^{-3}).
(d) The fraction of the volume occupied by a gas at s.t.p. which consists of molecules.

25–16 O-M Estimate the total mass of the Earth's atmosphere.

Quantitative Problems

Molar Mass and the Equation of State

†25–17 The pressure of a fixed mass of gas at 300 K is 0.15 MPa. The pressure is now increased to 0.40 MPa, and the density is trebled. What is the final temperature?

[2.7×10^2 K]

†25–18 Helium is monatomic and has a relative atomic mass of 4.0. Calculate

(a) the molar mass of helium
(b) the molar volume at s.t.p.
(c) the density of helium at s.t.p.

Assume R, p_0, T_{ice} [(c) 0.18 kg m^{-3}]

†25–19 The density of chlorine at s.t.p. is 3.16 kg m^{-3}. What is its mean relative molecular mass?

Assume V_m

†25–20 When a superconducting electromagnet is inadvertently quenched, 0.80 kmol of helium are boiled off. What volume does this attain at room temperature, 290 K?

Assume V_m [19 m^3]

25–21 Nitrogen is diatomic and has a relative atomic mass of 14. A sample of the gas occupies 2.0 m^3 at a pressure of 0.15 MPa and a temperature of 300 K. Calculate

(a) the amount of nitrogen in the sample,
(b) the molar mass of nitrogen, and
(c) the mass of the nitrogen.

Assume R [(c) 3.4 kg]

25–22 A sample of oxygen consists of 3.6×10^{24} molecules. Given that oxygen is diatomic and has a relative atomic mass of 16, calculate

(a) the amount of oxygen in this sample,
(b) the molar mass,
(c) the total mass of the sample,
(d) the mass of one oxygen molecule,
(e) the molar volume of oxygen at s.t.p.,
(f) the total volume of this sample at s.t.p., and
(g) the number density of oxygen molecules at s.t.p.

Assume N_A, R [(g) 2.7×10^{25} molecules m^{-3}]

25–23 Solid lead has a relative atomic mass of 207 and density 1.13×10^4 kg m^{-3}. Calculate the following for lead near room temperature:

(a) the molar mass,
(b) the mass of an atom,
(c) the molar volume,
(d) the number density of atoms,
(e) the number of atoms in a volume of 1.0 mm^3, and
(f) the mass of 1.0 mm^3.

Assume N_A [(d) 3.3×10^{28} molecules m^{-3}]

25–24 A car tyre of capacity 1.5×10^{-2} m^3 contains air at atmospheric pressure. Air at 300 K is now pumped into it until a gauge records 0.20 MPa. When disconnected the gauge reads zero. Calculate

(a) the amount of extra air pumped in, and
(b) an approximate value for its mass. (Take $M_r = 30$.)

Assume R [(a) 1.2 mol (b) 36 g]

25–25 Connected containers. Two containers of volumes V and $4V$ are connected by a capillary tube. Initially both are at 280 K, and contain air at a pressure 0.10 MPa. The larger is now warmed to 350 K, and the smaller cooled to 210 K.

(a) Why may we *not* apply $pV/T = $ constant to either container *separately*?
(b) Write down expressions for the *total* amount of gas present both before and after the changes.
(c) Calculate the final pressure in the system.

[(c) 0.11 MPa]

The Dynamics of Gas Molecules

25–26 Ten particles have the following speeds (v/m s^{-1}):

1.0 3.0 4.0 5.0 5.0 5.0 6.0 6.0 7.0 9.0.
Calculate

(a) the most probable speed of these particles
(b) the average or mean speed
(c) the r.m.s. speed.

[(a) 5.0 m s^{-1} (b) 5.1 m s^{-1} (c) 5.5 m s^{-1}]

25–27 The following table shows the number of particles (N) with a particular speed (v):

N	6	8	10	12	9	5
v/m s^{-1}	10.0	20.0	30.0	40.0	50.0	60.0

Draw a histogram to represent these figures, and calculate

(a) the most probable speed of these particles
(b) the average or mean speed
(c) the r.m.s. speed.

[(a) 40.0 m s^{-1} (b) 35.0 m s^{-1} (c) 38.1 m s^{-1}]

25–28 A beam of hydrogen molecules, each of mass 3.3×10^{-27} kg and travelling at 1.6 km s^{-1}, strikes a wall at $\pi/3$ rad to the normal. If 2.0×10^{20} molecules s^{-1} are incident on an area of 1.2×10^{-4} m^2 of wall, calculate

(a) the average normal force exerted by the beam on the surface if all the molecules are absorbed by the surface,

(b) the average normal force exerted if all the molecules rebound elastically, and

(c) the pressure exerted on the wall in each case.

In (b) you should assume (unrealistically) that each molecule obeys the law of reflection.

[(a) 0.53 mN   (b) 1.1 mN   (c) 4.4 Pa, 8.8 Pa]

*25–29   Molecular effusion. Hydrogen gas ($M_r = 2.0$) at s.t.p. surrounds an evacuated container which has a volume 5.0×10^{-3} m^3. A small hole of area 2.0×10^{-14} m^2 is made in the container's wall. Calculate

(a) the r.m.s. speed of the hydrogen molecules,

(b) the number of hydrogen molecules entering the container every second,

(c) the time for 1.0×10^{-6} mol of hydrogen molecules to enter the container, and

(d) the pressure in the container after this time.

State any assumptions you need to make in deriving your answers.

Assume N_A, R

[(a) 1.8 km s^{-1} (b) 1.6×10^{14} s^{-1} (c) 3.7 ks   (d) 0.45 Pa]

Deductions from $p = \frac{1}{3}\rho\overline{c^2}$

25–30   Dalton's Law of partial pressures. A container holds 66 g of carbon dioxide ($M_r = 44$), 12 g of helium ($M_r = 4$) and 39 g of benzene ($M_r = 78$) at a pressure 0.25 MPa and temperature 320 K. Calculate

(a) the partial pressures of the three gases, and

(b) the volume of the container.

Assume R [(a) benzene 25 kPa   (b) 5.3×10^{-2} m^3]

25–31 A container of volume 0.10 m^3 is full of air at a temperature 300 K and pressure 1.5 MPa. Consider the air to be 80% nitrogen and 20% oxygen by mass, where the relative molecular masses are 28 and 32 respectively. Calculate

(a) the amount of gas in the container,

(b) the total mass of gas present, and

(c) the partial pressures of the two gases.

Assume R

[(a) 60 mol   (b) 1.7 kg   (c) 1.2 and 0.27 MPa]

25–32   Graham's Law of diffusion. The rate of diffusion of a gas through a very small hole is given by $\rho\sqrt{RT/2\pi M_m}$ where ρ is the density of the gas, T is the temperature and M_m is the molar mass.

(a) In what unit is this rate of diffusion expressed?

(b) How do the rates of diffusion of different gases at the same pressure and temperature depend upon (i) their densities, and (ii) their relative molecular masses?

(c) How will this rate of diffusion of oxygen compare with that of hydrogen under the same conditions? ($M_r = 32$ and 2 respectively.)

[(c) 4.0 times as great]

25–33   Composition of the Earth's atmosphere. Calculate the r.m.s. speeds of hydrogen, helium, nitrogen and oxygen at 300 K. The relative molecular masses are 2.0, 4.0, 28 and 32 respectively. Compare these speeds with the escape speed from the Earth. What bearing do your answers have on the composition of the Earth's atmosphere? (*The temperature of the upper atmosphere is* $\sim 10^3$ K.)

Assume R [0.2, 0.1, 0.05 and 0.04 v_e respectively]

Transport Phenomena

*25–34   Mean free path. The average distance that a gas molecule travels between collisions with other gas molecules is known as its **mean free path** λ, and is given by $\lambda = 1/\sqrt{2}\,n\pi r_0^2$, where n is the number density of molecules and r_0 is the molecular diameter.

(a) How would you expect λ to vary with density and with pressure at a particular temperature?

(b) The quantity πr_0^2 is known as the **collision cross-section** of a molecule. How does this compare with the *geometrical* cross-section, and what do you think is meant by this quantity?

(c) Calculate the value of λ for carbon dioxide molecules at s.t.p. if $r_0 = 0.42$ nm. How does your value of λ compare with r_0?

(d) What would be the approximate value of λ at densities of 10^{-16} kg m^{-3} (best laboratory vacuum), and 10^{-20} kg m^{-3} (interstellar space)?

(e) Estimate the times between collisions for the three mean free paths that you have calculated, and the corresponding collision frequency.

Assume N_A, V_m [(c) 48 nm   (d) 1 Gm, 10 Tm]

*25–35   Viscosity – transport of momentum. The viscosity η of a gas may be written $\eta = \frac{1}{3}\rho\overline{c}\lambda$ where ρ is the density of the gas, $\overline{c}$ is the mean speed of the gas molecules and λ is the mean free path.

(a) How would you expect the viscosity of a gas to vary with (i) pressure and (ii) temperature?

(b) How does a measurement of the viscosity of a gas lead to an estimate of its molecular diameter?

(c) The viscosity of nitrogen at s.t.p. is 17 μPa s, and its density is 1.3 kg m^{-3}. If the relative molecular mass of nitrogen is 28, estimate the diameter of the nitrogen molecule. Say what assumptions you have made.

Assume R [(c) 0.3 nm]

*25–36   Thermal conductivity – transport of energy. The thermal conductivity k of a gas is given by $k = \frac{1}{3}\rho\overline{c}c_V\lambda$, where ρ is the density of the gas, $\overline{c}$ is the mean speed of the gas molecules, c_V is the specific heat capacity at constant volume and λ is the mean free path.

(a) Derive a relation between k and the viscosity η (Qu. 25–35), and discuss the dependence of thermal conductivity upon pressure and upon temperature.

(b) *Estimate* the diameter of a helium molecule ($M_r = 4$) given that $k = 0.14$ W m^{-1} K^{-1} and $c_V = 3.1$ kJ kg^{-1} K^{-1} at s.t.p. Again state your assumptions clearly.

$$[(b)\ 0.1\ \text{nm}]$$

*25–37   Diffusion – transport of molecules. The diffusion coefficient D for a gas is given by $D = \frac{1}{3}\lambda\bar{c}$, where λ is the mean free path and $\bar{c}$ is the mean speed of the gas molecules. D is a measure of the rate of diffusion of mass across a given area for a particular mass concentration gradient.

(a) How would you expect D to depend upon pressure and upon temperature?

(b) Find a relation between D and η for a gas, and sketch a graph of $D\rho$ plotted against the pressure p.

(c) *Estimate* the value of D for air at s.t.p.

(d) Suggest a direct method of studying D.

$$[(c)\ \sim 2 \times 10^{-5}\ \text{m}^2\ \text{s}^{-1}]$$

26   Ideal Gases: Thermal Behaviour

Questions for Discussion

26–1 Explain what is meant by saying that all natural processes are *irreversible*. Include a brief description of the following natural phenomena: a waterfall, lightning and the weathering of rocks. Is energy 'lost' in an irreversible change?

26–2 When a gas expands by a given volume does it do more work at constant pressure or at constant temperature?

26–3 Is it possible to change the state of an ideal gas so that p and V are related by $pV^{\frac{1}{2}} = $ constant?

If so, suggest what the corresponding p–V graphs would look like.

26–4   The first law. The law can be written $\Delta Q = \Delta U + \Delta W$, where ΔQ is the heat supplied to a substance, ΔU is the increase in its internal energy, and ΔW is the external work done by the substance.

(a) What information does the law provide about the energy involved in a given process?

(b) What situation or process is described by writing (i) $\Delta Q = 0$ (ii) $\Delta U = 0$ (iii) $\Delta W = 0$?

(c) Use the law to discuss the transformations of energy which occur when water freezes.

26–5 Suppose that U is the balance of a bank account, ΔQ the money paid in as cash, and ΔW the money drawn out by cheque.

(a) Write an equation which relates an initial balance U_i, a final balance U_f, ΔQ and ΔW.

(b) If the bank manager could keep records of only *one* of these three variables, which would he choose, and why?

(c) To what extent does the analogy between this situation and the *first law of thermodynamics* hold good?

26–6 Some water is placed in a vacuum flask, and is shaken up.

(a) Does the water temperature rise?

(b) Has the internal energy of the water been altered?

(c) Has any heat been given to the water?

(d) Has any work been done on the water?

26–7 Consider the equation $C_m = \Delta Q / \mu\Delta T$, in which C_m is the molar heat capacity of a gas, ΔQ is the heat added to an amount μ [mol] of gas, and ΔT is the resulting change in temperature. Describe situations in which the value of C_m could be (a) plus infinity (b) zero (c) minus infinity.

What advantage is obtained by specifying a process at (i) constant volume (ii) constant pressure?

26–8 Discuss the relative differences between the two principal molar heat capacities for solids, liquids and gases. Which of these heat capacities is measured experimentally for solids and liquids?

26–9 When a gas expands adiabatically it may do work. What is the source of energy which enables this work to be done?

26–10 Discuss what happens when (a) a dry gas, (b) a saturated vapour, expands adiabatically.

26–11 Hot air rises. Why then does the temperature decrease when one climbs a mountain?

*26–12 A molecule of a particular substance consists of two atoms joined to one another as a dumb-bell. What might be the theoretical molar heat capacity at constant volume when the substance is (a) gaseous (b) solid?

*26–13 Describe and explain how $C_{V,m}$ for a diatomic gas varies with temperature.

*26–14 Explain what is meant by saying that *entropy* gives a measure of the disorder of a system. Express your answer in terms of the laws of thermodynamics.

*26–15 Two gases at different temperatures are isolated from their surroundings, and separated from each other by a partition through which heat exchange can take place.

(a) Is the total entropy more likely to increase or to decrease?
(b) What would have to happen if the entropy were to decrease?
(c) What would be the final result if the entropy were to increase?

*26–16 In a real heat engine, some of the heat taken in is converted into work, and the rest is rejected as heat at a different temperature. What would happen with an ideal heat engine?

In a real refrigerator, work is required to transfer heat from a low-temperature to a high-temperature reservoir. What would happen with an ideal refrigerator?

*26–17 Why do high-performance internal combustion engines use high compression ratios?

26–18 O-M Estimate the change in the temperature of the air in a car cylinder when a piston moves quickly up the length of the cylinder. (Ignore combustion processes.)

Quantitative Problems

External Work and Molar Heat Capacity

†26–19 The heartbeat of an athlete in action is measured to be 1.5 beats s^{-1}. If the heart pumps 7.5×10^{-5} m^3 of blood at each beat against an average pressure of 16 kPa, at what rate is it working?   [1.8 W]

26–20 When 5.0×10^{-6} m^3 of water is boiled away at standard pressure, 8.3×10^{-3} m^3 of steam is formed. Calculate

(a) the work done by the water in expanding
(b) the increase in the internal energy of the water.

The specific latent heat of vaporization of water is 2.3 MJ kg^{-1}.

Assume p_0, ρ_{H_2O} [(a) 0.84 kJ   (b) 11 kJ]

26–21 The temperature of 4.0 mol of oxygen is increased from 280 K to 300 K at a constant pressure of 0.50 MPa. Take $C_{p,m}$ for oxygen as 29 J mol^{-1} K^{-1}, and calculate

(a) the heat supplied,
(b) the volume change,
(c) the work done by the gas in expanding, and
(d) the value of $C_{V,m}$ for oxygen.

How does your value of $(C_{p,m} - C_{V,m})$ compare with R?

Assume R

[(a) 2.3 kJ   (b) 1.3×10^{-3} m^3 (c) 0.66 kJ   (d) 21 J mol^{-1} K^{-1}]

†26–22 10.0 mol of oxygen are heated from 280 K to 295 K. Calculate the molar heat capacities of oxygen over this temperature range if the gas is in

(a) a rigid container, in which case 3.12 kJ of energy are supplied
(b) a freely extensible container, when 4.38 kJ are required.

Calculate the difference between (a) and (b), and explain the significance of the result.

[(a) 20.8 J mol^{-1} K^{-1} (b) 29.2 J mol^{-1} K^{-1}]

*26–23 Use the information given in the previous question to calculate the increase in

(a) the molar internal energy U_m
(b) the molar enthalpy H_m.

[(a) 0.312 kJ mol^{-1} (b) 0.438 kJ mol^{-1}]

*26–24 If the ratio of the principal molar heat capacities of a gas is 1.4, calculate the molar heat capacity at constant pressure.

Assume R [29 J mol^{-1} K^{-1}]

Isothermal and Adiabatic Changes

*26–25 *Cloud chamber.* The gas and vapour in a cloud chamber are at 290 K, and undergo a rapid expansion which can be treated as approximately reversible and adiabatic. The *expansion ratio* is 1 : 1.3. Take $\gamma = 1.4$, and calculate the final temperature. (*In a cloud chamber condensation then takes place on the trail of ions left by the ionizing radiation.*)

[2.6×10^2 K]

*26–26 1 mol of nitrogen ($\gamma = 7/5$) and 1 mol of helium ($\gamma = 5/3$) are mixed and contained in a volume of 15×10^{-3} m^3 at 250 K. The mixture is then compressed reversibly and adiabatically to a volume of 5.0×10^{-3} m^3. Calculate, for the mixture,

(a) the effective value of γ
(b) the final temperature
(c) the ratio of the initial to the final pressure.

[(a) 23/15   (b) 4.5×10^2 K   (c) 1 : 0.19]

*26–27 An ideal monatomic gas ($\gamma = 5/3$) expands reversibly from a volume V_1 and pressure p_1 to a volume V_2. Calculate the work done by the gas if the change takes place

(a) isothermally
(b) adiabatically
(c) so that pV^2 is constant.   [(c) $p_1V_1(V_2 - V_1)/V_2$]

*26–28 Calculate the values of dp/dV at p_1, V_1 for parts (a), (b) and (c) of the previous question. What do these values represent? What deductions can you make from them?

[(c) $-2p_1/V_1$]

*26–29 A diatomic gas ($\gamma = 1.4$), initially at standard pressure, expands reversibly to five times its original volume of 2.0×10^{-3} m^3. If the change takes place at 273 K (isothermally), calculate

(a) the final pressure of the gas, and
(b) the work done by the gas during the expansion.

What would be the values of (a) and (b) if the change took place adiabatically?

Assume p_0, R [(a) 20 kPa   (b) 0.33 kJ; 11 kPa, 0.24 kJ]

*26–30 A vacuum pump is used to lower the pressure inside a container of volume 0.20 m³ to 0.10% of its original value. If 1.0 × 10⁻⁴ m³ of gas for which $\gamma = 1.4$ are expelled at each stroke, and the pump is operating at 12 strokes s⁻¹, calculate the time required under isothermal conditions. What time would be needed if the conditions were adiabatic? Explain the physical reason for the two times being different.

[1.2 ks, 0.82 ks]

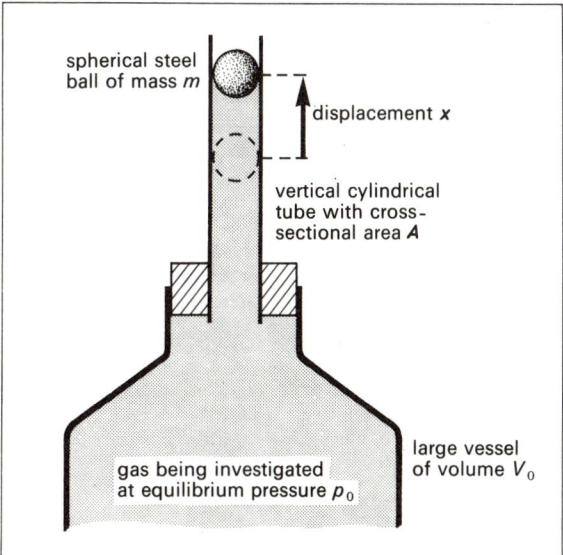

For Qu. 26–31

*26–31 **Rüchardt's method for** γ. Refer to the diagram. The ball and cylindrical tube must be accurately shaped so that the gap between the two is everywhere small. If the ball is displaced, the elasticity of the gas will cause a restoring force to act on it. Use the expression $-V(\mathrm{d}p/\mathrm{d}V)_{\mathrm{ad}} = \gamma p$ for the adiabatic bulk modulus of an ideal gas to show that the period T of oscillation of the ball is given by $T = 2\pi \sqrt{(mV_0/\gamma p_0 A^2)}$.

On what assumptions is your argument based? (*This is a useful method of finding γ for the gas under test.*)

*26–32 Use the result of the last question to find γ for the gas contained in the vessel. The frequency of oscillation of the ball is 0.88 Hz, $V_0 = 12 \times 10^{-3}$ m³, $p_0 = 0.11$ MPa, $A = 5.0 \times 10^{-4}$ m², and the mass of the ball is 0.12 kg.

[1.6]

Equipartition of Energy

*26–33 A stationary sound wave is set up in nitrogen ($M_{\mathrm{m}} = 28 \times 10^{-3}$ kg mol⁻¹) at a temperature of 310 K. When the frequency is 2.0 kHz, the distance between adjacent nodes is 90 mm. Use this information to find the atomicity of nitrogen.

Assume R

*26–34 An oxygen molecule has a total mass of 5.3 × 10⁻²⁶ kg, and the separation of the two atoms is 0.12 nm. What is the average angular speed of rotation of a molecule about each of the three axes of symmetry at 300 K? (*Treat the atoms as point masses.*)

Assume k [4.7 Trad s⁻¹ (twice), and zero]

*26–35 Use the principle of equipartition of energy to predict a value of γ for the following gases: (a) argon (b) nitrogen (c) carbon dioxide (collinear) (d) water vapour (not collinear). State your assumptions.

*26–36 An air-filled box contains a galvanometer mirror suspended on a thread which has a torsion constant c. The average total energy of oscillation is constant and equal to $\frac{1}{2}I\omega^2 + \frac{1}{2}c\theta^2$, where I is the moment of inertia of the mirror for torsional oscillations, and ω is the angular velocity at angular displacement θ.

(a) How many degrees of freedom has this oscillating system?
(b) How is the average total energy of oscillation related to the temperature?
(c) If $c = 0.10$ pN m rad⁻¹, calculate the r.m.s. value of θ at 300 K.
(d) What is the cause of these random movements?
(e) If the box is now highly evacuated, and remains at the same temperature, calculate the new r.m.s. value of θ.

What is the cause of these random movements in this new situation?

Assume k [(c) 0.20 mrad]

Heat Engines

*26–37 **Carnot refrigerator**. The efficiency of a reversible Carnot cycle engine is given by $\dfrac{Q_1 - Q_2}{Q_1} = \dfrac{T_1 - T_2}{T_1}$, where Q_1 is the heat transferred into the engine at temperature T_1, and Q_2 the heat transferred out of the engine at temperature T_2. An ideal refrigerating machine operating in a reversible Carnot cycle at 330 K transfers 8.0 kJ of heat from a room at 278 K. Calculate how much work must be done on the machine for this to take place.

[1.5 kJ]

*26–38 A reversible Carnot cycle engine operates between temperatures of 1000 K and 250 K. If 1.5 kJ of heat are transferred to the engine at 1000 K in one cycle, find

(a) the efficiency of the engine
(b) the heat transferred from the engine at 250 K.

[(a) 75%   (b) 0.38 kJ]

*26–39 A steam engine uses pressurized steam at 470 K, and exhausts it at 373 K. What would be the efficiency of the machine if it were ideal? Explain why the actual efficiency will be lower than this.

[21%]

27   Change of Phase

Questions for Discussion

27–1 We say that heat is the transfer of energy caused by a temperature difference; yet when we supply energy (other than work) to melting ice, no change in temperature results. Does this mean that strictly the latent *heat* of fusion should be given some other name?

27–2 What is the *triple point* of a substance? Why has the triple point of water, rather than its standard (or normal) *melting* point, been chosen as a standard fixed point to define the ideal-gas temperature scale?

27–3 What are the relative sizes of the specific latent heats of fusion, vaporization and sublimation of a substance? Comment on the significance of your answers.

27–4 How is the molar latent heat of vaporization of a substance related to the binding energy between a pair of molecules of the substance? Estimate the magnitude of the binding energy between water molecules.

27–5 Can a given species of glass be said to have a particular melting point at standard pressure? What about a metal, such as zinc? What difference in the *structure* of these materials enables us to make this distinction?

27–6 How does the behaviour of a saturated vapour differ from that of an unsaturated vapour?

27–7 How does the latent heat of vaporization of perspiration form the basis for the temperature regulation of the body?

27–8 Why do small droplets appear on the side of a sherry glass above the liquid level?

27–9 O-M How much would it cost to vaporize a snowman using an electrical heater?

27–10 O-M Estimate the speed at which a lead bullet should hit its target so that it just melts on impact.

27–11 O-M Estimate the electric power of an underground heating system which is designed to prevent a rugby football pitch from freezing.

Quantitative Problems

Molar Latent Heat

†27–12 The relationship between molar latent heat of sublimation $L_{s,m}$ and the interaction p.e. ε between an atom (or molecule) and its neighbour is given by $L_{s,m} = \frac{1}{2} N_A n \varepsilon$, where n is the number of nearest neighbours. Calculate $L_{s,m}$ for hydrogen ($\varepsilon = 4.0 \times 10^{-22}$ J) and krypton ($\varepsilon = 25 \times 10^{-22}$ J). Consider the structures to be close-packed so that $n = 12$. (For the purposes of this calculation ignore the external work.)
Assume N_A [hydrogen, 1.4 kJ mol^{-1}]

*27–13 Trouton's Rule. Systems in different phases have different entropies – the more disordered the phase, the higher the entropy. Thus when there is a change of phase there is also a change of entropy, ΔS. **Trouton's Rule** states that

$$\Delta S_{v,m} = \frac{L_{v,m}}{T} \approx 88 \text{ J mol}^{-1} \text{ K}^{-1},$$ where $L_{v,m}$ is the molar latent

heat of vaporization and T is the standard boiling point of the liquid. Use this rule to estimate the value of $L_{v,m}$ for (a) benzene ($T = 353$ K), and (b) ethyl ether ($T = 307$ K).

Specific Latent Heat

†27–14 A lead bullet at 320 K is stopped by a sheet of steel so that it reaches its melting point of 600 K and completely melts. If 80% of the k.e. of the bullet is converted into the internal energy of the bullet, calculate the speed with which the bullet hits the steel sheet. For lead, the specific heat capacity is 0.12 kJ kg^{-1} K^{-1} and the specific latent heat of fusion is 21 kJ kg^{-1}.

[0.37 km s^{-1}]

†27–15 A piece of metal of mass 25 g is suspended in a constant temperature enclosure at 290 K. The enclosure is then filled with dry water vapour at 373 K and 0.40 g of steam condenses on the piece of metal. Calculate the specific heat capacity of the metal given that the specific latent heat of vaporization of water is 2.2 MJ kg^{-1}.

[0.42 kJ kg^{-1} K^{-1}]

27–16 The Bunsen ice calorimeter. Refer to the diagram. The instrument is prepared by pouring some ether into the test tube and blowing air through it: this produces the mantle of ice. It is then left until the temperature of the instrument is 273 K and the mercury thread is steady.

When 3.0 g of a substance at 338 K were placed inside the test tube, the mercury thread moved 60 mm. Use the following information to calculate the specific heat capacity of the

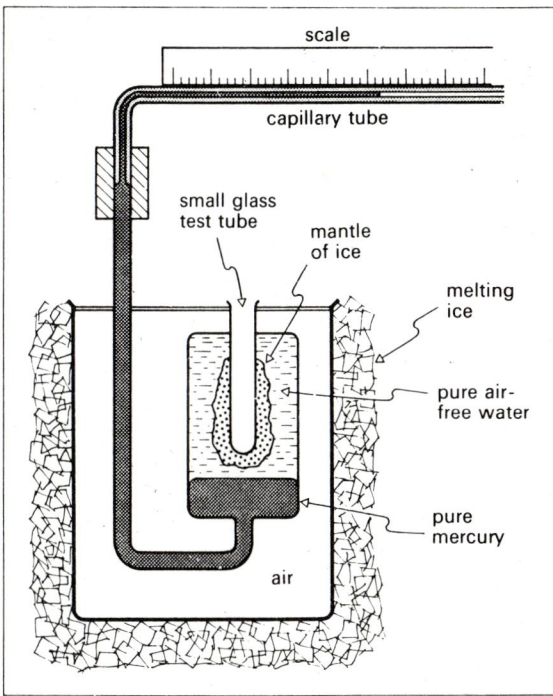

scale

capillary tube

small glass test tube

mantle of ice

melting ice

pure air-free water

pure mercury

air

*For Qu. **27–16***

substance. Density of ice at 273 K = 920 kg m^{-3}, specific latent heat of fusion of ice = 0.34 MJ kg^{-1}, cross-sectional area of the capillary thread = 0.50 mm^2. What is the sensitivity of the calorimeter in mm J^{-1}?

Assume ρ_{H_2O} [0.60 kJ kg^{-1} K^{-1}, 0.51 mm J^{-1}]

27–17 *The Henning method.* After a steady state has been reached, 20 g of the boiling liquid evaporates and condenses in 0.18 ks. A current of 4.0 A flows through the heating coil and a p.d. of 15 V is applied across it. If the heat loss through the jacket is 5.0% of the heat supplied by the coil, calculate

(a) the power loss, and
(b) the specific latent heat of vaporization of the liquid.

[(b) 0.51 MJ kg^{-1}]

27–18 A 90 W electric heater was used to raise the temperature of 0.20 kg of camphor contained in a calorimeter. The temperature rose steadily to 176 °C and then remained constant for 0.11 ks before further increase. At 190 °C the heater was

switched off. The temperature started to decrease, but remained constant at 176 °C for twelve times as long as before. Calculate

(a) the heat loss during the first time interval when the temperature was 176 °C, and
(b) the specific latent heat of fusion of camphor.

[(b) 46 kJ kg^{-1}]

27–19 0.10 kg of a hot liquid is placed in a container of negligible heat capacity, and a cooling curve is plotted. The gradient of the curve is −35 mK s^{-1} at the temperature of the phase change, and this takes 0.90 ks. After solidification the gradient again becomes negative. The specific heat capacity of the liquid is 2.0 kJ kg^{-1} K^{-1}.

(a) Calculate the specific latent heat of fusion of the solid.
(b) Will the graph's gradient immediately after the phase change equal its value immediately before? Discuss.
(c) If 0.20 kg of liquid had been used under otherwise identical conditions, outline the changes you would have expected in the given information.

[(a) 63 kJ kg^{-1}]

Vapours

†27–20 A movable piston keeps air in a cylindrical space which is kept saturated with water vapour at 281 K and standard atmospheric pressure. The s.v.p. of water at this temperature is 1.1 kPa. Calculate the new pressure when the volume is
(a) doubled, or (b) halved, both isothermally.

Assume p_0 [(a) 51.0 kPa   (b) 201 kPa]

27–21 The space inside a closed vessel contains air and is kept saturated with water vapour at 300 K under a pressure of 100 kPa. The s.v.p. of water at 300 K is 4.00 kPa, and at 324 K it is 12.3 kPa. Calculate the pressure in the vessel

(a) at 324 K
(b) at 373 K, the standard (normal) b.p. of water.

Will the water boil before or after 373 K?

Assume p_0 [(a) 116 kPa   (b) 220 kPa]

27–22 A space contains 3.00×10^{-3} m^3 of air and is saturated with water vapour at 373 K. It is cooled to 293 K, at which temperature the s.v.p. of water is 2.3 kPa. Calculate the volume of the air after cooling if the total pressure has been kept at 133 kPa.

Assume p_0 [5.8×10^{-4} m^3]

27–23 Measurements on samples of air from a room at 290 K show that the water content is 11.2 g m^{-3}. If the s.v.p. of water at 290 K is 1.93 kPa, estimate the relative humidity in the room. The density of water vapour at s.t.p. is 0.806 kg m^{-3}.

[78%]

28 Real Gases

Questions for Discussion

28–1 A gas undergoes a rapid adiabatic expansion into a vacuum. Is there any change of temperature if the gas is (*a*) ideal, or (*b*) real?

28–2 Sketch an isothermal curve for a real gas below its critical temperature to illustrate the changes which occur in the transformation from vapour to solid.

28–3 It is sometimes said that attractive forces exerted on one another by the molecules of a real gas are operative at such large distances that they cause the gas to exert a smaller pressure on the container walls than it would if it were ideal. Can you reconcile this statement with the observed fact that the pressure exerted on the walls is independent of their nature (i.e. of the intermolecular forces exerted by the walls)?

28–4 A real gas consists mainly of a mixture of monomers (mass m) and dimers (mass $2m$). How does the average momentum of a dimer compare with that of a monomer? (*Hint: apply the principle of equipartition of energy.*)

28–5 Suggest a physical explanation for the *van der Waals* pressure correction coefficient a being temperature-dependent. Would you expect b to depend on temperature? If so, to what extent?

28–6 What physical meaning can be attached to the constants a and b in the *van der Waals* equation?

28–7 Is it possible for the p.e. possessed by a gas by virtue of its intermolecular forces to *increase* when its volume is increased?

28–8 Find a simplified form of the *van der Waals* equation which can be used for large values of the molar volume.

28–9 Sketch some curves to show the form of typical *van der Waals* isotherms. Use one of the curves to explain what is meant by (*a*) superheating, and (*b*) supercooling.

28–10 O-M What fraction of the total volume available to a gas at standard conditions is covolume? What fraction of the observed pressure is the **internal pressure** (a/V_m^2)?

Quantitative Problems

Critical Phenomena

*28–11 The molar volume, pressure and temperature of a real gas at its critical point can be expressed by $V_{c,m} = 3b$, $p_c = a/27b^2$ and $T_c = 8a/27Rb$ respectively, where a, b and R are constants. Evaluate the quantity $RT_c/p_cV_{c,m}$, which is called the **critical coefficient**. What can you deduce from its value?

*28–12 The observed values of the critical coefficients of many real gases are the same, but appreciably greater than the calculated theoretical value. For carbon dioxide the value is 3.5. If $T_c = 304$ K and $p_c = 7.3$ MPa, calculate the following for carbon dioxide ($M_r = 44$):—

(*a*) the molar critical volume,
(*b*) the specific critical volume, and
(*c*) the critical density.

Assume R   [(*c*) 4.5×10^2 kg m^{-3}]

*28–13 *Critical temperature.* If the k.e. of a molecule in the equilibrium position is greater than the magnitude of the (negative) p.e. ε at this point, then we may assume that it will always be able to escape from its neighbours. Take this thermal energy to be kT, and use it to predict temperatures above which it will be impossible to liquefy the following gases: helium ($\varepsilon = 8.0 \times 10^{-23}$ J), hydrogen (4.0×10^{-22} J) and nitrogen (1.3×10^{-21} J). How do your answers compare with experimental values for their critical temperatures?

Assume k

Equations of State

28–14 *Calculation of molecular diameter.* For hydrogen $b = 2.7 \times 10^{-5}$ m^3 mol^{-1}. Use this value, that of N_A, and the assumption that b is four times the volume of a mole of spherical hydrogen molecules to find a value for the molecular 'diameter' r_0.

[0.28 nm]

28–15 In the *van der Waals* equation of state $\left(p + \dfrac{a}{V_m^2}\right)(V_m - b)$
$= RT$ the numerical values of the constants for nitrogen are $a = 0.14$ and $b = 3.9 \times 10^{-5}$ when each is expressed in its SI unit.

(*a*) What are the SI units of a and b?
(*b*) Calculate the pressure of 2.00 mol of gas at 200 K contained in a volume 6.00×10^{-3} m^3.
(*c*) What would be the pressure if the gas were ideal?

Assume R   [(*b*) 0.546 MPa   (*c*) 0.554 MPa]

28–16 Use the information in the previous question to calculate the pressure required to change the volume isothermally to 1.00×10^{-3} m³

(a) if the gas is ideal

(b) assuming that the gas obeys the *van der Waals* equation.

[(a) 3.32 MPa   (b) 3.05 MPa]

*28–17 By comparing the virial equation $pV_m = A + Bp + Cp^2 \ldots$ with the *van der Waals* equation $(p + a/V_m^2)(V_m - b) = RT$, show that $A = RT$, $B = b - a/RT$ and $C = ab/R^2 T^2$. At the *Boyle* temperature T_B, $B = 0$ and C is negligible. Calculate T_B for nitrogen given that $a = 0.14$ N m⁴ mol⁻² and $b = 3.9 \times 10^{-5}$ m³ mol⁻¹. (*Remember to work the units as*

well as the numbers in your calculation. Compare your answer with the experimental value for T_B, 323 K.)

Assume R [4.3×10^2 K]

*28–18 *The Dieterici equation of state.* This equation describes the observed behaviour of real gases better than the *van der Waals* equation, particularly for heavier and more complex gases. Use the equation

$$p(V_m - b) = RT \exp(-a/RTV_m)$$

to show that if b/V_m and a/RTV_m are small then the *Dieterici* gas obeys *Boyle's Law* very accurately. Show that the temperature T_B at which the terms containing a and b cancel each other's effects is given by $T_B = a/bR$. This is the *Boyle* temperature. (*Hint:* $e^x \approx 1 + x$ when x is small.)

29 Thermal Conduction

Questions for Discussion

29–1 By what physical mechanism is heat conducted through (a) silver (b) wood (c) air (d) glass (e) glycerine?

29–2 How would you expect the thermal conductivities of metals, non-metals, liquids and gases to vary with temperature? Give reasons for your answers.

29–3 In what ways is the rate of heat flow in thermal conduction analogous to the rate of flow of charge in electrical conduction?

29–4 Explain in detail the difference between the transfer of energy by a sound wave and that by the conduction of heat. Do these two processes have anything in common?

29–5 How would you measure the thermal conductivity of the material of a copper wire of cross-sectional area about 1 mm²? State any difficulties that you would encounter and precautions that you would take in attempting to achieve accurate results.

29–6 What are the thermal effects of the 'scale' deposit left when hard water is heated in (a) a saucepan, and (b) an electric kettle?

29–7 Sketch a graph of the possible variations in temperature as one moves from a point 0.2 m inside a room at 300 K through a glass window to a point 0.2 m outside, where the temperature is 270 K.

29–8 Why is there an optimum spacing for the gap between the two panes of glass in a double-glazing installation? Would it be worth evacuating this space?

29–9 Discuss whether the rate of heat loss would be increased if some very poor lagging were wrapped round a steam delivery pipe of small diameter.

29–10 *The Wiedemann-Franz Law.* The thermal and electrical conductivities of metals are related by the expression $\lambda/\sigma = 3(k/e)^2 T$, where k is the Boltzmann constant, e is the electronic charge and T is the temperature. Discuss the significance of this relationship, and evaluate λ/σ at room temperature.

*29–11 *Electron gas.* The movement of electrons accounts for the major part of the thermal conductivity of a metal. The thermal conductivity λ can be written $\lambda = \frac{1}{3} n \bar{v} l c_e$, where n is the number density of electrons, $\bar{v}$ is their mean thermal speed, l is their mean free path and c_e is the heat capacity of an electron. If $c_e \approx 1.8 \times 10^{-25}$ J K⁻¹, estimate the value of l at room temperature. Comment on the magnitude of your answer.

*29–12 *Phonons.* The principal mechanism for transfer of heat through a non-metal is by lattice vibrations, which travel through the material as extremely high frequency acoustic waves called **phonons**. These particle-like packets of waves possess both energy and momentum, and travel with the speed c of a sound wave through the substance. The thermal conductivity of a substance is given by $\lambda = \frac{1}{3} c l (C/V)$, where C

is the heat capacity of a body of the substance, V is its volume and l is the mean free path of a phonon. Estimate the value of l for a typical poor conductor.

*29–13   The hot end of an unlagged uniform metal bar is maintained at a temperature θ_0 in an environment at 273 K. The temperature θ of a point distant x from this end is given by $\theta = \theta_0 e^{-wx}$. Sketch a graph to show how the temperature excess varies with x. Upon what physical quantities is w likely to depend? How would your graph be modified if the bar were lagged?

29–14   O-M Estimate the power required to keep the inside of *Concorde* at the right temperature for passenger comfort.

29–15   O-M What is the rate of heat loss from an average house in Winter time?

29–16   O-M Radioactive decay processes within the Earth generate internal energy which is eventually conducted out through the crust. Given that the mean temperature gradient is about 10 mK m^{-1} in the surface rocks, estimate the energy lost by the Earth each day by this process.

Quantitative Problems

Parallel Lines of Heat Flow

†29–17   When the heat flow through unit area of a sheet of insulating material of thickness 3.0 mm is 8.0 kW m^{-2}, the temperature drop across the sheet is 100 K. Calculate the thermal conductivity of the material.

[0.24 W m^{-1} K^{-1}]

†29–18   *Conduction through windows.* The temperature of a room is maintained at 292 K when the outdoor temperature is 274 K. The glass windows in the room have a total area of 6.0 m^2 and a uniform thickness of 4.0 mm. Use this information to calculate the power required to maintain this temperature difference assuming that the only heat loss is through the glass which has thermal conductivity 0.80 W m^{-1} K^{-1}. Why is your answer unrealistically large?

[22 kW]

†29–19   Steam is produced at a rate of 0.55 kg s^{-1} in a steel boiler whose plates are 6.0 mm thick. The thermal conductivity of steel is 48 W m^{-1} K^{-1}, and the specific latent heat of vaporization of water is 2.1 MJ kg^{-1}. Calculate the temperature drop across the boiler plates if the area through which heat is conducted from the furnace is 8.0 m^2.

[18 K]

†29–20   *The effect of an insulating lining.* If the inside of the boiler in the previous question becomes coated with a non-metallic deposit to a thickness of 1.0 mm, calculate the temperature difference between the outside of the metal plates and the exposed surface of the coating if steam is to be produced at the same rate. The thermal conductivity of the coating is 0.40 W m^{-1} K^{-1}. Compare your answer with that of the previous question.

[3.8 × 10^2 K]

29–21   A cylindrical rod of cross-sectional area 5.0 mm^2 is made by joining a 0.30 m rod of silver to a 0.12 m rod of nickel. The silver end is maintained at 290 K and the nickel end at 440 K. Given that the thermal conductivities of silver and nickel are 0.42 kW m^{-1} K^{-1} and 91 W m^{-1} K^{-1} respectively, calculate

(a) the temperature of the join under steady conditions, and
(b) the rate of conduction of heat down the rod.

State any assumptions you need to make. (*You will find that a slide rule gives more reliable results if you work in terms of temperature differences from* 290 K. *Why?*)

[(a) 343 K   (b) 0.37 W]

29–22   *Power loss during conduction.* A uniform bar of cross-sectional area 80 mm^2 is made of metal of thermal conductivity 0.32 kW m^{-1} K^{-1}, and has one end kept at a steady high temperature. When steady conditions have been reached, the temperature gradients at two cross-sections along the bar are 4.0×10^2 K m^{-1} and 1.8×10^2 K m^{-1}. Calculate the average power loss from the surface of the bar between these two cross-sections.

[5.6 W]

29–23   0.60 kg of ice is placed in a wooden box which is 15 mm thick and which is covered with a 5.0 mm layer of insulating material. Calculate how long it would take all the ice to melt given that the temperature of the surroundings is 300 K, and that the effective area of the box is 1.8 m^2. The thermal conductivities of the wood and the insulation are 0.15 W m^{-1} K^{-1} and 30 mW m^{-1} K^{-1} respectively, and the specific latent heat of fusion of water is 0.34 MJ kg^{-1}. If you think that your answer disagrees with your experience then suggest a possible explanation.

[1.1 ks]

29–24   *The Searle bar method.* A heating coil is wrapped round one end of a metal bar and, when the current is 1.8 A, the p.d. across the coil is 20 V. The cross-sectional area of the bar is 1.2×10^{-3} m^2 and the average temperature gradient along the bar is measured as 75 K m^{-1}. Cooling water passes round the other end of the bar and, when the rate of flow is 3.0 g s^{-1}, the temperature increase is 2.5 K. Calculate

(a) the thermal conductivity of the metal
(b) the percentage power loss.

Assume c_{H_2O} [(a) 0.38 kW m^{-1} K^{-1} (b) 12%]

29–25 *The Lees disc method.* A glass disc is clamped between two brass discs of the same diameter, one of which is kept at 380 K by an electrical heater. When conditions are steady the temperature of the other brass disc is 368 K. If the current is switched off, its initial rate of cooling is 0.18 K s^{-1}. The brass discs have specific heat capacity 0.39 kJ kg^{-1} K^{-1} and each has mass 0.30 kg. Calculate the thermal conductivity of the glass if the disc has area 1.2×10^{-2} m^2 and thickness 5.0 mm. What assumptions do you make in your calculation?

[0.73 W m^{-1} K^{-1}]

29–26 *Effect of an air layer.* When measuring the thermal conductivity λ_x of a material X using the *Lees* disc method, a layer of air of thickness 1.0% of that of the disc becomes trapped between one face of the disc and the adjacent plate of the apparatus. If the thermal conductivity of the air is 0.050 times that of X, calculate the % error in the value of λ_x found by assuming that the disc is in thermal contact with the plate.

[17%]

Problems Involving Integration

*29–27 *Growth of ice on ponds.* When heat is removed from the water in a pond which is at 273 K by the cold air above it, an ice layer grows. If the thickness of the layer is x at time t, show that the rate of increase of thickness is given by $dx/dt = \lambda\Delta\theta/\rho lx$, where λ is the thermal conductivity of the ice, ρ the density, l the specific latent heat of fusion and $\Delta\theta$ is the temperature difference between the air and the water. Derive an expression for the time taken for the layer to grow to a thickness y. Why is this treatment approximate only? Under what conditions might the approximations be justified?

*29–28 Use the method of Qu. **29–27** to find the increase in thickness of ice on a pond over a period of twelve hours if the initial thickness was 80 mm. $\lambda = 2.2$ W m^{-1} K^{-1}, $\Delta\theta = 10$ K, $\rho = 9.2 \times 10^2$ kg m^{-3}, $l = 0.33$ MJ kg^{-1}.

[33 mm]

*29–29 *Radial flow between coaxial cylinders.* Show that the radial rate of outward flow of heat under steady-state conditions in a material which is between two coaxial cylinders of length l is given by $dQ/dt = 2\pi\lambda l(\theta_1 - \theta_2)/\ln(r_2/r_1)$. The inner cylinder has radius r_1, held at constant temperature θ_1, and the outer cylinder has radius r_2 held at constant temperature θ_2. Neglect end-effects. (*Hint: consider a cylindrical shell of thickness δr, and integrate between r_1 and r_2.*)

*29–30 A long wire is heated electrically at the linear rate of 4.0 kW m^{-1} and runs along the axis of an insulating cylinder. The wire has radius 0.20 mm and is at a temperature of 1.8×10^3 K. The insulating cylinder has radius 50 mm and the outer surface is at 3.0×10^2 K. Use the result proved in the previous problem to find the thermal conductivity of the insulating material.

[2.3 W m^{-1} K^{-1}]

*29–31 Use the result of Qu. **29–29** to calculate the rate of condensation of steam within every metre length of an iron delivery pipe which has internal and external radii of 40 mm and 50 mm. Assume that both the steam and (unrealistically) the inside of the pipe are at 373 K, and consider an air temperature of 288 K. The specific latent heat of vaporization of water is 2.3 MJ kg^{-1}, the thermal conductivity of iron is 65 W m^{-1} K^{-1} and the heat emitted from the surface of the pipe is 30 J m^{-2} for every kelvin excess temperature. (*Hint: the external temperature of the pipe will not be the same as the air temperature.*)

[0.35 g s^{-1}]

30 Thermal Radiation

Questions for Discussion

30–1 Compare typical values for the *speeds* at which heat is transferred by the three processes of conduction, convection and radiation. For the conduction process consider only the mechanism involving lattice vibrations (not the movement of electrons).

30–2 Name substances which are (*a*) adiathermanous but transparent (*b*) diathermanous but opaque (*c*) diathermanous

and transparent. Consider the terms transparent and opaque to refer to the visible spectrum.

30–3 A thermal radiation detector based on the photoelectric effect or on photoconductivity may fail to respond to radiation in the middle of the infra-red. Does this imply that these wavelengths are producing no heating effect? Explain.

30–4 What is the *microscopic* origin of the black-body radiation in a cavity?

30–5 Do you consider the term *black-body radiation* to be an appropriate one for radiation emitted (e.g.) through a small hole in the side of a furnace? What would you see if you looked into such a hole?

30–6 To what extent is a black body of radiation theory analogous to an ideal gas in kinetic theory?

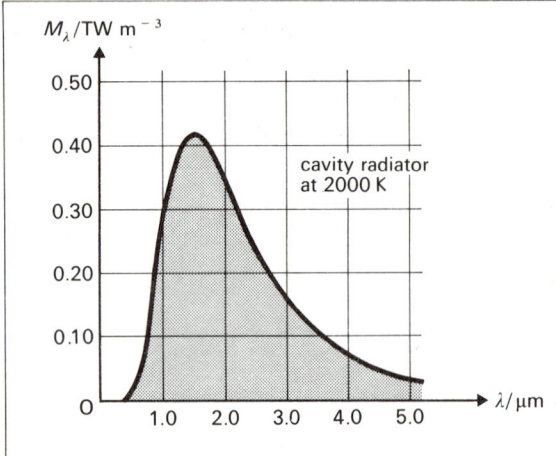

For Qu. **30–7**

*****30–7** Refer to the diagram. Molybdenum has an average emissivity of 0.21. Sketch a graph which shows the approximate variation in its spectral radiant exitance M_λ with wavelength.

*****30–8** If a piece of white porcelain with a dark pattern is heated to a high enough temperature then the bright and dark parts of the original pattern may be reversed when it is viewed by emitted light. Does this illustrate *Kirchhoff's Law*? Discuss.

30–9 Estimate the spectral emissivity of glass for the waveband 0.50–0.51 μm.

30–10 A particular surface is effectively a perfect reflector of a specified waveband. What is its spectral emissivity for that waveband?

30–11 Why does the Earth's atmosphere absorb a high proportion of the energy radiated *by* the Earth, but only a small proportion of that incident *on* the Earth from the Sun?

*****30–12** *Net loss of energy by radiation.* A body of area A and at temperature T shows black-body behaviour. It is placed in an enclosure whose walls are at temperature T_0, and which have a total emissivity ε. Write down

(*a*) the power radiated by the object,

(*b*) the power absorbed by the object from that emitted by the walls by (*i*) radiation (*ii*) scattering and reflection.

Hence show that the net power loss is $\varepsilon A\sigma(T^4 - T_0^4)$.

30–13 How would your answer to Qu. **30–12** be modified if

(*a*) the body had an area which was very small compared to the walls, *or*

(*b*) the walls also showed black-body behaviour?

30–14 A light bulb with a tungsten filament is about 10% efficient. How could its performance be improved?

30–15 Can a total radiation pyrometer measure the *real* temperature inside a furnace, or does it measure only the black-body temperature?

30–16   O-M What is the initial rate of cooling of a tungsten filament when an electric lamp is switched off?

30–17   O-M Use some order-of-magnitude calculations to establish whether a domestic radiator warms a room more by radiation or by convection.

30–18   O-M A spherical ball of lead of mass 10 kg is placed at the principal focus of a parabolic reflector whose axis is aimed at the Sun. Estimate the minimum effective area of cross-section that the reflector must have to bring the lead to its melting point.

Quantitative Problems

*****30–19** The interior of a furnace, which can be treated as an equal temperature enclosure, is at 3.0×10^3 K. How much energy is lost by radiation if a door of area 0.40 m² is opened for 30 s?

Assume σ   [55 MJ]

*****30–20** The total radiant exitance of tungsten at 2.45×10^3 K is 0.50 MW m⁻². Calculate

(*a*) that of a black body at the same temperature, and hence

(*b*) the total emissivity of tungsten at that temperature.

What does this enable you to say about its absorptance?

Assume σ   [(*b*) 0.25]

*****30–21** A body at 3.0×10^3 K is radiating energy. By how much must the temperature be changed to reduce the power output by 1%?

[8 K]

*****30–22** *Irradiance.* What is the irradiance (the incident radiant flux on an area A divided by A) within a *cavity* whose interior walls are maintained at 2.0×10^3 K if they have a total absorptance of (*a*) 1.0 (*b*) 0.50?

Assume σ   [(*a*) 0.91 MW m⁻²]

*30–23 A *small* hot body of total emissivity 1, surface area 8.0×10^{-2} m^2 and heat capacity 20 J K^{-1} is placed in a room whose walls are at 300 K. Calculate its rate of fall of temperature when it is at 600 K. Explain the significance of the body being *small*.

Assume σ [28 K s^{-1}]

*30–24 A tungsten filament of total emissivity 0.32 has a diameter 0.10 mm, and a length 0.25 m. At what temperature should it operate if it is rated at 100 W? Say what assumptions you make.

Assume σ [2.9×10^3 K]

*30–25 *Storage of liquid helium.* Liquid helium at 4 K is usually stored in a *Dewar* (or vacuum) flask whose outside walls are at 78 K, since they are cooled by liquid nitrogen. How much greater would the rate of heat gain by radiation be if the *Dewar's* outer walls were at room temperature (say 300 K)? Give an order-of-magnitude answer only.

[2×10^2]

*30–26 In a particular *Dewar* flask helium, whose standard boiling point is 4.2 K, is kept in a container whose total absorptance is 0.34, and whose surface area is 8.0×10^{-3} m^2. The container is surrounded by walls maintained at 78 K by liquid nitrogen. The liquid helium has a specific latent heat of vaporization of 25 kJ kg^{-1}. Calculate the rate at which

(*a*) the inner container radiates energy,
(*b*) the inner container absorbs energy, and
(*c*) the helium boils away.

Assume σ [(*c*) 2.3×10^{-7} kg s^{-1}]

*30–27 *The Solar constant.* The Sun, whose radius is 700 Mm, can be considered to radiate as a black body of temperature 5.80×10^3 K. Its mean distance from the Earth is 150 Gm.

The value of the Solar constant measured at the Earth's *surface* when the sky is cloudless is 1.35 kW m^{-2}. Calculate

(*a*) the total power radiated by the Sun,
(*b*) the power received by an area 1.00 m^2 placed normal to this uninterrupted radiation at the edge of the Earth's atmosphere, and
(*c*) the power absorbed before this radiation reaches the corresponding 1.00 m^2 on the Earth's surface.

What becomes of the loss in (*c*)?

Assume σ

[(*a*) $3.9(5) \times 10^{26}$ W (*b*) 1.40 kW m^{-2} (*c*) 5×10^1 W m^{-2}]

*30–28 The radius of the Earth's orbit is 2.2×10^2 times the Sun's radius, and the black-body temperature of the Sun is 6.0×10^3 K. Suppose that the Earth receives radiation from the Sun only, and that it can be treated as a black body.

(*a*) What, under these conditions, would you expect the mean temperature of the Earth's surface to be? (Ignore any internal generation of energy.)
(*b*) Where are the assumptions you have made most likely to break down in practice? What, for example, is the net effect of the oceans?
(*c*) Indicate qualitatively how the temperature of a planet is likely to be related to its radius and its mean distance from the Sun.

[(*a*) 2.9×10^2 K]

*30–29 *Wien's displacement law.* (*a*) The tungsten filament of an electric light bulb acquires an equilibrium temperature of 2.5×10^3 K. At what wavelength would the graph plotting its spectral radiant exitance against wavelength show a maximum value? To what part of the spectrum does this correspond?

(*b*) At what surface temperature would an astronomical body show a corresponding maximum at a wavelength of 50 nm?

Assume $\lambda_{max} T = 2.9$ mm K   [(*a*) 1.2 μm   (*b*) 5.8×10^4 K]

VI Geometrical Optics

31   Principles of Geometrical Optics

Questions for Discussion

31–1 *The nature of light.* We can view the propagation of electromagnetic waves in three ways: (1) as straight rays, (2) as waves, and/or (3) as photons of discrete energy.

(*a*) To what extent are these ideas mutually compatible? (*Think in terms of some correspondence principle.*)
(*b*) Give examples of situations in which the explanation offered by one view is obviously superior to that of the others.

31–2 Is it true to say that there is no interaction between visible light and a transparent object through which it is transmitted?

31–3 How is it that an eclipse of the Sun is sometimes *total*, and sometimes *annular*?

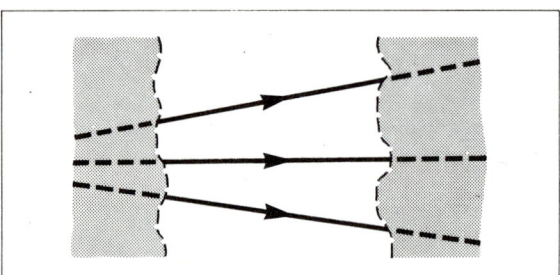

For Qu. **31–4**

31–4 Refer to the diagram. Could the three rays shown

(*a*) come from an object
(*b*) come from a real image
(*c*) come from a virtual image
(*d*) form an image on a photographic plate?

Justify each answer by an appropriate complete diagram.

31–5 Is it possible to view a *point* object with a point plane mirror? If not what size of mirror is needed, and why?

31–6 Is reflection *always* independent of wavelength? (Could we ever distinguish wavelengths by some reflection process?)

31–7 A plane mirror is said to produce **lateral inversion** in the image (that is, it is reversed from side to side). Is it also reversed from top to bottom?

31–8 A small object is placed 80 mm in front of a plane mirror. Where would you place your eye to view it with maximum clarity?

31–9 The image of the setting Sun viewed by its reflection in the surface of the sea is often elongated in the line of vision, but not sideways. Why?

31–10 Many stars appear white. To what extent is this evidence that electromagnetic waves of all colours travel through a vacuum at the same speed?

31–11 Discuss how the corpuscular (particle) theory of light can account for *partial* reflection and transmission. (*Examine in detail the possible behaviour of a single particle at an interface.*)

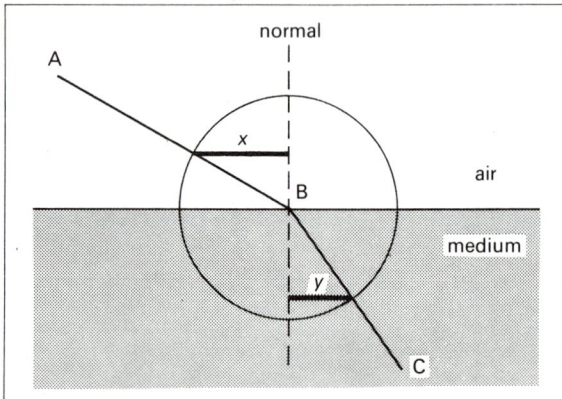

For Qu. **31–12**

31–12 Refer to the diagram. Prove that the value of x/y is the relative refractive index for light passing from air into the medium.

31–13 How does an object which is transparent, and virtually invisible, such as an eddy of warm air above a convector heater, cast a detectable shadow?

31–14 A parallel beam of light enters a block of glass obliquely from air. How is its intensity altered, other than by absorption and partial reflection?

31–15 Refer to the diagram on the next page, which shows a light ray incident on a series of parallel-sided air layers. Complete the path of the light ray, and deduce whether total reflection is possible at the interface AB. Name some natural phenomena which result from this effect.

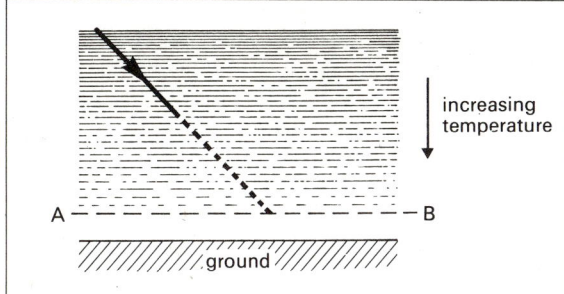

For Qu. 31–15

31–16   An observer stands at the edge of a pool filled to a uniform depth. Why does the far end appear more shallow than that near his feet?

31–17   It is possible for a transparent material to cause a refracted ray to lie in a plane other than that defined by the incident ray and the normal. What physical properties does such a material have?

31–18   Is the persistent red colour of the setting Sun a refraction effect?

31–19   Why do stars twinkle? Consider, in formulating your explanation, both the optics of the problem, and also the physiology of the eye.

31–20   Is it possible for a light ray to enter a transparent homogeneous sphere, and *then* to suffer total internal reflection?

31–21   A cube of transparent material of critical angle 44° is placed over a mark on a piece of dry paper. Can the mark be seen through the vertical sides? What might happen if the paper were damp?

31–22   Refer to the *Pulfrich refractometer* of Qu. **31–35**. For what ranges of values for n_g and n_l do you think the instrument will be useful?

31–23   Consider this definition of an **image**: the point image of a point object is that *single* point through which *all* rays which leave the point object, and which are intercepted by an optical system, are made to pass, or appear to pass. Use *this* criterion to decide which of the following systems can produce an image:

(*a*) a plane mirror
(*b*) a spherical mirror of (*i*) large (*ii*) small aperture
(*c*) a parabolic mirror, using a distant point object (*i*) off the axis (*ii*) on the axis
(*d*) a uniformly dense optical medium in which an immersed point object is viewed from a rarer medium at nearly normal incidence
(*e*) a lens bounded by spherical surfaces.

*31–24   Discuss, without detailed analysis, the formation of a rainbow. Your answer should include a sketch to indicate ray behaviour, and to show the direction along which the observer should look.

Quantitative Problems

Reflection

†**31–25**   When *Boys* measured *G* he observed a light spot which was reflected from the glass beam onto a screen 7.00 m away. The *total* deflection of the spot was 184 mm. Through what angle was the end of the quartz fibre twisted?

[13.1 mrad]

†**31–26**   A tall man is 2.0 m high. Draw a plane mirror an arbitrary distance from the man, and trace the paths of rays from the top of his head and his feet into his eyes.

(*a*) What is the smallest length of mirror that would enable him to see his whole image?
(*b*) What is the length of this image?
Would it be seen with equal clarity for all man-mirror distances? Where would *you* choose to put the mirror?

[(*a*) 1.0 m]

Refraction

†**31–27**   The following is a set of measurements of the angles made by a light beam with the normal in air and water respectively.

$\theta_a/°$	10.0	20.0	30.0	40.0	50.0	60.0	70.0	80.0
$\theta_w/°$	7.75	15.0	22.1	28.8	35.1	40.6	45.3	48.2

(*a*) Draw a graph of θ_a (*y*-axis) against θ_w.
(*b*) Using the same axes, draw a graph of $\sin \theta_a$ against $\sin \theta_w$, and comment.
(*c*) From your graph find the relative refractive index of water, and a value for the critical angle for these media.

[(*c*) 1.33, 48.6°]

†**31–28**   *Wave speeds.* The relative refractive indices for light of wavelengths 405 nm and 770 nm passing into vitreous silica are respectively 1.470 and 1.454. Calculate the speeds of the two wavelengths, and the difference of their speeds. What colours are represented by these wavelengths? Which wave travels faster?

Assume c   [$\Delta c = -2.(2)$ Mm s^{-1}]

†**31–29**   *Approximations.* At small angles $\sin \theta \approx \theta$, where θ is expressed in radians. Light is incident at an angle 10.0° from air to a plane surface of relative refractive index 1.50. Calculate exact and approximate values for the angle of refraction, and comment on your answer.

31–30 A fish looks straight upwards through a water-air interface, and sees a fly apparently 40 mm directly above. At what point should the fish aim in order to catch the fly? $_a n_w = 1.33$. (*Draw a large ray diagram from which to relate the real and apparent heights from first principles.*)

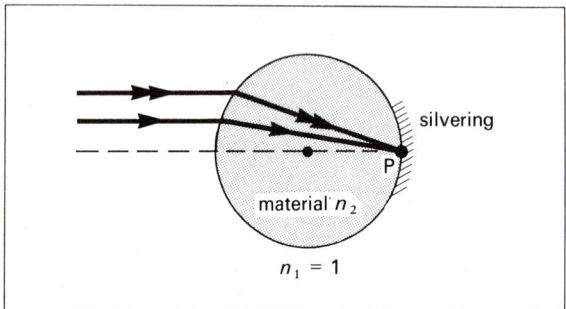

For Qu. **31–31**

31–31 Refer to the diagram. For what value of n_2 will a parallel beam of *paraxial* rays be focused at P? What would be the effect of silvering the back surface as shown? (**Scotchlite** *is a material which uses this principle. It has tiny glass spheres embedded in a silver background, and is used (e.g.) for road signs.*)

[2.0]

*31–32 A student is measuring the *internal* diameter of a cylindrical glass tube with a travelling microscope. Instead of viewing the plane end of the tube, and obtaining the true answer d, he observes through the curved side, and finds the apparent diameter D. How are d and D related to n, the refractive index of the glass? Why must one assume that the internal diameter is much less than the external diameter?

[$D = nd$]

Total Internal Reflection

†31–33 (*a*) The critical angle for diamond is 27°. What is its refractive index?

(*b*) Potassium iodide has a refractive index 1.667, and that of aniline is 1.586. Calculate the critical angle for an interface between the two media. Which way across an interface would a ray have to be travelling if it were to be totally reflected?

(*c*) At a particular temperature and for a particular light frequency polystyrene has a refractive index 1.586. How would it appear when totally immersed in aniline, *and why*?

[(*a*) 2.2   (*b*) 72°]

31–34 *The air cell.* A liquid is poured into a rectangular perspex tank. In the liquid is placed a parallel-sided air film which is sealed between a pair of microscope slides (an **air cell**). A ray of light is shone through the tank and cell, and the source is observed on the far side.

(*a*) When sodium light is used the observer can see the source while the cell is rotated about a vertical axis through an arc of 96.0°, but not outside these limits. Calculate the refractive index of the liquid.

(*b*) If the sodium source were replaced by a white-light source, what would the observer see near the cut-off positions?

(*c*) Show that the refractive index of the material of the microscope slide does not affect the measurement in any way.

(*The solution of this problem is greatly simplified by the symmetrical statement of Snell's Law: $n \sin \theta_n = constant$.*)

[1.35]

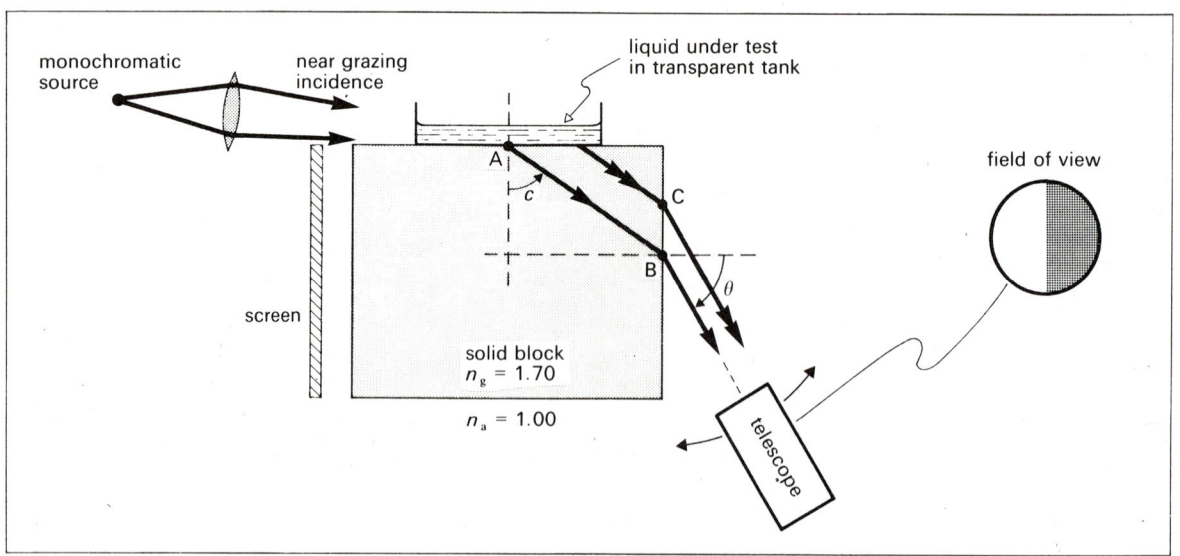

For Qu. **31–35**

31–35 *Pulfrich refractometer.* Refer to the diagram. The telescope is rotated until the field of view assumes the appearance shown.

(*a*) In one particular measurement the angle θ was found to be 60.0°. Calculate the refractive index of the liquid under test.

(*b*) Explain carefully why the beam of light emerging at C does not cause the region to the right of the crosswire to appear bright.

[1.46]

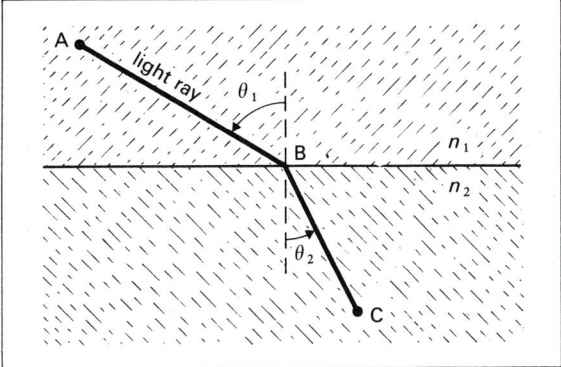

For Qu. **31–36**

Fermat's Principle

*31–36 *Fermat's principle. The principle states that the time taken by light to travel from a point A to a point B has a stationary value (which is usually a minimum).*

(*a*) Show that when an object gives rise to an image by reflection in a plane mirror, Fermat's principle is obeyed if the angles of incidence and reflection are equal.

(*b*) Refer to the diagram. Show that the principle is obeyed provided $n_1 \sin \theta_1 = n_2 \sin \theta_2$ (*Snell's Law*).

(*c*) An ellipse can be drawn using a pencil, a taut piece of cotton and two fixed pins (the foci). Use this information and *Fermat's principle* to show that light rays radiating from a source placed at one focus must be focused, after reflection, at the other.

32 The Prism and Thin Lens

Questions for Discussion

32–1 (*a*) Under what conditions may we use

$$n = \sin \tfrac{1}{2}(A + D_{min})/\sin \tfrac{1}{2}A$$

with prisms of small angle?

(*b*) Can the equation $D = (n - 1)A$ be applied to minimum deviation?

32–2 Why is it unwise to refer to a given lens in general as 'diverging' or 'converging'? (*Is the statement necessarily valid at all times?*)

32–3 Refer to the diagram. Complete it by drawing in the lens, its principal foci and two further rays from O to I.

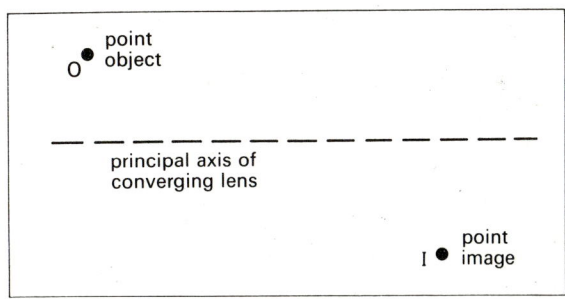

For Qu. **32–3**

32–4 Discuss these conflicting statements.

(*a*) Light from the Sun is parallel, and therefore it forms a real point image at the principal focus of a converging lens.
(*b*) The Sun is an extended object which therefore produces an extended image.

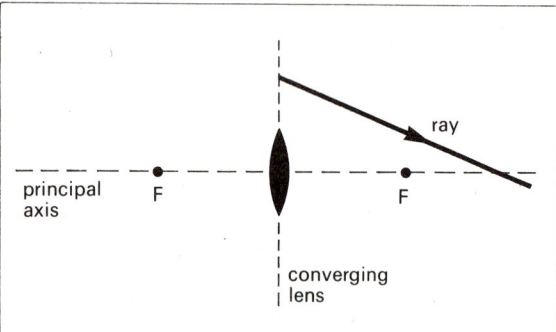

For Qu. **32–5**

32–5 Refer to the diagram. Construct the complete path (before refraction) of the ray shown.

32–6 Show that the size of the image of the Sun formed by a converging lens is proportional to its focal length. What properties of a converging lens would make it a good burning glass?

32–7 Rewrite the lens equation $1/u + 1/v = 1/f$ in the form $xy = c^2$ (that of a rectangular hyperbola), and sketch the curve of u plotted against v.

(*a*) What are the equations of the asymptotes?
(*b*) What information is given by drawing the line $u = v$?

32–8 Indicate *three* conditions that must be observed in situations to which the lens-makers' formula can be applied with some accuracy.

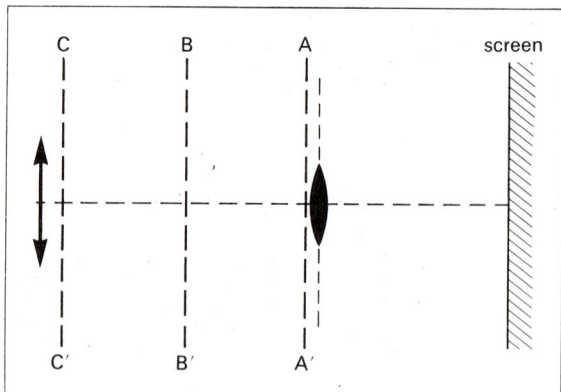

For Qu. **32–9**

32–9 Refer to the diagram, in which the double arrow is an extended object which forms an image on the screen. Ignore diffraction effects, and discuss what changes would be observed on the screen in the following circumstances.

(*a*) A large opaque card is moved (*i*) upwards from the bottom, and then (*ii*) downwards from the top, along the planes AA′, BB′ and CC′ respectively.
(*b*) A circular opaque disc of diameter equal to the lens radius is inserted along AA′ until it is coaxial with the lens. (*This answer should be quantitative.*)

32–10 A *thin* lens whose surfaces have different radii of curvature forms an image on a screen. What changes would be observed if the lens were turned to present the other face to the incident light?

32–11 What condition must be satisfied for a concave (diverging) lens, or a plane mirror or a convex (diverging) mirror to give rise to a real image?

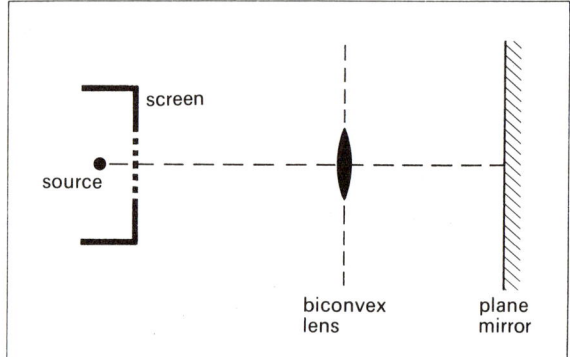

For Qu. **32–12**

32–12 Refer to the diagram. Suggest three different ways in which it may be possible for a real image to be formed adjacent to the object.

32–13 How would you measure the focal length of a thick lens if it was mounted in such a way that the exact location of its optic centre was uncertain?

*32–14 *Longitudinal magnification.* *Transverse* magnification ($m = v/u$) refers to objects placed *across* the principal axis. *Longitudinal* magnification refers to objects placed *along* the axis, and has the form $\Delta v/\Delta u$. Show that for a *very short* object the longitudinal magnification has the same size as m^2.

32–15 **O-M** *Depth of focus.* A miniature camera whose lens has focal length 50 mm is focused for infinity. If the diameter of the aperture is 10 mm, find the smallest distance away from the camera at which an object may be placed so that its image is acceptably sharp. Ignore diffraction effects and all lens aberrations.

Quantitative Problems

Prisms

†**32–16** *Minimum deviation.* A ray passes symmetrically through a prism, and is deviated by 48°. The prism has a refracting angle of 60°. Draw a representative ray diagram, and calculate the following:

(a) the angle between the normal and the ray in the glass
(b) the deviation at each refraction
(c) the angle between the normal and the ray in the air
(d) the refractive index of the prism material.

[(d) 1.62]

†**32–17** *A small-angled prism.* A ray of light is incident at an angle 3.0° on a prism of refracting angle 5.0° and refractive index 1.50. Calculate

(a) the first angle of refraction
(b) the second angle of incidence (in the glass)
(c) the second angle of refraction (in the air)
(d) the total deviation of the ray.
(*Use Snell's Law in the form* $\theta_a \approx n_g \theta_g$, *since the angles are small.*)

(e) What would the deviation have been for a ray incident at an angle of 7.5° on the first surface?

[(d) 2.5°]

32–18 An isosceles triangular glass prism of $n_g = 1.50$ has two sides of 100 mm, and the third is 10 mm. A ray of light makes an angle of incidence ~ 0.1 rad on a long face. Through what angle is it deviated?

[50 mrad]

32–19 *The biprism.* An isosceles glass prism has $A = 178.0°$ and $n_g = 1.60$. A parallel beam of monochromatic light is incident on the prism normal to the long face. What is the angle between the two refracted beams?

[1.2°]

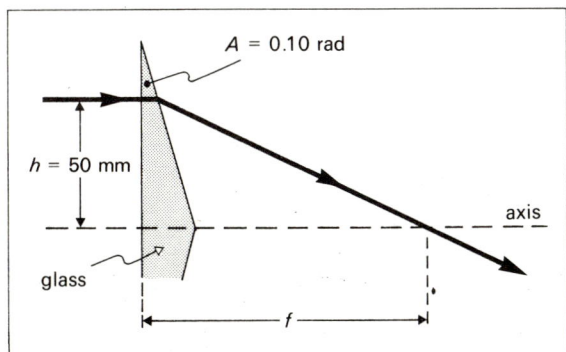

For Qu. **32–20**

32–20 Refer to the diagram, which shows a thin prism of refractive index 1.50.

(a) Calculate the quantity marked f.
(b) How could you make f independent of h if (i) the prism were to retain its present shape, but could be inhomogeneous (its refractive index could vary from point to point), or (ii) there were to be no variation of n, but the shape of the solid body could be altered?
Give qualitative answers.

[(a) 1.0 m]

Objects and Images

†**32–21** A beam of parallel rays of light makes an angle 0.10 rad with the principal axis of a converging lens of 150 mm focal length. Find the position of the image (two coordinates) and its size.

†**32–22** A converging lens has a focal length 200 mm. Find the position, nature (real or virtual), orientation (upright or inverted) and size of the image of a 10 mm object placed (a) 100 mm (b) 200 mm (c) 400 mm (d) 800 mm from the optic centre. If you use a lens equation then each answer should be illustrated by an appropriate *ray* diagram which shows how the image is formed.

†**32–23** Repeat Qu. **32–22** (a) and (c) for a diverging lens of the same focal length.

†**32–24** Choose a focal length for a projection lens which is to give an image 1.4 m wide from a 35 mm transparency in a room 6 m long. (*Your answer should be made a round number.*)

[0.15 m]

†**32–25** *Camera focusing.* (a) In a typical miniature camera the lens focal length is fixed at 50 mm, and the focusing range extends from infinity down to 1.0 m. What range of movement between lens and film is necessary?

(b) A life-sized photograph of an insect is needed. What would the film/lens separation be? (*This would be achieved by the use of extension tubes.*)

[(a) 2.6 mm]

32–26 *The virtual object.* An object 10 mm high is placed 400 mm in front of a converging lens of focal length 200 mm. A second converging lens of the same focal length is placed in the further focal plane of the first.

(a) Where would the first lens have formed an image?
(b) What is the size and position of the final image?

[(b) 5.0 mm high, 100 mm from second lens]

32–27 *Newton's formula.* A gauze 12 mm in diameter was placed in front of a lens enclosed in a tube. A real image of diameter 48 mm was located 0.80 m beyond the principal focus on the image-side of the lens. Calculate its focal length.

[0.20 m]

32–28 Two bright slits are separated by a distance s, and placed 1.0 m from a white screen. A lens placed between the slits and screen gives an image of the slits in which they are separated by 40 mm. The same lens, placed in a different position, gives a second image in which the slit separation is 2.5 mm.

(a) What is the value of s?

(b) Calculate the focal length of the lens.

(*This sort of procedure can be used to find the separation of the two virtual slit images in a Fresnel biprism experiment.*)

[(a) 10 mm   (b) 0.16 m]

Lens Construction

†32–29 Calculate four possible values for the focal length of a lens made to the following specification: the material has $n = 1.50$, and the spherical surfaces have radii of curvature 0.20 m and 0.40 m. Draw sketches which illustrate the lenses.

[± 0.27 m, ± 0.80 m]

32–30 A thin planoconvex lens of diameter 60 mm is to have a focal length 0.40 m, and is made of glass of refractive index 1.50. If the lens is very thin at its edge, what is its thickness at the centre?

[2.2 mm]

32–31 The real thickness of the centre of an equiconvex lens is measured by travelling microscope to be 3.0 mm. The lens diameter is 50 mm, and its focal length 0.20 m. Calculate

(a) its radii of curvature, and

(b) the refractive index of the lens material.

[(a) 0.21 m   (b) 1.52]

32–32 A biconvex glass lens has a focal length in air of 0.20 m, and the glass has $n_g = 1.50$. It is submerged in a solution of potassium iodide for which $n = 1.80$. Think carefully about (i) the sign convention for the radii of curvature, and (ii) the term that is used to take account of the two different refractive indices, and calculate the new focal length.

[0.60 m, diverging]

32–33 *Camera lens aperture.* The **aperture** of a camera lens is usually quoted by giving its diameter relative to its focal length. Thus

$$aperture = f/5.6.$$

(a) The correct exposure for a particular subject is 40 ms at an aperture $f/16$. What aperture should be used if a moving subject makes it desirable to use a shutter time (usually called 'speed') of 2.5 ms?

(b) Would a different relative aperture have to be used if the nature of the subject required the use of a lens of double focal length?

(c) Suggest why $f/12.5, f/9, f/6.3, f/4.5$ is a series of apertures often marked on camera lenses.

[(a) $f/4.0$]

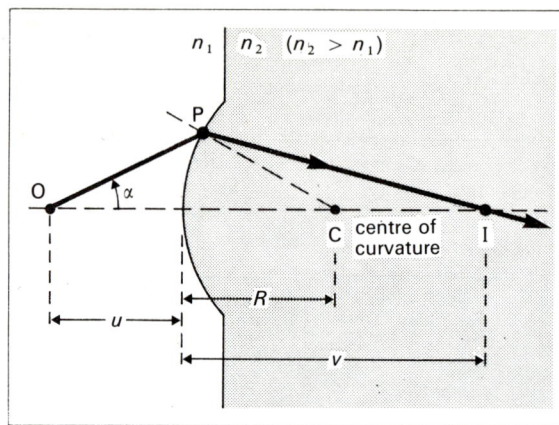

For Qu. **32–34**

*32–34 *Refraction at a single spherical surface.* Refer to the diagram. Assume that the ray shown is paraxial (that α is small), and apply the symmetrical form of *Snell's Law* to the refraction at P to prove

$$\frac{n_1}{u} + \frac{n_2}{v} = \frac{n_2 - n_1}{R}.$$

(a) What is the focal length of this surface? (*Let* $u \to \infty$.)

(b) What is the relationship between an object's true depth (u) and apparent depth (v) when viewed across a plane interface? (*Let* $R \to \infty$. *The equation derived here must simplify in agreement with the well-known result.*)

Lens Combinations

32–35 *Principle of the supplementary lens.* A camera whose lens has a focal length 50 mm is focused on an object 2.0 m away. An object is now placed on the 0.50 m mark, and an image is obtained on the film in the same position by placing a **supplementary lens** over that of the camera lens. What is the power of this lens?

[1.5 rad m^{-1}]

32–36 *Principle of the telephoto lens.* In a camera a converging lens of focal length 175 mm was mounted 75 mm in front of a second lens which was diverging, and of focal length 150 mm. The camera was used to photograph a distant object.

(a) Where would an image have been formed by the converging lens used on its own?

(b) Where is the final image formed by the combination?

(c) What is the magnification produced by the diverging lens?

(d) What focal length converging lens would have to be used (on its own) to give an image of the same size?

Illustrate your answer with ray diagrams. (*Note that the increased weight involved in (d) might lead to mechanical distortion of the camera body.*)

[(c) 3.00   (d) 525 mm]

32–37 *A liquid lens.* Glycerol ($n = 1.47$) is poured onto a horizontal plane mirror, and then a thin equiconvex glass lens ($n_g = 1.60$) is placed on top, so that the glycerol takes the form of a thin planoconcave lens. The glass lens's radii of curvature are 300 mm. A bright object placed above the lenses forms an image coincident with itself a distance f above the mirror.

(a) Calculate the focal length of the glass lens.

(b) What values of R would be substituted into the lens-makers' formula for the liquid lens?

(c) What is the value of f?   [(c) 0.41 m]

The Boys Image

32–38 (a) A planoconvex lens is placed with its curved surface facing a bright source. When the lens-source separation is 400 mm, a faint image appears alongside the source. Calculate the focal length of the lens.

(b) When the lens is turned so that the source is nearer the plane surface, the faint image is formed when the separation is 133 mm. Calculate the refractive index of the lens material.

[(b) 1.50]

33 Spherical Mirrors

Questions for Discussion

33–1 A small spherical object is placed so that its centre of curvature coincides with that of a concave mirror. Describe the shape, size and position of the image.

33–2 Discuss the advantages and disadvantages of using a diverging (convex) mirror as a driving mirror in a motor car.

33–3 Is it possible for a diverging (convex) mirror to produce an image which is (a) enlarged, or (b) real? Discuss.

33–4 An image is to be formed by a single spherical mirror of a bright distant object. Discuss the relative brightnesses of images formed by mirrors of

(a) focal length f and aperture a

(b) focal length f and aperture $2a$

(c) focal length $2f$ and aperture a

(d) focal length $2f$ and aperture $2a$.

Quantitative Problems

†33–5 A converging (concave) spherical mirror has a radius of curvature of 400 mm. Find the position, nature (real or virtual), orientation (upright or inverted) and size of the image of a 10 mm object placed the following distances from the optic centre: (a) 100 mm (b) 200 mm (c) 400 mm (d) 800 mm.

If you use the mirror equation then each answer should be illustrated by an appropriate *ray* diagram which shows how the image is formed.

†33–6 Repeat Qu. **33–5** (a) and (c) for a diverging (convex) mirror of the same radius of curvature.

†33–7 A converging spherical mirror has a radius of curvature 400 mm.

(a) If your eye is 300 mm from the mirror, what is the location and nature of its image? Can the image be seen?

(b) If the eye is withdrawn until it is 600 mm from the mirror, could the image then be seen? Would the new image be different in nature from that in (a)?

33–8 (a) An object 20 mm high is placed 100 mm from a spherical mirror, and gives rise to an erect image which is 40 mm high. What is the mirror radius of curvature?

(b) If the erect image had been 10 mm high, what would then have been the radius of curvature?

Draw ray diagrams which illustrate your answers.

[(a) 400 mm, converging   (b) 200 mm, diverging]

33–9 An object forms images twice its own size when placed either 50 mm or 150 mm from the pole of a spherical mirror. What is the focal length of the mirror?

[0.10 m, converging]

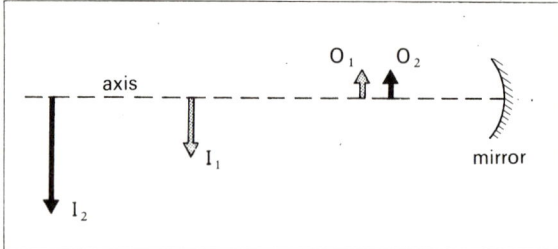

For Qu. **33–10** *(not to scale)*

33–10   Refer to the diagram. The linear magnification for position 1 is $2\times$, and that for position 2 is $4\times$. The distance from O_1 to O_2 is 40 mm.
Calculate

(*a*) the focal length of the mirror
(*b*) the distance from I_1 to I_2.

[(*a*) 0.16 m   (*b*) 0.32 m]

33–11   *Measurement of n.* Glycerol is placed in a concave spherical mirror of radius of curvature 0.50 m, and covered by a microscope slide with which it makes contact. An object coincides with its own image when it is 0.34 m from the mirror's pole. Consider arbitrary rays that are incident (say) 20 mm from the pole, and calculate

(*a*) their angles of incidence and refraction, and *hence*
(*b*) the refractive index of glycerol.
What is the significance of the microscope slide?

[(*b*) 1.47]

33–12   Calculate the radius of curvature of a diverging (convex) spherical mirror from the following information. An object was placed 200 mm in front of the mirror, and the image produced coincided with that of the same object formed by a plane mirror placed 150 mm from the object.

[400 mm]

33–13   *Measurement of f for diverging mirror.* A luminous object is placed 400 mm from a converging lens of focal length 200 mm. A diverging (convex) mirror is placed on the far side of the lens, and coaxial with it. When the mirror is moved backwards and forwards, a real image is formed coincident with the luminous object when the mirror is 100 mm from the lens. What is the focal length of the mirror?

[150 mm]

*33–14   *Longitudinal magnification.* An object 2.0 mm long is placed along the axis of a converging mirror of focal length 0.12 m. The mid-point of the object is 0.20 m from the pole.

(*a*) Draw a sketch diagram to show the position and orientation of the image.
(*b*) Calculate (*i*) the position of the image of the mid-point, and (*ii*) the length of the image.
(*Hint: refer to Qu.* **32–15**.)

[(*b*) (*ii*) 4.5 mm]

34   Lens and Mirror Aberrations

Questions for Discussion

34–1   *Spherical aberration for a lens.* In one proof of the lensmakers' formula we use the relationship

$$D = (n-1)\,A.$$

(*a*) Under what conditions does this equation hold?
(*b*) For which parts of a lens (inner or outer zones) are the conditions most likely to be flouted? Draw a diagram which illustrates your answer.
(*c*) A planoconvex lens and an equiconvex lens of the same focal length are available for focusing a parallel beam of light. Which would you choose, why would you choose it, and how would you use it?

34–2   Two equiconvex lenses are made of materials of refractive index 1.5 and 1.6 but they have the same power. Which will show less spherical aberration?

34–3   *Spherical aberration for a mirror.* Draw a large semicircle to represent a section through a hemispherical converging mirror. Trace accurately the paths of rays making up a beam of light parallel to the principal axis.

Discuss the effect of reducing the aperture of the mirror relative to the radius of curvature.

34–4 The spherical aberration of a concave mirror can be explained by saying that the inner and outer zones have different powers. Which zones have the greater? In the **Schmidt system** this aberration is corrected by placing a thin lens in front of the mirror. Discuss the form of this lens, and explain its action.

34–5 Is there a place at which a point object can be placed so that its image formed by a concave mirror is also a point (without spherical aberration)? (*Such a position would be called* **anastigmatic**.)

34–6 Why does an uncorrected converging lens *appear* to show no chromatic aberration when it is used as a magnifying glass? What is the significance of this in the design of optical instruments?

34–7 A converging glass lens has $f_{red} > f_{blue}$. What about a similar diverging lens?

34–8 Is it in principle possible (by appropriate grinding of surfaces) to make a lens free of *all* aberrations when it focuses monochromatic light?

Quantitative Problems

Spherical Aberration

34–9 A ray parallel to the principal axis is incident on a spherical mirror at an angle θ. Suppose f_θ is the focal length for this ray, and f that for paraxial rays.
  (a) Prove $\Delta f/f = (\sec\theta - 1)$ where $\Delta f \equiv (f - f_\theta)$.
  (b) Calculate the value of θ for which $\Delta f/f = 1\%$.
  (c) Calculate the aperture of a mirror for which $\Delta f/f \leqslant 1\%$, and $f = 1.0$ m.

[(a) 8.1°   (b) 0.56 m]

Chromatic Aberration

34–10 The **longitudinal chromatic aberration** of a lens is defined to be $(f_C - f_F)$, where C and F refer to the wavelengths of the C and F *Fraunhöfer lines* (emitted by hydrogen).

Use the lens-makers' equation, and the relationship $f_C f_F \approx f_D^2$ to prove

$$(f_C - f_F) = \left(\frac{n_F - n_C}{n_D - 1}\right) f_D$$

where D refers to the yellow D-line (emitted by sodium). (*The quantity* $\left(\dfrac{n_F - n_C}{n_D - 1}\right)$ *is called the* **dispersive power** *of the material.*)

34–11 The following measurements were made for a thin glass lens: $f_C = 1016$ mm and $f_F = 984$ mm. Calculate an approximate value for the dispersive power of the lens material.

[0.032]

34–12 *Calculation of longitudinal chromatic aberration.* The curved surface of a planoconvex lens has a radius of curvature of 400 mm, and is made of a crown glass for which $n_C = 1.515$ and $n_F = 1.520$. Calculate the value of $(f_C - f_F)$.

[7.5 mm]

*34–13 *Design of the achromatic doublet.* Two lenses of focal lengths f_1 and f_2 are made of different materials of dispersive powers ω_1 and ω_2, and put into contact. Use the result of Qu. **34–10** to show that when

$$\frac{\omega_1}{(f_1)_D} + \frac{\omega_2}{(f_2)_D} = 0,$$

then $f_C - f_F = 0$, i.e. that chromatic aberration is eliminated for this pair of focal lengths. (*Note that if one lens is converging, then we have proved that the other must be diverging since ω is positive for all ordinary optical materials.*)

*34–14 A converging achromatic doublet is to have a focal length f_D of 0.40 m, and the glasses available are flint ($\omega = 0.027$) and crown ($\omega = 0.020$).
  (a) Show that the converging component must necessarily be made of crown glass.
  (b) Calculate (using the result of Qu. **34–13**) the focal lengths of the two lenses.

(*Note that such a pair of lenses is usually constructed with a common face, and hence a common radius of curvature for a surface of each component.*)

[(b) Crown 0.10 m, flint 0.14 m]

35   Dispersion and the Spectrometer

Questions for Discussion

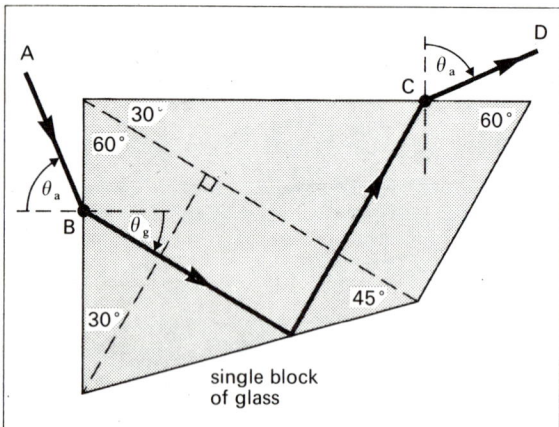

For Qu. 35–1

35–1   *The constant deviation prism.* Refer to the diagram.

(a) For what value of θ_g will the two angles marked θ_a be equal?

(b) What is then the angle between the incident and emergent rays?

(c) What would be observed along CD if a beam of white light was incident along AB, and the prism slowly rotated?

35–2   A parallel beam of light has two components of wavelength 656 nm (the *Fraunhöfer* C-line) and 486 nm (the F-line). It is made to pass through a potassium iodide solution for which $n_d = 1.80$. Would it be possible, by using a parallel-sided plate of crown glass for which $n_d = 1.54$, to separate the original beam into its two components?

35–3   It has been suggested that the variation in the refractive index of glass with wavelength can be represented by
$$n - 1 = A + Bf(\lambda).$$

Given that B is positive, examine a graph which plots n against λ, and suggest a simple form for $f(\lambda)$.

35–4   If the slit of a spectrometer is made wide, then the image is more easily focused since more energy passes through the image plane. Why is the slit usually very narrow?

35–5   Why are *line* spectra so called? What would be seen in the telescope of a spectrometer if a small circular aperture were placed in the focal plane of the collimating lens, and illuminated by sodium light?

35–6   What would be the difference between the line spectrum of a gas at low pressure, and that of the same gas at high pressure? (*Hint:* think of the Sun.)

35–7   A beam of light is shone over the top of an open beaker containing mercury. What would be seen on a white card held behind the beaker if the light came from (a) a tungsten filament (b) a sodium vapour lamp (c) a mercury vapour lamp?

35–8   While on the one hand CaF_2, $CaCl_2$ and $CaBr_2$ give different band spectra, they can all be made to produce a line spectrum in which the same lines appear. Suggest reasons for this. Under what conditions would this be observed?

35–9   Discuss the differences between the spectra of light emitted by a red-hot poker, a neon lamp, and an electric lamp enclosed by red translucent glass.

Quantitative Problems

Dispersion

35–10   *The achromatic prism.* A flint glass prism has a refracting angle of 8.0°, and is to be combined with a crown glass prism so that the combination is achromatic for the C and F *Fraunhöfer* lines.

glass	n_C	n_D	n_F
crown	1.517	1.519	1.524
flint	1.602	1.605	1.612

Use the information in the table to find

(a) the refracting angle of the crown glass prism

(b) the deviation produced by the combination for the D-line.

Discuss how many significant figures can be justified in your answers.

[(a) 1(1)°   (b) 1.(1)°]

35–11   *The direct vision prism.* The flint glass prism of the previous question is to be combined with a different crown glass prism so that the combination gives zero deviation for the D *Fraunhöfer* line. Calculate

(a) the refracting angle of the crown glass prism

(b) the angular dispersion of the C- and F-lines.

[(a) 9.3°   (b) 0.01(5)°]

The Spectrometer

35–12 *Width of slit image.* In one setting of a particular spectrometer the slit had a width 1.0 mm, the collimating lens a focal length 250 mm, and the telescope objective a focal length 400 mm. The slit was illuminated by monochromatic light.

(a) What width image could one obtain of the slit in the focal plane of the eyepiece if no prism were placed on the table?

(b) What width would the image have if the light were deviated through its minimum deviation by a prism before being intercepted?

(c) Discuss qualitatively the effect on the image width of passing the rays through the prism at some setting other than that of minimum deviation.

(d) If white light were used what meaning would you give to the term *slit image*?

[(b) 1.6 mm]

35–13 When a parallel beam of white light has passed through a prism the C and F *Fraunhöfer* lines have suffered an angular dispersion of 0.20°. A pure spectrum is then produced by focusing the light with an achromatic lens of focal length 400 mm.

(a) Where is the image formed?

(b) What is the *linear* separation of the C- and F-line images?

(c) These real images are viewed through an eyepiece of focal length 50 mm, so that the final (virtual) images are formed at infinity. What is the *angular* separation of the C- and F-line images?

[(b) 1.4 mm   (c) 28 mrad = 1.6°]

35–14 A hydrogen discharge tube is set up in front of the very narrow slit of a spectrometer collimator, and the parallel beams of light pass through a prism for which $n_C = 1.517$ and $n_F = 1.524$, where C and F refer to the red and blue *Fraunhöfer* lines. The prism has a refracting angle of 60°, and is arranged so that the helium d-line (in the yellow) gives minimum deviation.

(a) Why do the C- and F-lines show a deviation which is effectively a minimum?

(b) Calculate the deviations of the C- and F-lines, and hence their angular dispersion.

(c) The telescope objective of this spectrometer has $f = 200$ mm, and is achromatic. What is the linear separation of the slit images formed by the C- and F-lines?

[(b) 0.6(0)°   (c) 2.1 mm]

35–15 A spectrometer whose collimator and telescope lenses have focal lengths 200 mm and 250 mm respectively is used to display the spectrum from a mercury vapour lamp. The prism used has a refracting angle of 60.0°, and refractive indices 1.610 and 1.590 for the blue and green lines.

(a) Find the angular dispersion between the two lines, assuming that both are effectively showing minimum deviation.

(b) What is the linear separation of the centres of the blue and green images of the slit?

(c) What width of slit would have an image of this width if monochromatic light only were used?

(d) What is the maximum permissible width of slit if the blue and green images are not to overlap?

Ignore diffraction effects.

[(a) 34 mrad   (b) 8.4 mm   (c) 6.7 mm]

Miscellaneous

35–16 *Measurement of refractive index.* In an experiment to measure the refractive index of a glass prism, the following measurements were made using a spectrometer. The angles quoted refer to readings on the vernier scale.

(a) *Measurement of A:*
  Telescope position (1) 170° 16′ ± 02′
               (2) 290° 20′ ± 02′.

(b) *Measurement of* D_{min}:
  Reading without prism = 100° 00′ ± 02′,
  Reading using prism and light of the
    helium d-line   = 140° 32′ ± 02′.

Calculate the refractive index of the glass at this wavelength, together with a value for the experimental uncertainty in n. Suggest a way in which the accuracy of the measurement of D_{min} could be improved.

[1.538 ± 0.003]

35–17 *Measurement of dispersive power.* In an experiment to measure the dispersive power of a glass prism, the following measurements were made using a spectrometer. The angles quoted refer to readings on the vernier scale.

(a) *Measurement of A:*
  Telescope position (1) = 150° 04′
               (2) = 270° 04′.

(b) *Measurement of* D_{min}:
  Reading without prism = 200° 00′.
  Reading using prism: C-line = 238° 30′
                D-line = 238° 45′
                F-line = 239° 16′.

Calculate the dispersive power of this glass.

[0.01(7)]

36 Optical Instruments

Questions for Discussion

36–1 Is it possible to obtain an enlarged retinal image of a distant object by observing a photographic print? Your discussion should include orders of magnitude for the focal length of the camera lens.

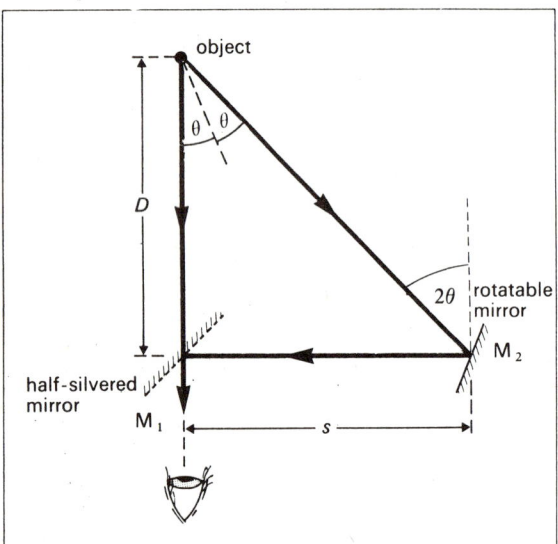

*For Qu. **36–2** (not to scale)*

36–2 *The rangefinder.* Refer to the diagram.
(*a*) Find an approximate relationship between D, s and θ for when $D \gg s$.
(*b*) Suggest a precise way of measuring θ which could be linked to a dial on the instrument.
(*c*) How may the sensitivity of the instrument be increased, so that (for example) one might distinguish between objects for which $D = 100$ m and 105 m?
(*d*) Estimate the value of θ for a human being (i.e. binocular vision) watching a cricket match.

36–3 Explain *qualitatively* why a magnifying glass of small focal length enables finer detail to be studied than does one of long focal length.

36–4 When long- and short-sighted people use the same magnifying glass, they obtain different angular magnifications. Who benefits the more?

36–5 (*a*) What is the focal length of a magnifying glass for which the maximum angular magnifying power is 2?
(*b*) Is there a way of using the lens which gives a retinal image of a small accessible object more than twice the size obtainable with the naked eye?

36–6 Discuss the difference between the nature of the objects viewed by a telescope and by a microscope that makes it desirable for the objective of the former to have a long focal length, while that of the latter should be short.

36–7 When an astronomical telescope produces an image of a *point* object (such as a distant star), then ideally the image is a point too. What advantage is there in using a telescope to observe such stars?

36–8 In one form of dew-point hygrometer the thermometer must be read from the far side of the room (at a range of say 5 m). The maximum distance at which it can be read clearly by the naked eye is 0.25 m. Devise a method of making the measurement using lenses of focal lengths 0.10 m and 50 mm.

36–9 A *Galilean telescope* is similar to an astronomical telescope, but the eye-lens is diverging instead of converging. Trace the paths of three rays through such an instrument. What advantages and disadvantages do you think it has compared to the astronomical telescope?

36–10 O-M What percentage of the incident light in a *Cassegrain* reflecting telescope is intercepted by the back of the convex mirror?

Quantitative Problems

Principles

†36–11 *Visual angle.* Which of the following two objects would give rise to the larger image on the retina of the human eye?
(*a*) A man of height 2.0 m at a range of 0.20 km, or
(*b*) the Moon, which subtends an angle of 9 mrad at the Earth's surface?

†36–12 *Simple magnifying glass.* An object 10 mm high is to be viewed through a simple magnifying glass of focal length 50 mm.

(1) *Image at infinity*
(*a*) Where should the object be placed?

(b) What angle does the final image subtend at an eye placed very close to the lens?

(c) What angle would the object subtend at the near point, 250 mm from the eye?

(d) What angular magnification is obtained?

(2) *Image at near point*
Repeat Qu. (a) – (d) of (1).

Is there any advantage to be obtained in forming the final image (i) at infinity, (ii) at the near point, (iii) closer than the near point? Discuss.

[(1) (d) 5 ×, (2) (d) 6 ×]

36–13 *Visual acuity.* The **visual acuity** of the eye is the least angular separation of rays coming from point objects that are just distinguishable, and for many people has the value 0.5 mrad.

(a) What is the least separation of point objects that can be distinguished at a range of 0.10 km?

(b) At what range can one clearly distinguish two point objects whose separation is 0.50 mm?

[(a) 50 mm   (b) 1.0 m]

The Compound Microscope

†**36–14** A compound microscope has an objective lens of focal length 40 mm and an eye-lens of focal length 60 mm separated by 250 mm. An object 2.0 mm long is set up 50 mm from the objective lens.

(a) Calculate the position and linear magnification of the image formed by the objective lens.

(b) Calculate the position and linear magnification of the final image.

(c) What is the length of the final image?

(d) What angle does the final image subtend at an eye placed close to the eye-lens?

(e) What angle would the object subtend at the distance of most distinct vision (250 mm)?

(f) What is the magnifying power of the microscope?

[(c) 48 mm]

36–15 A particular compound microscope has an objective lens of focal length 20 mm, and the eye-lens has a focal length 60 mm.

(a) Where should the intermediate real image be formed if the final virtual image is to be 0.24 m from the eye-lens?

(b) An object placed 24 mm from the objective gives a final image in the desired position. What is the separation of the lenses?

(c) What angular magnification is obtained by the observer if 240 mm is his distance of most distinct vision?

[(a) 48 mm from eye-lens   (b) 168 mm   (c) 25 ×]

36–16 A compound microscope has an eye-lens of focal length 62.5 mm placed 150 mm from the objective lens. The instrument gives a final image at the distance of most distinct vision (250 mm), and it is found that in this adjustment the magnifying power is 50 ×. What is the focal length of the objective lens?

[9.1 mm]

The Telescope

†**36–17** *Angular magnification.* A surveyor looks through the telescope of a levelling instrument, and would like to see the measuring stick (which is 0.12 km away) in the same detail as he would if it were 3.0 m away.

(a) What angular magnification does he need?

(b) The eyepiece of his telescope has a focal length 5.0 mm. What is that of the objective lens?

[(b) 0.20 m]

36–18 An astronomical telescope consists of a pair of converging lenses of focal lengths 1.00 m and 100 mm separated (originally) by 1.10 m. It is used for viewing a distant object.

(a) Calculate the magnifying power when it is used in normal adjustment.

(b) What happens to the magnifying power if the eyepiece is moved slowly toward the objective lens?

(c) What, for the normal eye, is the nearest point at which it is worth while forming the final image if it is to be observed for some time?

(d) What is the objective-eyepiece separation for this adjustment?

(e) What is the new value of the magnifying power?

Discuss the two answers (a) and (e), and the fact that astronomical telescopes of this kind are usually used in normal adjustment.

[(a) 10 ×   (e) 14 ×]

36–19 The Moon, whose diameter is 3.5 Mm and whose distance from the Earth is 0.38 Gm, is viewed through an astronomical telescope in normal adjustment. The objective lens focal length is 5.0 m, and that of the eyepiece is 0.10 m. What is the angular diameter of the image?

[0.46 rad]

36–20 Mars was photographed through a telescope whose objective had a focal length of 10 m. The diameter of the image was measured to be 1.0 mm.

(a) Where was the photographic plate placed?

(b) What was the angular diameter of Mars when the photograph was taken?

(c) Suppose that (in the absence of a photographic plate) the real image formed by the objective was viewed through an eyepiece of 50 mm focal length. What would be (i) the angular diameter of the final image in normal adjustment, and (ii) the magnifying power of the telescope?

(d) What would be the visual angle of the photographic image placed at the distance of most distinct vision?

[(b) 0.10 mrad   (c) (ii) 200 ×   (d) 4.0 mrad]

36–21 *The Cassegrain telescope.* (a) A concave mirror of large aperture and focal length 10.0 m is used to form an image of a planet which subtends an angle 0.12 mrad at the mirror. Find the diameter and position of the image.

(b) Before this real image can be formed, the convergent beam of light is intercepted by a small convex mirror which has its pole 9.00 m from that of the concave mirror. As a result a different real image is formed at the pole of the concave mirror (where a small aperture is cut). What is the diameter of this image? (*Think in terms of the linear magnification given by the small mirror.*)

(*c*) This real image is now viewed by an eyepiece of focal length 100 mm, and a virtual image is formed at infinity. What angle does the second real image subtend at the eyepiece?

(*d*) What is the total magnifying power of the system?

It would be helpful to illustrate each stage of your answer by appropriate sketch diagrams.

[(*a*) 1.2 mm   (*b*) 11 mm   (*c*) 0.11 rad   (*d*) 900 ×]

The Eye-Ring

36–22   *The eye-ring.* The objective lens of an astronomical telescope has a diameter 30 mm and focal length 0.50 m. The eyepiece has a focal length 0.10 m, and the instrument is in normal adjustment. Calculate

(*a*) the lens separation
(*b*) the position of the eye-ring

(*c*) the diameter of the eye-ring

(*d*) the ratio $\dfrac{\text{diameter of objective}}{\text{diameter of eye-ring}}$

(*e*) the magnifying power.

Your answers to (*d*) and (*e*) should be the same. Show that this is not coincidence. What pupil diameter would enable the observer to use the instrument to optimum advantage?

(*The aperture of the objective lens is sometimes called the* **entrance pupil**. *The eye-ring, sometimes called the* **Ramsden circle**, *is the* **exit pupil**.)

[(*b*) 0.12 m beyond eye-lens   (*c*) 6.0 mm   (*d*) 5.0]

36–23   An astronomical telescope in normal adjustment has an objective lens of diameter 50 mm, and a magnifying power of 10. Draw a diagram showing the paths through the instrument of rays incident parallel to the principal axis which are refracted at the extreme outer zones of the objective lens, and deduce an appropriate diameter for the eye-ring.

[5.0 mm]

VII Wave Properties of Light

37 The Nature of Electromagnetic Radiation

Questions for Discussion

37–1 Describe three *separate* phenomena which lead you to think that light is propagated by a wave motion. Explain whether these phenomena are ever shown by *particles* in classical physics.

37–2 Is it possible for a plate of satin-finish aluminium to act as a *specular* (regular) reflector of electromagnetic waves? Justify your answer by giving orders of magnitude.

37–3 Waves are transmitted from a radar source, reflected from a small distant object at range r, and then received near the source. How does the intensity of the returning waves depend upon r?

37–4 Is electromagnetic radiation *always* emitted by accelerated charge?

37–5 A synchrotron causes charged particles to describe circular paths at high speed. What is the cause of **synchrotron radiation**? What effect does it have on the particles? Do the electrons of atoms show a similar effect?

37–6 Electric dipole radiation is plane-polarized, and we can consider atoms and molecules to radiate in the same way as a dipole. Why then is the light from (say) a laboratory helium source unpolarized?

37–7 Devise apparatus to demonstrate that a beam of light carries linear momentum. Explain carefully whether air molecules affect the way in which the apparatus functions. (*Look up, for example, the* **radiometer effect**.)

37–8 What experimental evidence could *you* offer a sceptical non-scientest that light travels at a non-infinite speed?

37–9 Suggest possible sources of error in the *Michelson* measurement of the speed of light *in vacuo*. (Do not exclude natural phenomena.)

*37–10 *Radiation pressure.* Consider an electromagnetic wave approaching a metal surface. Draw the relative orientations of the vectors E, B and c, and show that if E sets a free charge into motion, then B will cause it to experience a force along the direction of c (whether the charge is positive or negative).

How would you expect the radiation pressure to change if the wave were incident on an absorbing surface?

*37–11 *Energy density in electromagnetic wave.* The energy density of an electric field is $\frac{1}{2}\varepsilon_0 E^2$ (Qu. **46–25**), while that of a magnetic field is $\frac{1}{2}B^2/\mu_0$ (Qu. **57–21**). Show that the two energy densities associated with an electromagnetic wave have the same size.

37–12 **O-M** Estimate values for
(a) the frequency of visible light waves
(b) the greatest frequency that you know for any macroscopic phenomenon
(c) the rotational frequency of an electron in the hydrogen atom in its ground state.
What does this suggest about the origin of visible light waves?

37–13 **O-M** The eye has a threshold sensitivity to light of about 10^{-17} W. Calculate the minimum number of photons per second required for the eye to see something.

37–14 **O-M** How long does a light pulse take to travel
(a) through a window-pane
(b) across a room
(c) to the Moon and back
(d) from the Sun to the Earth
(e) from the nearest star to the Earth?

37–15 **O-M** *Galileo* attempted to measure c using human observers. What length of baseline is required if their total reaction time is to be 1% of the transit time of the light signal?

Quantitative Problems

(*For questions on the Doppler effect, see* Chapter **15**.)

Wave and Quantum Aspects

†37–16 The dissociation energy of carbon monoxide is about 1.8 aJ for each molecule. Calculate the minimum frequency of the electromagnetic radiation that could separate the carbon atom from the oxygen atom.
Assume h $[2.7 \times 10^{15}$ Hz$]$

†37–17 A sodium lamp is rated at 20 W for visible light, and it gives out light of wavelength 590 nm. How many photons does it emit in each second?
Assume c, h $[6.0 \times 10^{19}]$

†**37–18**   The energy changes ΔE involved in many gaseous ionization processes are in the range 10^{-18} to 10^{-16} J. Use the relationship $\Delta E = h\nu$ to discover which part of the Sun's electromagnetic spectrum is largely responsible for the existence of the Earth's ionosphere.

Assume h

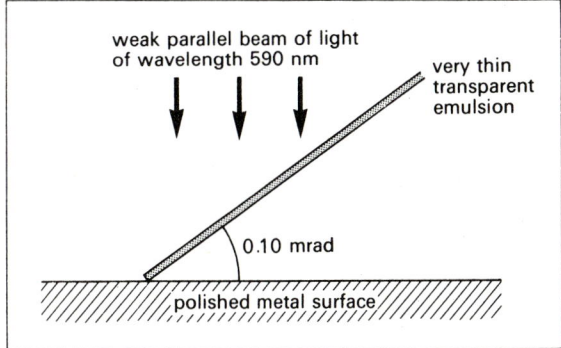

For Qu. **37–19**

37–19   *Stationary waves.* Refer to the diagram, which shows *Wiener*'s experiment to detect stationary light waves. What is the distance d which separates the centres of adjacent regions of darkness on the exposed film?

[3.0 mm]

37–20   The momentum p of a photon is given by $p = h/\lambda$.
(*a*) Why do you think c fails to appear in the equation?
(*b*) What is the momentum of a typical visible light photon (say $\lambda \sim 550$ nm)?
(*c*) What photon intensity (rate of arrival normal to unit area) would result in a radiation pressure of 10 μPa (about ten times that produced by the Sun at the Earth's surface)? Assume that all photons carry the same momentum.

Assume h   [(*c*) 8.3×10^{21} photons s^{-1} m^{-2}]

The Speed of Light

†**37–21**   A particular oscilloscope used as a timing device can be made to record time intervals down to 12 ns. What is the shortest distance that it could measure using a radar echo technique?

Assume c

†**37–22   *Essen's measurement of c.*** (*a*) We can measure the speed of sound by applying $c = \nu\lambda$ to measurements of ν and λ. Why can we not use this method directly for an accurate measurement of the speed of *visible light*?
(*b*) It is believed that all electromagnetic waves travel at the same speed *in vacuo*. *Microwaves* were found by *Essen* (1950) to resonate at a frequency 9.50 GHz in a cavity from whose dimensions the free-space wavelength was found to be 31.5 mm. Calculate a value for c. (*Essen's experiment was carried out with an experimental uncertainty of about 1 in 10^5.*)

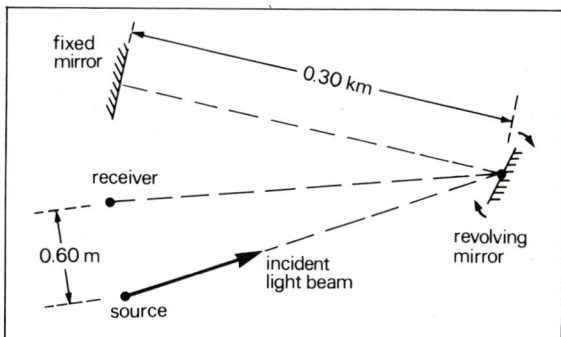

For Qu. **37–23**

37–23   Refer to the diagram. What is the lowest angular speed of rotation of the mirror which would enable the receiver to detect reflected light?

Assume c   [0.50 krad s^{-1}]

37–24   *Measurement of c using a Kerr cell.* A *Kerr cell* can be used as a non-mechanical shutter for light beams. The chopping or **modulating** of the light beam can be done at very high frequency by using the electrical oscillations of a vacuum-tube circuit.

In one experiment it was observed that when the length of the light path was 250 m, the pulses took one quarter of a cycle to travel from one *Kerr* cell to a second (with which it was synchronized) if the modulation frequency was 300 kHz. Use these figures to calculate a value for c.

(*In 1941 Anderson was able to use frequencies of* 19 MHz, *which should be compared with* 50 kHz, *the maximum available to Fizeau with his 1849 apparatus. The higher frequency enables a much smaller light path to be used.*)

[300 Mm s^{-1}]

37–25   *Stellar aberration.* Refer to the diagram, which is exaggerated. It shows an Earth-bound telescope being used to observe a star at two instants 6 months apart. Calculate the

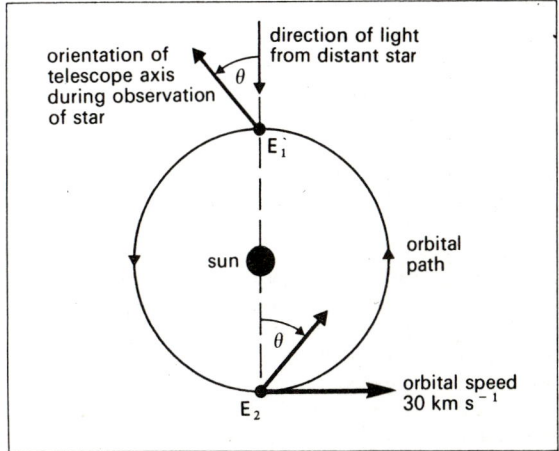

For Qu. **37–25**

angle 2θ. (*Hint: the velocity of light coming from the star combines vectorially with the Earth's orbital velocity. It should be noted that according to relativistic theory vector addition to the velocity of light gives the same magnitude in all cases.*)

Assume c   [0.20 mrad]

*37–26 The magnitudes of the quantities E, B and c for an electromagnetic wave are related by $c = E/B$. What is the value of E for a wave travelling along the positive z-direction at the instant when $B = 1.0$ μT along the positive y-direction? Calculate the unit as well as the number in your working.

Assume c

*37–27 *Radiation pressure.* Suppose a wave has an energy density W/V, and intensity I.

(*a*) Show that the relation $I = (W/V)\,c$ is dimensionally consistent.

(*b*) The electromagnetic radiation pressure p caused by the Sun at the Earth's surface is given approximately by $p = (W/V)$ if we assume that the radiation is incident as a beam, and is totally absorbed.

If $I = 1.4$ kW m^{-2}, then what is the value of this radiation pressure?

How does your answer compare with atmospheric pressure?

Assume c   [(*b*) 4.7 μPa]

38   Interference

Questions for Discussion

38–1 Which of the following pairs of wave sources are *coherent*?

(*a*) Two loudspeakers activated by separate sinusoidal audio-frequency oscillators of the same frequency.

(*b*) A narrow slit illuminated by white light, and its virtual image formed by (*i*) a stationary black glass plate, or (*ii*) a moving black glass plate.

(*c*) The virtual images of a narrow slit illuminated by sodium light, formed by a prism of angle close to π rad.

(*d*) A pair of narrow slits illuminated directly by a laboratory sodium source.

(*e*) A pair of separate lasers emitting light of different frequencies.

38–2 Is it possible for a pair of non-interacting atoms to emit coherent light waves? If so, what is the maximum duration of such emission?

38–3 The duration of emission of a typical visible light photon is ~ 1 ns, and the period of vibration of the associated wave is ~ 1 fs. It is proposed to try and photograph the interference pattern from two *incoherent* slit sources. What shutter speed would you like to choose?

38–4 When microwaves are reflected from a plane metal sheet, the measurement of angles of incidence and reflection is complicated by the maxima and minima of an interference pattern. Do such difficulties exist for experiments using visible light? – discuss.

38–5 When is it permissible to add the *intensities* (rather than the *amplitudes*) of the disturbances produced by a pair of superposed wavetrains?

38–6 What part does *diffraction* play in a *Young* double-slit experiment?

38–7 What would be the effect on a *Young* fringe system of
(*a*) gradually increasing the width of the single source slit
(*b*) immersing the apparatus in water?

38–8 Discuss reasons for the minima of intensity in a *Young* double-slit experiment not being zero.

38–9 Why does the contrast of a double-slit fringe pattern fall away as one moves from the centre of the pattern?

38–10 What is the function of the *single* slit in an optical double-slit experiment?

38–11 A soap film is formed on a wire frame, and then held up in a vertical plane and illuminated by white light. Explain the changes that will be observed as time passes.

38–12 A thin film of mica has been cleaved from a larger crystal. How would you set about an accurate determination of its thickness?

38–13 What would be the effect on a *Newton* ring system of

(a) reducing slowly the thickness of the air-film between the lens and the flat plate (assuming they are initially separated) (b) replacing the flat plate by a curved surface identical to that of the lens? (Give a quantitative answer.)

38–14 A *Newton* ring fringe system is established in the usual way using a planoconvex lens and an optically flat plate of the same refractive index 1.5. Describe qualitatively the effect on the system formed by reflection of

(a) replacing the air film by a layer of liquid of refractive index 1.6, and *then* (b) replacing the plate by one of refractive index 1.7.

(Hint: $\lambda_n = \lambda_0/n$)

38–15 Why does a *Newton* ring fringe system show better contrast when viewed by reflected light than by transmitted light?

38–16 A *Newton* ring fringe system is formed using sodium light. As the lens-plate separation is steadily increased from zero, the contrast of the fringe system fluctuates. Why?

Quantitative Problems

Visible Light Wavetrains

†**38–17** *Optical path.* Light of free-space wavelength 600 nm is incident on a film of material whose thickness is 80 µm, and of refractive index 1.50.

(a) What is the wavelength within the material? (b) How many complete waves are fitted into the thickness of the film? (c) What length of vacuum would contain the same number of waves? (*This quantity is referred to as the* **optical path** *for the material concerned.*)

[(c) 0.12 mm]

†**38–18** What is the wavelength in glass ($n_g = 1.5$) for an electromagnetic radiation of frequency 5.0×10^{14} Hz?

Assume c [0.40 µm]

†**38–19** An atom emitted light of wavelength 500 nm for a time period 0.40 ns. How many complete vibrations were contained in the wavetrain?

Assume c [2.4×10^5]

Double-Source Experiments

38–20 In a *Young* double-slit experiment a pair of slits 0.40 mm apart were placed 600 mm from the plane of observation. A travelling eyepiece was moved 7.5 mm in traversing the distance from fringe m to fringe $(m + 10)$.

(a) What was the wavelength of the light being used? (b) What would have been the fringe width if (independently)
   (i) the slit separation had been 0.20 mm,
   (ii) the slit-screen distance had been 800 mm, and
   (iii) red light with $\lambda = 700$ nm had been used?

[(a) 0.50 µm]

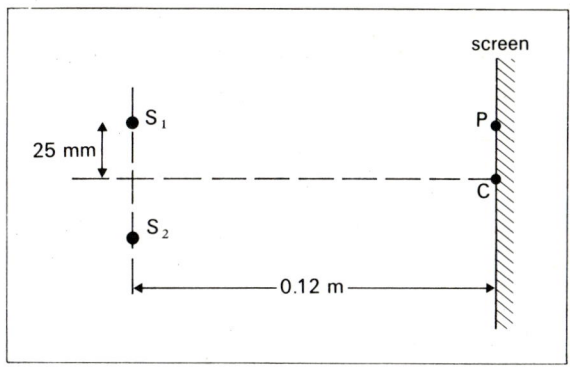

For Qu. **38–21**

38–21 Refer to the diagram. S_1 and S_2 are identical *phase-linked* sources giving sinusoidal waves of wavelength 20 mm, but they have a phase difference $\Delta\phi$. A maximum of intensity is observed at P, but there is no maximum between C and P. Find

(a) the value of $\Delta\phi$, and (b) the relative intensity at C.   [(a) π rad   (b) zero]

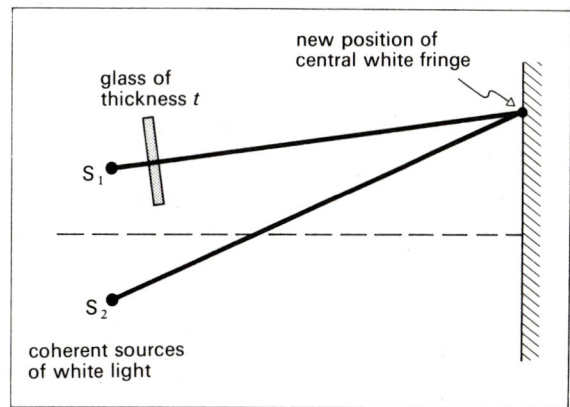

For Qu. **38–22**

38–22 *Optical path difference.* Refer to the diagram, in which $S_1S_2 = 0.40$ mm, and the slit-screen distance is 0.40 m. The glass has a refractive index 1.50.

(a) What, in terms of t, is the *extra* optical path introduced by the addition of the glass? (b) If this causes the central white fringe to be displaced a distance 10 mm, then what is the value of t?

[(b) 20 µm]

†**38–23** *The Pohl interferometer.* A very small light source S is set up close to a mica sheet of thickness 10 µm. Waves to be superposed come from the virtual sources that are the images of S in the front and back surfaces of the mica sheet. What is the separation of these sources? (*Your answer should be compared to the sort of slit separation obtainable in a double-slit experiment. The plane of observation is perpendicular to the line joining the sources, and shows large concentric circular fringes.*)

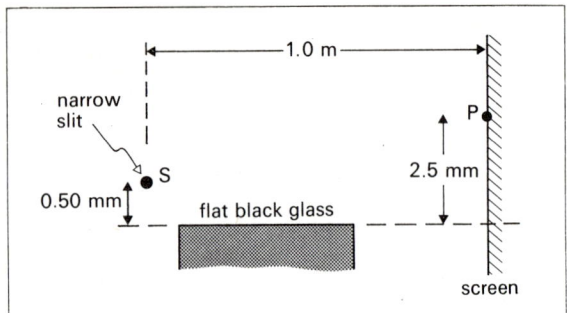

For Qu. **38–24**

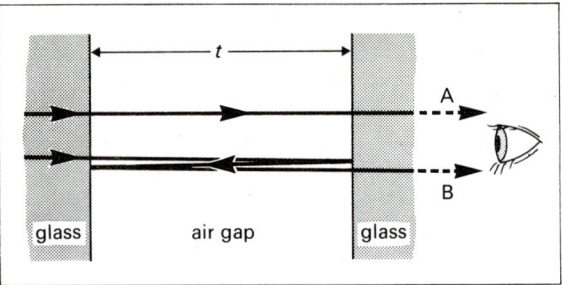

For Qu. **38–28**

*38–24　**The Lloyd mirror.** Refer to the diagram, which is not to scale. If the slit is illuminated by light of wavelength 500 nm, then what would be seen at P by an observer using a travelling eyepiece?

*38–25　**The Fresnel biprism.** Parallel monochromatic light from a spectrometer collimator was passed normal to the base of a prism of very large refracting angle ($\lesssim \pi$ rad), and the angle between the parallel beams emerging from the two halves of the biprism was measured by the spectrometer telescope to be 5.0 mrad. What would be the linear separation of the two virtual images formed by this prism if a narrow slit were placed 0.10 m from it? Draw a diagram to justify your answer. How does the brightness of the two images compare with that of a double-slit arrangement?

[0.50 mm]

*38–26　**Fringe visibility.** In a double-slit experiment, when one slit is covered, the centre of the viewing screen receives light of intensity I_0 from the other. When their roles are interchanged the intensity becomes $4I_0$.

(a) Calculate the relative *amplitudes* of the disturbances produced by the two slits separately.
(b) What are the maximum and minimum amplitudes on the screen when the two slits illuminate it together?
(c) The **fringe visibility** V is defined by

$$V = (I_{\max} - I_{\min})/(I_{\max} + I_{\min}),$$

where $I_{\max}$ and $I_{\min}$ are the intensities at adjacent bright and dark fringes. Calculate the visibility for (i) this system, and (ii) a system of perfect contrast in which the two slits give the same intensities.

[(c) (i) 0.80]

Films of Constant Thickness

38–27　**Division of amplitude.** For nearly normal incidence the (intensity) **reflection coefficient** R for light passing across an interface between media of refractive indices n_1 and n_2 is given by

$$R = \{(n_2 \sim n_1)/(n_2 + n_1)\}^2.$$

(a) What percentage of light energy is *transmitted* across a typical air-glass interface (for which $n_2 \approx 1.5$ and $n_1 = 1.0$)?
(b) What is then the relative *amplitude* of the reflected beam?

[(a) 96%　　(b) 0.20]

*38–28　Refer to the diagram.

(a) Write down, in terms of t and λ, the optical path difference between the waves travelling along paths A and B. Ignore refraction effects.
(b) What is the condition that such rays should be superposed destructively?
(c) What would an observer see for $t = 1.20$ μm if he used monochromatic light of wavelength
　　(i) 533 nm　(ii) 500 nm　(iii) 480 nm?

[(c) (i) cancellation]

*38–29　An observer shines a parallel beam of white light through a film like that of Qu. 38–28, and examines the transmitted light through a prism spectrometer. He finds the otherwise continuous visible spectrum to be crossed by only two dark bands. The centres of these bands correspond to wavelengths 606 nm and 433 nm.

(a) What is the thickness of the film?
(b) Explain what he sees on either side of the centre of each dark band, and why.

[(a) 0.76 μm]

*38–30　A draining soap film appears dark when viewed by reflected light of wavelength 600 nm. Suggest possible values for its thickness.

*38–31　**Coating (blooming) of a camera lens.** A camera lens made of flint glass is to be coated with magnesium fluoride ($n_c = 1.38$) in an attempt to reduce energy loss by reflection.

(a) What is the wavelength within the coating for yellow light of free-space wavelength 555 nm?
(b) What minimum thickness of coating will cause the waves reflected at the top and bottom surfaces to be superposed destructively? What assumption is necessary in this calculation? Will the destruction necessarily be complete?
What will then be the appearance of the lens by reflected light?

[(b) 0.10 μm]

*38–32　Refer to the previous question. The reflected intensity for a particular wavelength is reduced to zero when the amplitudes of the superposed waves are equal. For a lens in air this occurs when $n_c^2 = n_g n_a$.

(a) What value of n_g would produce completely destructive superposition? Find out from tables whether such glasses are made.

(b) What has now happened to the energy that is usually reflected?

Films of Varying Thickness

*38–33 *Newton rings.* A *Newton* ring system is established in the usual way using monochromatic light of wavelength 625 nm and a lens of radius of curvature 1.00 m. The lens is in contact with the glass plate.

(a) What is the radius of the dark ring corresponding to the tenth order of interference?

(b) What would be the behaviour of the fringe system if the plate were gradually lowered a vertical distance 0.78 μm while leaving the lens undisturbed?

What do you think is the smallest movement that you could *detect* by this method? How could you modify this arrangement to make it more sensitive?

[(a) 2.5 mm]

*38–34 *Wedge fringes.* The following experiment was carried out to measure the diameter of a hair. Two optically flat glass plates were placed in contact at one edge, and the hair then sandwiched between the plates a distance 100 mm from, and parallel to, the line of contact. The wedge of air was illuminated normally with sodium light wavelength 589 nm, and the reflected light revealed straight parallel interference fringes at the location of the wedge, and separated by 1.0 mm. Calculate the diameter of the hair. Would a micrometer screw gauge be as precise? Would it necessarily be less *accurate*?

[29 μm]

39 Diffraction

Questions for Discussion

39–1 Is there any meaningful distinction between diffraction and interference?

39–2 Give one reason why we may not see objects in such clear detail when the light is very bright.

39–3 Why is diffraction more of a nuisance in a telescope than in a camera?

39–4 Why is ultra-violet light sometimes used to illuminate objects under the microscope?

39–5 Is there any advantage in taking a photograph of a star through a *blue* filter, even if the telescope objective lens is achromatic? If so, how do you reconcile your answer with the fact that landscape photographs are often taken in *infra-red* light to achieve greater clarity?

39–6 When a diffraction grating is used at normal incidence it sometimes happens that the third order violet spectrum overlaps the second order red spectrum. Does this *always* happen? Your answer should include a calculation.

39–7 What advantages does a *reflection* grating have over a transmission grating?

39–8 Is there any way of distinguishing a photograph of a grating spectrum from that of a prism spectrum?

39–9 Can a *prism* spectrometer be used to make a satisfactory determination of wavelength? Your answer should include some orders of magnitude.

39–10 What would be seen on the screen of a *Young* experiment if one of the double slits were to be covered by an opaque screen?

39–11 Intense white light is shone through a narrow slit onto a white screen, and the latter shows coloured effects. What has caused the colours to separate?

*39–12 (a) Sketch the theoretical intensity-location pattern available from a double-slit experiment using slits of negligible width.

(b) Sketch the intensity-location pattern for monochromatic light diffracted under *Fraunhöfer* conditions through a single slit of small but finite width.

(c) Use the pattern of (b) to **modulate** the theoretical pattern of (a), and hence obtain a *realistic* sketch of the pattern available in a double-slit experiment. Your diagram must show clearly how the intensity variation is controlled by (i) the separation s of the double slits, and (ii) the width a of each slit.

*39–13   O-M A photograph is to be taken of a printed page using a pinhole camera. Suggest an optimum diameter for the pinhole if the time of the exposure can be set at any value.

39–14   O-M At what distance can an observer distinguish the headlamps of a *Land Rover* (separation ∼ 1 m) from that of a motor cycle? Mention in your answer any assumptions that you make about the human eye.

*39–15   O-M Estimate, *by calculation*, the greatest distance at which the human eye can read the headlines of a newspaper.

Quantitative Problems

The Diffraction Grating

†39–16 Find the largest grating spacing that can just give a fourth-order spectrum for sodium light of wavelength 589 nm incident normally on the grating.

[2.4 μm]

†39–17 When red light of wavelength 700 nm was passed normally through a grating it showed a first-order maximum at a deviation of effectively 90°.

(a) Calculate the grating spacing, and the angular dispersion that would be given over the whole of the visible spectrum (say 400 nm upwards).
(b) In what direction(s) would a detector respond to radiation of wavelength 750 nm?

[(a) 0.70 μm, 55°]

39–18 A helium lamp gives out waves of wavelength 587 nm and 668 nm which are incident normally on an optical transmission grating of 500 lines mm^{-1}.

(a) What will be the *angular* separation of these lines in the first-order spectrum?
(b) If an image of the spectrum is formed by a lens of focal length 200 mm, what will be the *linear* separation of the centres of the two line images?

[(a) 2.4°   (b) 8.6 mm]

39–19 The visible spectrum covers the wavelength range 400 to 700 nm approximately. White light passes through a grating of width 20 mm on which 12×10^3 lines have been ruled. What are the angular dispersions of the first- and second-order spectra? Comment on the relative advantages of these two spectra for measurement and observation purposes.

[11° and 28°]

*39–20 Find, *without* consulting five-figure tables, the angular separation of the sodium doublet (wavelengths 589.00 nm and 589.59 nm) given in the first-order spectrum by a grating of spacing 1.00 μm. Think carefully about the number of significant figures that you should use in your answer.

[0.7(3) mrad]

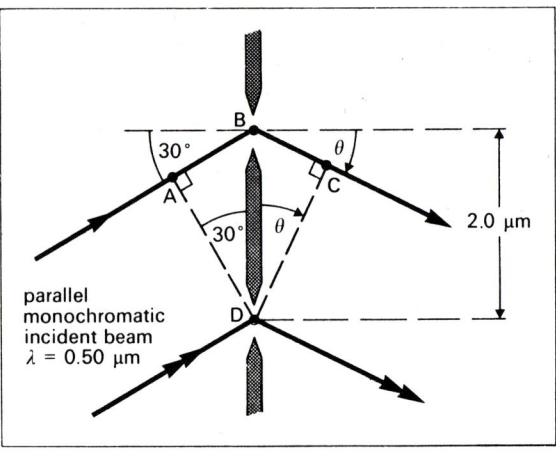

For Qu. **39–21**

*39–21   *Oblique incidence on a grating.* Refer to the diagram, which shows light incident on a diffraction grating at non-normal incidence.

(a) Write down, in terms of θ, the path difference between waves which traverse the path ABC and those on the wave-front at D.
(b) Write down the condition for the waves emerging from the grating to be superposed constructively.
(c) In which directions will an observer see
  (i) the zeroth-order spectrum, and
  (ii) the first-order spectra?

[(c) (i) − 30°   (ii) − 14° and − 49°]

*39–22 The dispersion D of a diffraction grating is defined by $D = d\theta/d\lambda$. Show that for a given grating spacing s and a given viewing angle θ the dispersion is proportional to the order m of the spectrum being examined.

*39–23   *Resolving power of a grating.* Suppose two wavelengths of difference Δλ can just be distinguished. Then for a first-order spectrum the resolving power R of a grating is given (according to *Rayleigh*'s approximate criterion) by

$$R \equiv \lambda/\Delta\lambda = N,$$

where N is the *total* number of lines on the grating, and λ is the mean wavelength.

How many lines must a grating have if an observer is to be able to separate the D-lines of sodium (589.00 nm and 589.59 nm) in the first-order spectrum? How would your answer be modified if the second-order spectrum could be used? Are there any restrictions on the grating spacing? Would you be able to distinguish the two D-lines with the gratings that *you* have used?

[1.0×10^3 lines]

Other Fraunhöfer Diffraction

*39–24 Plane waves of wavelength 600 nm are incident normally on a long narrow slit of width 0.40 mm. At what distance from the slit should a viewing screen be placed to demonstrate the first (central) dark fringes separated by 5.0 mm?

[1.7 m]

*39–25 A parallel beam of monochromatic light of wavelength 550 nm falls normally on a long slit of width a. The diffraction pattern is observed on a screen placed 1.0 m away, where it is found that the distance between the dark fringes which flank the central maximum is 2.0 mm. Find a.

[0.55 mm]

*39–26 In an attempt to demonstrate the wave nature of X-rays an observer proposes to diffract them through a single slit. An angular width of 0.1 mrad can just be detected, and it is thought that the wavelength involved is about 0.1 nm. What is the maximum slit width that could be used?

[1 μm]

*39–27 A parallel monochromatic beam of wavelength 600 nm falls on a circular lens of diameter 50 mm and focal length 1.0 m. Calculate the diameter of the first dark ring in the 'point' image, and use your answer to decide whether the naked eye could detect the diffraction effects. (*Do not forget that because here we have a circular aperture, we must include the factor* 1.22.)

[29 μm]

Resolving Power

*39–28 *The Rayleigh criterion.* (*a*) Parallel light of wavelength 600 nm is incident on a lens parallel to its axis, and is brought to a fringed focus in the focal plane. What is the angle subtended by the innermost dark fringe with the axis if the lens has a diameter 20 mm?

(*b*) At what angle may a second parallel beam enter the lens if the centre of its fringed focus is to lie on or outside the innermost dark fringe of the first?

(*c*) Sketch the intensity patterns of the two beams separately, and then superpose them to decide whether the eye could (in principle) distinguish *two* separate 'point' images. (*According to the* **Rayleigh criterion** *the patterns are distinguishable if the condition of* (*b*) *is satisfied.*)

[(*a*) 37 μrad]

*39–29 *Circular apertures.* Following the *Rayleigh criterion* we can just distinguish two point objects which subtend an angle θ if $\theta \geqslant 1.22 \, \lambda/D$, where λ is the wavelength of the electromagnetic radiation used for the detection, and D is the diameter of a circular aperture or receiving aerial.

(*a*) What is the value of θ for a human eye of pupil diameter 4 mm using light of wavelength 600 nm?

(*b*) When it receives blue light of wavelength 400 nm the *Mount Palomar* telescope has a resolving power such that $\theta = 0.1$ μrad. Make an estimate of its diameter.

(*c*) Suppose that the *Jodrell Bank* radiotelescope has a receiving aerial of diameter 70 m. What is the greatest wavelength that it can receive and still be able to separate objects that subtend an angle 3.7 mrad? (*This particular wavelength is emitted by hydrogen, and is much used in radioastronomy.*)

[(*a*) 0.2 mrad   (*b*) 5 m   (*c*) 0.21 m]

40   Polarization

Questions for Discussion

40–1 Can two non-zero perpendicular vectors combine to give a zero resultant? Under what conditions will plane-polarized light show interference phenomena?

40–2 You are handed an unmounted *Polaroid* sheet. No other marked polarizer is available. How would you set about determining the preferred direction?

40–3 The preferred direction of a *Polaroid* sheet is marked with a straight line. Is there any point in adding an arrowhead to a particular end of the line?

40–4 The intensity of unpolarized light entering an ideal *Polaroid* sheet is I_0. What is that of the transmitted light?

40–5 *Double refraction.* When a beam of monochromatic light enters a calcite crystal *normal* to the surface, the transmitted energy is split into two beams called the **ordinary** and **extraordinary**. The ordinary ray is transmitted normally.

(*a*) What can be deduced about the speed of light in crystalline calcite?

(*b*) What does *this* information allow you to say about the laws of refraction?

40–6   *The Nicol prism.* Suggest a method of separating *completely* the o- and e- waves referred to in Qu. **40–5**. (*Use a phenomenon which depends in a very definite way on the wave speed.*)

40–7   When light passes separately through two polarizers they appear transparent. What becomes of this transmitted energy when they are crossed?

40–8   Draw a sketch graph of the variation of transmitted intensity as an analyser receiving plane-polarized light is rotated through 2π rad.

40–9   An analyser is available. How could you use it to distinguish between (*a*) partially plane-polarized, (*b*) completely plane-polarized and (*c*) unpolarized light?

40–10   Why, when making accurate measurements in the *Malus* experiment, must one use a *parallel* beam of *monochromatic* light?

40–11   Can a material have a polarizing angle of 42^a?

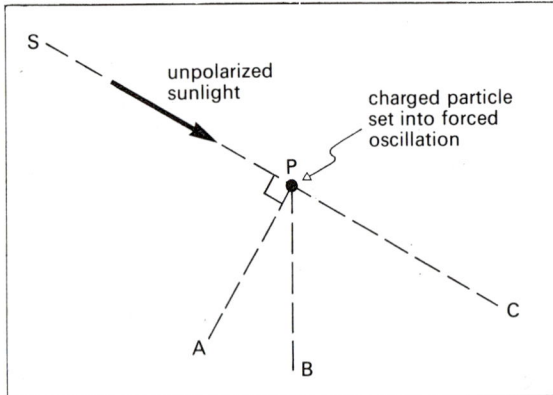

For Qu. **40–12**

40–12   *Polarization by scattering.* Refer to the diagram.
(*a*) Into which directions can the incident electromagnetic wave set the charged particle oscillating as a dipole?
(*b*) Discuss the nature of the polarization of the electromagnetic wave which is radiated (**scattered**) from P to an observer viewing along the directions PA, PB and PC.

40–13   *The colour of the sky.* Suppose the natural frequency of oscillation of the electrons in air molecules to be $\sim 10^{16}$ Hz. Use this information and the ideas of Qu. **40–12** to predict the apparent colour of the sky for observers along PA, PB and PC respectively.

40–14   Suppose that while the Sun is setting in the West you observe the plane-polarized light which is scattered from the South. What would be the plane of vibration? How would you test your answer with an analyser?

40–15   In a fine mist clearer photographs of distant objects can be taken by infra-red light than by visible light, yet when the fog is thick neither can be used. Why?

40–16   How should the preferred directions of *Polaroid* sunglasses be orientated to reduce glare
(*a*) from the surface of a river
(*b*) from sunlight when the Sun is vertically overhead?

40–17   O-M At what time of day is the light reflected from a still lake surface in *England* in May most nearly plane-polarized?

Quantitative Problems

40–18   *Brewster's Law.* (*a*) An unpolarized parallel beam of monochromatic light is incident on a plane air-glass interface of relative refractive index 1.50. Calculate the angle of incidence such that the refracted ray should be perpendicular to that which is partially reflected. (*The angle of incidence is then called the* **polarizing angle**.)
(*b*) What would be observed in (*a*) if the incident beam had been *plane-polarized* with the electric vibrations confined to the plane defined by the incident beam and the normal to the interface?
(*c*) Discuss the nature of the reflected ray in (*a*).

[(*a*) 56°]

40–19   When light is incident from air onto a particular glass the polarizing angle is 57°. What would it be for light incident onto the same interface from the glass side?

40–20   Calculate the polarizing angles for the red and blue light of the C and F *Fraunhöfer* lines incident on an air-water interface for which $n_C = 1.332$ and $n_F = 1.338$. Do you think that this sort of polarization phenomenon is likely to show coloured effects?

40–21   *Malus's Law.* A bright source of light is viewed through two sheets of *Polaroid* whose preferred directions are initially parallel.
(*a*) Through what angle should one sheet be turned to reduce the *amplitude* of the observed electric field vibration to half its original value?
(*b*) What effect does this have on the transmitted intensity?
(*c*) Through what angle should one sheet be turned to reduce the transmitted intensity to half its original value?
Assume there is no absorption of vibrations parallel to the preferred directions.

[(*a*) $\pi/3$ rad   (*c*) $\pi/4$ rad]

*****40–22**   Microwaves are transmitted at intensity I_0 through a grid of vertical metallic rods, and are then blocked by another whose rods are horizontal. What would be detected beyond the horizontal grid if a third grid were to be placed between the first two with its rods inclined at $\pi/4$ rad to each?

[$\frac{1}{4}I_0$]

VIII Sound Waves

41   Sound Waves

Questions for Discussion

41–1   Give a *physical* explanation as to why the passage of a sound wave gives rise to a molecular displacement which has a phase shift of $\pi/2$ rad relative to variations in air pressure.

41–2   A point source radiates sound energy isotropically at power P.
(a) What is the intensity at a point distant r from the source?
(b) How does the *amplitude* of molecular vibration depend on r?

41–3   The average energy density of a sound wave is the energy per unit volume of the medium which carries the wave. Use the method of dimensions to find how the energy density of a wave is related to its intensity and speed. There are no other variables, and the dimensionless constant is 1.

41–4   A microphone is connected to a c.r.o. so that the pattern on the screen illustrates the waveforms of different sounds. Discuss what would be seen when
(a) the sound from a recorder is compared to a handclap
(b) the sound from a tuning fork is compared to that from one an octave higher
(c) the waveform of a gong is observed over a long time interval
(d) a loaded tuning fork is sounded simultaneously with an otherwise identical unloaded fork
(e) notes of the same frequency and intensity are sounded on different instruments.

41–5   *Harmonic analysis of a waveform.* The spectrum of a piano note was analysed and found to have three main sinusoidal components. Their frequencies were 256 Hz, 512 Hz and 768 Hz, and their relative amplitudes 1.0, 0.25 and 0.50 respectively. Sketch these waveforms on the same axes, and then superpose them to show the waveform of the original sound.
(*In practice a sound would contain many more harmonics.*)

41–6   *Anisotropic medium.* The speed of sound through the oceans depends upon depth. In some places it is least at 1.3 km, becoming greater in both shallower and deeper regions. An explosion is set off at a depth 1.3 km. Draw sketches to show the resulting wavefronts and the paths taken by the sound energy.

41–7   Why, when sound waves try to pass from air into water, do they experience such a large reflection coefficient?

41–8   Comment on the statement: 'Sound waves would be isothermal if their frequency were lower – there would then be enough time for heat flow to occur'.

41–9   Sound waves are propagated through a gas at a speed $c = \sqrt{(\text{adiabatic bulk modulus}) / (\text{gas density})}$. Transverse waves travel along strings at a speed $c = \sqrt{(\text{tension}) / (\text{mass per unit length})}$. Rewrite this latter equation using the density of the string. What quantity for the string is analogous to (and has the same dimensions as) the bulk modulus of the gas?

*41–10   What is meant by *impedance matching*? (*See for example Qu.* **56–16**). Give two examples of its application in *acoustics.*

41–11   How does a car silencer operate? (*Think in terms of an acoustic filter.*)

41–12   Would you expect the quality of sound from a *large* loudspeaker to depend upon direction?

41–13   Would it be possible to construct a diffraction grating for sound waves? If so, estimate a suitable grating spacing for a wavelength 0.5 m.

41–14   Some Cathedrals and Abbeys have rectangular loudspeakers which are much longer than they are broad. Explain why they have this shape, and suggest how they should be orientated for use in these buildings.

41–15   When a tuning fork is held over a resonance tube, and the air column's length varied, it is found that the intensity of a loud sound dies away more rapidly than that of a quiet sound. Consider the phase relation between the vibrations of the fork prong and the air column to discuss in terms of *work*, and then in terms of *energy*, why this should be so.

41–16   A tuning fork is held in the resonance position over a tube containing an air column. What is the effect on the length of air column required for resonance of
(a) saturating the air with water vapour
(b) doubling the air pressure
(c) doubling the common (*Celsius*) temperature
(d) replacing the air by helium at the same temperature and pressure?

41–17   When a wind instrument is played the temperature of the air column inside it is increased. What effect does this have on its fundamental frequency? Is there any similar effect for a stringed instrument?

41–18   When a double-bass string is plucked vigorously, the note which is heard has a quality which changes with time. Suggest reasons for this.

41–19 In the *Melde* experiment the part of the wire into which the vibrator passes energy has a *smaller* amplitude at resonance than adjacent parts. Why? Can this amplitude be zero when there is a standing wave on the wire?

41–20 O-M A gas fails to behave as a continuum to sound waves at normal pressures if the frequency becomes very high (Qu. **41–26**). At what pressure would you expect to detect a similar effect for a frequency of 1.0 kHz? Would the sound be audible?

Quantitative Problems

(*For questions on the Doppler effect, see* Chapter **15**.)

Pressure, Displacement and Intensity

†**41–21** *Inverse square law.* A point source of sound gives out waves whose energy density is 25 nJ m^{-3} in a region 4.0 m from the source. What would be the energy density another 16 m away?

[1.0 nJ m^{-3}]

†**41–22** A sound source transmits energy at the same rate in all directions, and can just be detected 0.50 km from the source, where its intensity is 1.0 pW m^{-2}. What is the power of the source? How does this compare with the power output of a small loudspeaker? Why does the answer appear to contradict experience?

[π μW]

41–23 *The decibel.* The intensity I of a sound is expressed in W m^{-2} in the SI, but sometimes it is quoted relative to an arbitrary intensity $I_0 = 1.0$ pW m^{-2} (about the minimum detectable). A *logarithmic* scale is chosen because the ability of the ear to detect changes of intensity depends upon the intensity to which it is already subjected.

Thus   *intensity in decibels* $= 10 \lg (I/I_0)$.

Calculate, in decibels, the intensities of

 (*a*) a normal speaking voice, 1.0 μW m^{-2}
 (*b*) the threshold of pain, 1.0 W m^{-2}.

(*Note that* 1 bel = 10 decibels.)

[(*a*) 60 db = 6 bel   (*b*) 120 db = 12 bel]

***41–24** Sound waves of frequency 0.40 kHz are passing through air of density 1.3 kg m^{-3} at a speed 0.34 km s^{-1}. The pressure amplitude of sounds which can just be tolerated is 30 Pa, and that of sounds which an observer can just detect is 20 μPa. Calculate, for both sounds,

 (*a*) the wave intensity, and
 (*b*) the displacement amplitude.

[(*a*) 1.0 W m^{-2}, 0.45 pW m^{-2} (*b*) 27 μm, 18 pm]

Speed, Frequency and Wavelength

†**41–25** *Depth sounding.* A ship sends a sound pulse to the sea floor at a speed 1.5 km s^{-1}, and the reflected signal is detected after a time delay of 80 ms. How deep is the sea at this point? How could such a small time interval be measured with precision?

[60 m]

†**41–26** Sound waves are propagated through a gas so long as the wavelength is large relative to the mean free path of the gas molecules, which for air at room temperature is about 0.1 μm. Calculate a frequency above which sound waves could not propagate. Assume that the speed of sound is 0.34 km s^{-1}.

[3 GHz]

The Speed of Sound through Matter

†**41–27** *Solid.* The speed of a wave motion in general is given by an expression of the form

$$c = \sqrt{(\text{elasticity factor/inertia factor})}.$$

 (*a*) Confirm, by examining the units concerned, that the use of the *Young* modulus and density of a material is consistent with obtaining the speed of longitudinal waves through a rod.
 (*b*) For steel $E = 0.20$ TN m^{-2} and $\rho = 7.8 \times 10^3$ kg m^{-3}. Calculate the speed of sound through a steel rod.
 (*c*) How does this compare with the speed of sound through air at s.t.p.?

[(*b*) 5.1 km s^{-1}]

41–28 *Liquid.* Water has a density 1.0×10^3 kg m^{-3}, and longitudinal sound waves travel through it at 1.5 km s^{-1}. Calculate

 (*a*) the bulk modulus of water, and
 (*b*) the compressibility.

[(*a*) 2.2 GN m^{-2} (*b*) 4.4 × 10^{-10} m^2 N^{-1}]

†**41–29** *Refraction of sound.* Sound travels through air at 0.33 km s^{-1}, and through water at 1.5 km s^{-1}.

 (*a*) What is the relative refractive index for sound waves passing from air into water?
 (*b*) At what angle must the sound waves strike an air-water interface if no sound energy is to be transmitted into the water?

(*Compare this situation with the total reflection of light waves trying to pass from water to air.*)

[(*a*) 0.22   (*b*) 13°]

41–30 *Temperature variations.* The speed of sound through dry air at 273 K is 331 m s^{-1}. By how much will it change when the temperature becomes 274 K? Would the speed change by the same amount over the temperature range 300 K to 301 K?

[0.61 m s^{-1}]

41–31   Calculate the relative refractive index for sound waves travelling through air at 330 K and crossing an interface into cool air at 280 K.

[1.09]

*41–32   Degrees of freedom. Sound travels through hydrogen gas at 1.26 km s^{-1} when the temperature is 273 K. The molar mass of hydrogen molecules is 2.00×10^{-3} kg mol^{-1}.

(a) What is the value of $\gamma \; (= C_{p,\text{m}}/C_{V,\text{m}})$ at this temperature?
(b) How many degrees of freedom do the molecules have?
(c) Suggest what kinds of energy the molecules may have at this temperature.

Assume R   [(a) 1.4   (b) 5]

*41–33   Show that, for an ideal gas

$$\frac{\text{r.m.s. molecular speed}}{\text{speed of sound}} = \sqrt{(3/\gamma)}.$$

Superposition

41–34   Stationary waves. A tuning fork describes s.h.m. of frequency 0.68 kHz, and sends sound waves towards a wall at a speed 0.34 km s^{-1}. They are reflected normally without loss of energy. Discuss the variation in molecular displacement amplitude and air pressure amplitude in the region up to 1.0 m away from the wall. Illustrate your answers with appropriate diagrams.

41–35   Two-source interference in sound. Two loudspeakers are placed 3.0 m apart, and are energized by the same oscillator whose frequency is 0.66 kHz. They emit sound waves of speed 0.33 km s^{-1}.

(a) A detector microphone is moved along a line parallel to that joining the speakers, but 20 m away. What is the separation of points at which maximum intensity is detected near the centre of the pattern?
(b) What would be the effect on this separation distance of
  (i) increasing the source frequency
  (ii) increasing the separation of the loudspeakers
  (iii) moving the microphone further away
  (iv) raising the temperature of the air in the room?
(c) Would the pattern be detectable if both speakers were energized by the sound of an orchestra (rather than the oscillator)?

[(a) 3.3 m]

41–36   The Quincke tube. Refer to the diagram. In this method for measuring the speed of sound in air, a source of frequency 1.7 kHz was used. The sliding tube had to be moved 0.10 m between consecutive positions of minimum intensity at 300 K. Calculate the speed of sound at this temperature.

The temperature of the air in the tube was increased to 375 K. Calculate the new separation of the tube positions using the same source frequency.

[0.34 km s^{-1}, 0.11 m]

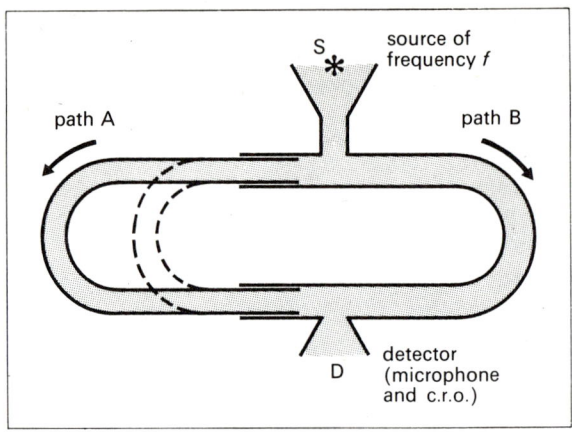

For Qu. **41–36**

†41–37   Beat frequency. An accurately calibrated tuning fork of frequency 256.0 Hz is sounded with middle C of a piano, and a beat frequency of 3.0 Hz is heard.

(a) What are the possible frequencies of the piano string?
(b) How could you (by experiment) obtain a unique answer?

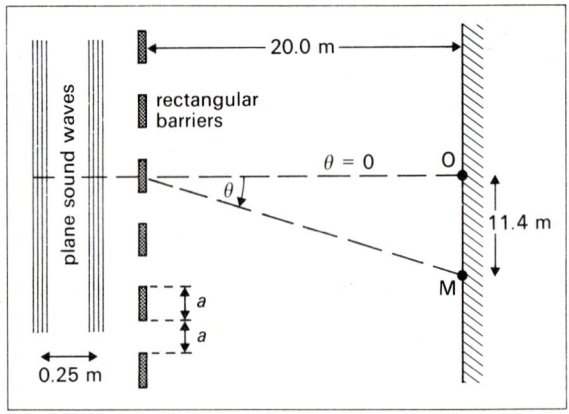

For Qu. **41–38**

41–38   The diffraction grating. Refer to the diagram, which is not to scale. Plane sound waves of wavelength 0.25 m approach a series of rectangular planks which are arranged parallel to each other, and separated by a distance a equal to their widths. When the microphone is moved from O to M, a distance of 11.4 m, the loudness of the received sound varies from a maximum, through a minimum, to a second maximum. What is the width of each plank?

[0.25 m]

*41–39   Diffraction at a slit aperture. Refer to the diagram. Plane sinusoidal waves of wavelength 2.0 m are incident on a slit-type opening of width 4.0 m. Calculate the smallest angle θ

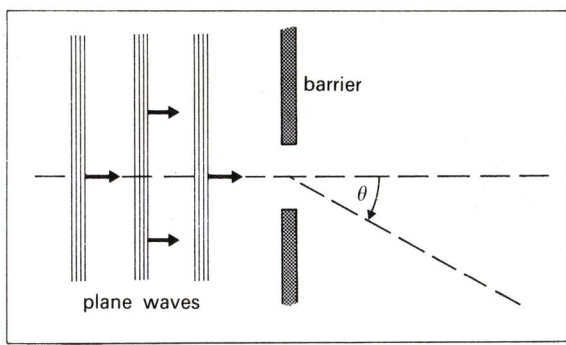

For Qu. **41–39**

41–44   *The resonance tube.* The following results were obtained using a resonance tube closed at one end. f is the tuning fork frequency, and l the shortest resonance length for that fork.

f/Hz	200	250	300	400	500
l/mm	402	318	260	190	149

(*a*) Plot these results so as to obtain a straight-line graph.
(*b*) Your graph should not pass through the origin. Explain why, and use the intercept to make some quantitative deduction about the tube.
(*c*) Find a value for the speed of sound at the air temperature used.

[(*c*) 33(8) m s^{-1}]

along which a minimum of sound intensity will be detectable some distance from the source.

Do you think this sort of situation would arise with the sound waves sent out by a person talking?

[$\pi/6$ rad]

Air Columns

†**41–40**   The length of the air column in a glass tube closed at one end can be adjusted over the range 0.50–2.0 m. At what lengths would there be resonance for a tuning fork of frequency 0.28 kHz? Assume the speed of sound to be 0.34 km s^{-1} and ignore end-corrections.

[0.91 m, 1.5(2) m]

41–41   The second overtones of an open pipe A and a closed pipe B have the same frequency at a given temperature. Calculate the ratios

(*a*) length of A to that of B
(*b*) fundamental frequency of A to that of B.
Ignore end-corrections.

[(*a*) 6:5   (*b*) 5:3]

41–42   Two identical pipes are taken, and one is stopped at one end, while the other is open at both ends. The same overtone is then excited in both pipes, and the open one resonates at 340 Hz, while the closed one resonates at 255 Hz. Take the speed of sound as 340 m s^{-1}, and calculate

(*a*) the length of the pipes, and
(*b*) the number of the overtone.
Ignore end-corrections.

[(*a*) 1.00 m   (*b*) 1]

41–43   In a typical resonance tube experiment a tuning fork of unknown frequency caused the air column to resonate when it had the lengths 115 mm, 365 mm, and 615 mm respectively. The speed of sound is 340 m s^{-1}.

(*a*) Calculate the tuning fork frequency.
(*b*) Estimate a value for the radius of the tube by evaluating the end-correction.

[(*a*) 680 Hz   (*b*) 1(7) mm]

41–45   *The dust tube.* (*a*) In a dust-tube experiment a brass rod is set into oscillation by rubbing. The rod's length is 1.0 m, its density is 8.0×10^3 kg m^{-3}, and the *Young* modulus for brass is 0.10 TN m^{-2}. Calculate (*i*) the speed of waves through the rod, (*ii*) the wavelength of the fundamental vibration of the rod if it is clamped at its mid-point, (*iii*) the frequency of the note given out by the rod.
(*b*) At resonance the lycopodium powder gathers into piles, with the nodes separated by 96 mm. Calculate a value for the speed of sound through the gas in the tube at this temperature.

[(*a*) 3.5 km s^{-1}, 2.0 m, 1.8 kHz   (*b*) 0.34 km s^{-1}]

Strings

†**41–46**   A horizontal string of mass 50 g and length 8.0 m is subject to a tension of 40 N. One end is given a brief transverse displacement. How long does it take the resulting disturbance to travel to the far end?

[0.10 s]

†**41–47**   A steel wire of density 7.8×10^3 kg m^{-3} is subjected to a tension of 12.5 N. Transverse waves travel along the wire at 40 m s^{-1}. What is the wire's cross-sectional area?

[1.0 mm^2]

41–48   The tension in a steel wire of cross-sectional area 0.10 mm^2 is 16 N. The density of steel is 7.8×10^3 kg m^{-3}. The wire is fixed at two points separated by 0.75 m, and then plucked at a point 0.19 m from one end. Calculate

(*a*) the mass per unit length of the wire
(*b*) the speed at which transverse waves are propagated along the wire
(*c*) the wavelength of the fundamental mode of vibration
(*d*) the fundamental frequency of vibration.

What other frequency will sound with an appreciable intensity on this occasion, and why?

[(*a*) 0.78 g m^{-1} (*b*) 0.14 km s^{-1} (*c*) 1.5 m
(*d*) 96 Hz;   0.19 kHz]

41–49 A string of tension 0.10 kN and linear mass density 2.5×10^{-3} kg m^{-1} is placed near an audiofrequency oscillator of limited range. It is found that the string resonates at 0.30 kHz, and that the *next* resonance is at 0.36 kHz. It is not known which overtones these are. What is the length of the string?   [1.7 m]

41–50 A wire resonates at 0.10 kHz. Its tension is increased by 69%. What is the new corresponding resonant frequency?

[0.13 kHz]

41–51 A heavy flexible rope hangs vertically. The tension at each cross-section is determined by the weight of that part of the rope hanging below. Let the distance of any cross-section from the bottom be h. Show that

(*a*) the speed of a transverse wave at height h is $\sqrt{gh}$, and

(*b*) the wave accelerates as it climbs the rope.

41–52 A string is stretched so that its new length is $(1 + e)l_o$, where l_o is its original length and $e \ll 1$. Find an expression for the ratio of the frequencies of the fundamental longitudinal and transverse vibrations. (Qu. **41–27** *will be helpful.*)

$[f_l/f_t = e^{-\frac{1}{2}}]$

IX Electrostatics

42   Charge and Coulomb's Law

Questions for Discussion

42–1 Suppose you thought you had discovered a third type of electric charge (in addition to positive and negative). What experimental procedure would you follow to test your hypothesis?

42–2 What is meant by the statement 'Electrostatic forces obey the principle of superposition'?

42–3 A rubbed cellulose acetate rod attracts small pieces of paper. After contact some adhere to the rod, and others are repelled. Why?

42–4 Why, if we use the SI, may we not allot a *defined* value (such as 1) to the permittivity constant ε_0?

42–5 How can an atomic nucleus be stable if it is composed of particles which are either neutral or carry *like* charges?

42–6 Discuss the experimental evidence for believing that charge is quantized. Is mass quantized?

42–7 Discuss the experimental evidence for the law of conservation of electric charge.

42–8 O-M Consider a typical 150 mm perspex ruler. Estimate how many nuclei it contains, and hence the number of electrons and their associated charge.

42–9 O-M What is the size of the electrostatic attractive forces exerted by neighbouring ions of an NaCl crystal on each other?

Quantitative Problems

Macroscopic

†42–10 *Maxwell's equations* dealing with electromagnetic fields suggest that $c = 1/\sqrt{\varepsilon_0\mu_0}$. Use the defined value of μ_0 and the experimental value of c to find a value for ε_0.

42–11 *Coulomb's torsion balance.* (a) In one experiment using a torsion balance two small spherical charges Q_1 (movable) and Q_2 (fixed to the beam) caused the beam to be deflected $\pi/4$ rad from the equilibrium position when the charges were placed 40 mm apart. When the separation was reduced to 20 mm, the deflecting force remaining perpendicular to the beam, the angle of deflection became θ. Calculate θ, assuming the truth of the inverse square law.

(b) Given that $Q_1 = Q_2 = 1.0$ nC, evaluate the torsion constant of the suspending fibre (in μN m rad^{-1}). Take the length of the moment arm to be 100 mm.
Assume ε_0 [(b) 0.72 μN m rad^{-1}]

42–12 Two small identical conducting spheres, each of weight 4.0 mN, share an electric charge, and are suspended from the same point by silk threads 1.0 m long. When they reach equilibrium their centres are separated by 20 mm. What was the size of the charge that they shared?
Assume $1/4\pi\varepsilon_0$ [$\pm$ 2.7 nC]

42–13 Two identical point charges repel each other with a force 0.1 mN. They are moved 5.0 mm further apart, and the repulsive force is reduced to 25 μN.

(a) How far apart were they originally?
(b) What was the size of the charges?
Assume $1/4\pi\varepsilon_0$ [(a) 5.0 mm   (b) $\pm$ 0.53 nC]

42–14 Two charges -0.20 nC and -0.80 nC are held 30 mm apart, and a 1.0 pC test charge moved along the line joining them. Where would it experience zero resultant force? (*This is a **neutral point** in the electric field.*)   [10 mm from -0.20 nC]

*42–15 (a) Two positive charges $+Q$ are held a distance d apart. A particle of mass m and charge $-Q$ is placed mid-way between them, and given a small displacement perpendicular to the line joining them. Show that the particle describes s.h.m. of period $\sqrt{\varepsilon_0 m\pi^3 d^3/Q}$.

(b) Calculate the period for a particle of charge $+Q$ displaced along the line joining the charges.

*42–16 *Method of images.* A point charge $+Q$ is placed a distance a from an infinite conducting plane on which it induces a charge $-Q$.

(a) Draw the electric field lines for this arrangement.
(b) Draw the field pattern associated with charges $+Q$ and $-Q$ separated by a distance $2a$.
(c) Compare (a) with (b), and evaluate the attractive forces that the plane and charge exert on each other.

Is there an analogous situation in gravitation?

Microscopic

42–17 A proton and an electron are separated by a distance x. Calculate

(a) the attractive electric force F_e each exerts on the other
(b) their attractive gravitational force F_g
(c) the ratio F_e/F_g, and comment on its value.

Does x appear in the answer? Comment.
Assume m_p, m_e, $1/4\pi\varepsilon_0$, G, e [(c) $\sim 10^{39}$]

42–18   Imagine two α-particles, each of mass 6.7×10^{-27} kg, to be placed 1.0 pm apart. They experience mutual electrostatic and gravitational interactions. If they were released, what would be their instantaneous accelerations?

Assume e, $1/4\pi\varepsilon_0$, G $[1.4 \times 10^{23}$ m s$^{-2}]$

42–19   Consider a model of the hydrogen atom in which the electron describes a circular orbit of radius 53 pm about a proton whose mass is very large in comparison.

(*a*) Calculate the centripetal electric force experienced by the electron.

(*b*) Why does the electron speed not increase?
(*c*) What is its orbital speed?
(*d*) What is the electron's acceleration?
(*e*) Is the acceleration constant?

Assume e, m_e, $1/4\pi\varepsilon_0$

$[(c)$ 2.2 Mm s^{-1} (d) 9.0×10^{22} m s$^{-2}]$

†**42–20**   When a mole of singly-charged ions is discharged at an electrode during electrolysis, then the charge transferred is called the **Faraday constant** *F*. What is its size?

Assume N_A, e $[96$ kC mol$^{-1}]$

43   The Electric Field

Questions for Discussion

43–1   *Electric field lines.* Sketch the patterns of field lines that result from the following arrangements of charges:

(*a*) an isolated negative point charge
(*b*) an isolated negatively charged sphere
(*c*) a pair of positive point charges placed close to one another
(*d*) a positive point charge placed close to a negative point charge
(*e*) a positive point charge placed close to a large earthed metallic plate
(*f*) a uniformly charged positive conducting plate held just above, and parallel to, a large earthed metallic plate.

43–2   Why do we *not* define *E* as numerically equal to the force acting on a charge of one coulomb?

43–3   Discuss the principles behind the calculation of the electric force exerted by one cubic charge distribution on a similar distribution placed relatively near. (*Refer also to the analogous situation in* Gravitation Qu. **20–1**.)

43–4   Under what circumstances will a charged particle follow the path of electric field lines?

43–5   A charged particle moves in a straight line through an electric field. Is it necessarily accelerated?

43–6   If, in finding the speed to which a charged ion had been accelerated, you obtained an answer 6 Gm s^{-1}, what conclusions would you draw?

43–7   An electric dipole has zero net charge: how can it establish an external field?

43–8   Can a *neutral* atom be made into an electric dipole? Can an ion?

43–9   Describe the trajectory of an electric dipole moving through

(*a*) a uniform electric field
(*b*) a non-uniform field.

Discuss the stability of different orientations of the dipole axis relative to the electric field lines.

43–10   *The oscilloscope.* What is the shape of the trajectory of an electron in an oscilloscope after it has left the deflecting field? Calculate the effect on the observed deflection of doubling the following dimensions:

(*a*) the vertical separation of the Y-plates
(*b*) the length of the deflecting plates
(*c*) the separation of the deflecting plates from the screen.

In each change all *other* factors are held constant.

***43–11**   Charges are placed at each end of an insulating rod of moment of inertia *I* so that the system has an electric dipole moment *p*. It is then placed in a uniform electric field *E*, and suspended from a torsionless fibre.

(*a*) Show that when the system is given a small angular displacement it executes s.h.m.

(*b*) Show that the time period is $2\pi\sqrt{I/pE}$, and hence that if this dipole is set oscillating in different fields then the frequency of oscillation is proportional to the square root of the field strength.

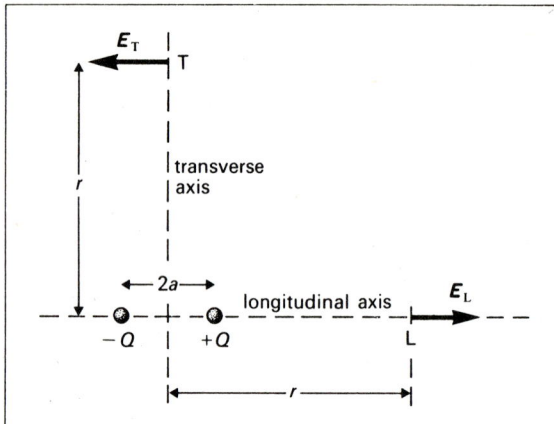

For Qu. 43–12

*43–12 Refer to the diagram, in which $r \gg 2a$. Show that, in magnitude, $E_L/E_T = 2$.

43–13 O-M A polar molecule might have positive and negative charges separated by roughly one molecular diameter. Estimate its electric dipole moment.

Quantitative Problems

Cause *E* and its Effect *F*

†43–14 What point charge should be placed at the centre of a sphere of radius 1.0 m so as to make the radial field at the surface 1.0 N C⁻¹?
Assume $1/4\pi\varepsilon_0$ [0.11 nC]

43–15 Calculate the value of E
(a) at the 'surface' of a gold nucleus ($r = 6.9$ fm)
(b) established by the proton at the most probable position of the electron in a hydrogen atom ($r = 53$ pm).
For (a) take $Z = 79$, and assume that the nucleus behaves as a point charge.
Assume e, $1/4\pi\varepsilon_0$
[(a) 2.4×10^{21} N C⁻¹   (b) 0.51 TN C⁻¹]

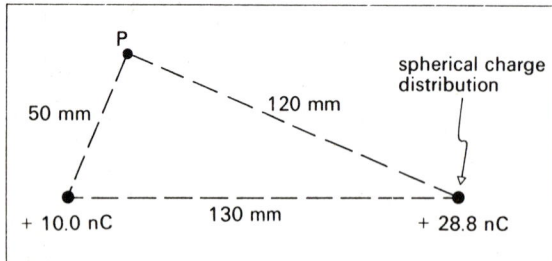

For Qu. **43–16**

43–16 Refer to the diagram. What is the magnitude of the resultant electric field at P? Show its direction on a sketch diagram.
Assume $1/4\pi\varepsilon_0$ [40 kN C⁻¹]

43–17 An electron is held motionless in the evacuated space between a pair of charged metallic plates of opposite sign.
(a) What is the field direction?
(b) What is the size of the field?
(c) Given that the unit N C⁻¹ ≡ V m⁻¹, and that the plates are separated by 10 mm, calculate the p.d. between them. Comment on your answer.
(*Note that this is a conceptual question, since one cannot observe an electron directly.*)
Assume e, m_e, g_0 [(b) 56 pN C⁻¹   (c) 0.56 pV]

43–18 *The cathode ray oscilloscope.* In an arrangement like that of the cathode ray oscilloscope, an electron enters a field of 0.40 kN C⁻¹ directed downwards with an initial horizontal velocity of 10 Mm s⁻¹. The length of the deflecting plates is 40 mm, and the fluorescent screen is 0.25 m away from the nearer edge of the plates. Calculate
(a) the time spent between the deflecting plates
(b) the upward acceleration
(c) the upward velocity component acquired by the electron
(d) the upward displacement of the electron while between the plates
(e) the final velocity direction
(f) the upward displacement of the electron where it hits the screen.
What size field would have given a deflection on the screen of 30 mm?
Assume e/m_e
[(a) 4.0 ns   (b) 70 Tm s⁻²   (c) 0.28 Mm s⁻¹   (d) 0.56 mm
(e) 28 mrad above horizontal   (f) 7.6 mm;   1.7 kN C⁻¹]

Electrical Energy

43–19 (a) An atom which has gained one electron is accelerated by electric forces in a uniform field of 0.30 kN C⁻¹ until it acquires an energy 0.48 aJ. Over what distance was it accelerated?
(b) What speed would the same ion acquire if it fell freely through the same distance in the Earth's gravitational field?
(c) What was the final speed of the ion in the electric field if its mass is 6.0×10^{-26} kg?
(*Note that the speed acquired in the electric field depends upon both m and Q, since there is no direct relationship (in general) between mass and net electric charge. The gravitational force $\propto m$.*)
Assume e, g_0 [(a) 10 mm   (b) 0.44 m s⁻¹   (c) 4.0 km s⁻¹]

43–20 An electron travelling in the direction of the field lines has its speed reduced in a uniform electric field from 1.0 Mm s⁻¹ to rest in a space of 40 mm. Calculate
(a) the k.e. lost by the electron
(b) the (negative) work done on the electron by the agent responsible for the field

(*c*) the electric force that acts on the electron

(*d*) the size of the field.

In what distance would a field of the same size (but oppositely directed) be able to arrest a proton of (*i*) the same speed (*ii*) the same energy?

Assume m_e, m_p, e [(*d*) 71 N C^{-1} (*i*) 73 m]

43–21 Consider an α-particle and a β-particle each travelling at 5.0 Mm s^{-1}. Calculate the minimum size of field required to bring each to rest in a distance 60 mm. The mass of an α-particle is 6.7×10^{-27} kg. Can a γ-ray be stopped by an electric field?

Assume e, m_e [(α) 4.4 MN C^{-1} (β) 1.2 kN C^{-1}]

The Electric Dipole

†43–22 *Electric dipole moment.* An electric dipole consists of two spherical charges ± 1.0 nC (Q) separated by 5.0 mm (2*a*). It is placed in a uniform electric field E of 20 kN C^{-1} so that the line joining the negative charge to the positive makes an angle $\pi/2$ rad with the electric field lines. Calculate

(*a*) the force exerted on each charge by the agent that establishes the external field

(*b*) the moment arm of the couple on the dipole, and hence its torque T

(*c*) the electric dipole moment p. (*Use both the expressions $p = 2aQ$, and $p = T_{max}/E$, and hence confirm that they are consistent.*)

[(*c*) 5.0 pC m]

*43–23 An electric dipole placed in a uniform electric field experiences a torque 0.70 mN m when it is perpendicular to the field lines. It is then held antiparallel to the field, and given an infinitesimal displacement. What rotational k.e. does it acquire?

[1.4 mJ]

*43–24 An electric dipole of moment 60 nC m is placed in its stable position in a uniform field, and then rotated through an angle $\pi/2$ rad. It is found that this causes the system to gain 12 μJ of energy.

(*a*) What is the field strength?

(*b*) How much more energy is needed to increase the angle of rotation by a further $\pi/6$ rad?

(*c*) What is the maximum p.e. that this system could possess? (Take the initial value of the p.e. as your reference zero.)

[(*a*) 0.20 kN C^{-1} (*b*) 6.0 μJ   (*c*) 24 μJ]

44   Gauss's Law

Questions for Discussion

44–1 Which, if either, is the more fundamental, *Coulomb's Law* or *Gauss's Law*?

44–2 Suppose we assume the truth of *Gauss's Law*. Does this mean that *Coulomb's Law* must be *exactly* an inverse square law?

44–3 Consider a closed surface which encloses zero net charge.

(*a*) What is the electric field flux through this surface?

(*b*) What can you say about the value of E at each point on the surface?

44–4 The value of E at each point on a closed surface is zero.

(*a*) What is the electric field flux through this surface?

(*b*) Can you say that there are no charges inside the surface? Discuss.

44–5 What is the electric field flux cutting closed surfaces drawn round

(*a*) a neutral atom (such as A)

(*b*) a negative ion (such as Cl$^-$)

(*c*) a positive ion (such as Na$^+$)

(*d*) a polar molecule (such as H_2O, which is an electric dipole of moment 6×10^{-30} C m)?

44–6 Use *Gauss's Law* to compare the following electric field strengths:

(*a*) that close to a large conducting plane sheet of area A which carries a charge Q (uniformly distributed over both faces), with

(*b*) the field between a pair of parallel plates of the same area A, each of which carries a *net* charge $\pm Q$, when they are placed relatively near.

(*Note that in* (*b*) *the effect of the second plate is to cause a redistribution of the charge on the original plate.*)

44-7   We often say $E = 0$ at all points inside a conductor in electrostatic equilibrium. This is not true – what do we *really* mean?

44-8   How does one achieve the complete removal of excess charge from a small conducting body?

44-9   Suppose that a neutral gold atom has a spherically symmetric charge distribution, and therefore zero external electric field. How can such an atom deflect an α-particle?

Quantitative Problems

Field Flux and Gauss's Law

†**44-10**   A uniform electric field of 40 N C^{-1} is directed horizontally due North. A cardboard sheet of area 0.50 m^2 is free to rotate about an E–W axis in the plane of the sheet. Calculate the electric field flux through the sheet when it is

(*a*) horizontal
(*b*) vertical
(*c*) inclined at an angle $\pi/3$ rad to the horizontal.

[(*a*) zero   (*b*) 20 N m^2 C^{-1} (*c*) 17 N m^2 C^{-1}]

44-11   A charge $+ 13.6$ nC is placed inside a cubical box.

(*a*) What is the field flux out of the box if its side is 100 mm and the charge is placed (*i*) at the centre (*ii*) just inside one corner?
(*b*) What is the field flux through any *one* face if the charge is placed at the centre?
(*c*) Why does *Gauss's Law* not lead to a simple method of calculating the field flux through one face when the charge is off-centre?
(*d*) Repeat (*a*) and (*b*) for a cubical box of side 200 mm.

Assume ε_0 [(*b*) $+ 0.26$ kN m^2 C^{-1}]

Radial Field Lines

†**44-12**   A solid metallic sphere of radius 0.20 m is given a charge 10 nC. Calculate the electric field strength at these distances from the centre:

(*a*) 0.10 m (*b*) 0.20 m (*c*) 0.40 m.

Would your answer to (*a*) be modified if instead of being solid the sphere had been an empty shell?

Assume $1/4\pi\varepsilon_0$ [(*b*) 2.2 kN C^{-1} (*c*) 0.56 kN C^{-1}]

44-13   A long straight wire is given a linear charge density 1.0 nC m^{-1}

(*a*) What is the field at a point 20 mm from the wire?
(*b*) How far from the wire is the field one quarter of this value?

Assume $1/4\pi\varepsilon_0$ [(*a*) 0.90 kN C^{-1}]

44-14   Consider a long wire of linear charge density λ about which an electron is orbiting in a circular path of radius a.

(*a*) Write down an expression for the electric field size at a point distance a from the wire.
(*b*) Calculate a value for the electron's orbital speed in terms of its specific charge e/m_e.

Does the electron's angular speed depend on its orbital radius?

$$\left[(b)\ \sqrt{\left(\frac{\lambda}{2\pi\varepsilon_0}\right) \times \left(\frac{e}{m_e}\right)}\right]$$

***44-15**   (*For the following hypothetical problem you will find useful the analogous gravitational situation in which a particle oscillates inside a tunnel bored through a diameter of the Earth, Qu. 20–48.*) A sphere of radius R is given a charge Q which is uniformly distributed through its volume, and a particle of charge $- Q_0$ and mass m is held at the opening of a diametrical tunnel and released.

(*a*) Write down the force acting on the particle in terms of its displacement r from the centre.
(*b*) Show that the particle describes s.h.m., and find the time period.

In the gravitational analogy the time period is independent of the mass of the particle, whereas here it is not. Discuss.

$$[(b)\ T = 2\pi\sqrt{4\pi\varepsilon_0 mR^3/QQ_0}]$$

Parallel Field Lines

†**44-16**   Consider a sphere of radius 1.0 m and a very large plane sheet, both made of conducting material and carrying a charge density 17.6 nC m^{-2}.

(*a*) Calculate the value of E just outside the surface of the sphere (*i*) using the result $E = \sigma/\varepsilon_0$, (*ii*) by assuming that the whole charge on the sphere behaves as though it were a point charge at the centre of the sphere. Are these results consistent?
(*b*) Discuss the differences in the variation of E as one moves away from the sphere and the plane sheet, and account for them in terms of the behaviour of electric field lines.

Assume ε_0, $1/4\pi\varepsilon_0$ [(*a*) 2.0 kN C^{-1}]

44-17   *Derivation of $E = \sigma/\varepsilon_0$.* A large metallic plate has a surface charge density $+ 8.85$ μC m^{-2}. Imagine a cylinder of cross-sectional area 1.0×10^{-4} m^2 and height 10 mm to be constructed with one plane end just inside the metal surface, and the other outside. Let the field strength at the second plane surface be E.

Calculate (in terms of E) the electric field flux through

(*a*) the first plane surface
(*b*) the second plane surface
(*c*) the curved surface.

Equate $\varepsilon_0 \times$ (the total field flux through the surface) to the net charge enclosed, and evaluate E. Does E depend on the height of the cylinder? Discuss.

Assume ε_0 [$E = 1.0$ MN C^{-1}]

44–18   An electron is placed at the centre of a large metal plate which carries a negative charge, and is accelerated towards a nearby positive plate at 2.0×10^{15} m s^{-2}. Both plates have charge densities of the same size. What is its value?

Assume ε_0, e/m_e [0.10 μC m^{-2}]

*44–19   (*In working this problem you will find Qu. 42–16 helpful.*) A point charge $+ 1.0$ nC is placed 50 mm from a very large conducting sheet whose surface charge density, when isolated, is $\boldsymbol{\sigma}$.

(*a*) Calculate the attractive force that results because of the charges induced on the conducting sheet.
(*b*) Calculate, in terms of σ, the repulsive force experienced by the point charge from the original charge distribution.
(*c*) What value of σ would enable the charge to experience zero resultant force? Discuss the stability of this equilibrium position.

Assume ε_0, $1/4\pi\varepsilon_0$

[(*a*) 0.90 μN   (*b*) (0.11 σ) kN m^2 C^{-1} (*c*) $+ 8.0$ nC m^{-2}]

45   Electric Potential

Questions for Discussion

45–1   Give examples of situations in which the potential of a charged body has a sign opposite to that of its charge.

45–2   Can we assign a value -240 V to the potential of the Earth? How would this affect observed potential differences?

*45–3   When calculating the potential established by a point charge, we choose $V = 0$ when $r = \infty$ m as our reference zero. Can we follow the same procedure for

(*a*) an infinitely long charged wire, and/or
(*b*) an infinitely large charged sheet?

Use $W = \int F \cdot dr = \int EQ_0 \cdot dr$ as the basis of your reasoning.

45–4   (*a*) What can you say about the electric field strength in a region over which the electric potential has everywhere the same value?

(*b*) If in a region $E = 0$ at every point, what can you say about the value of V? Give examples of situations which demonstrate your conclusion.

45–5   Sketch the *equipotential surfaces* which result from the following charge configurations:

(*a*) a point charge
(*b*) a spherically symmetric charge distribution
(*c*) a very large plane uniformly charged sheet
(*d*) a long uniformly charged cylinder
(*e*) an electric dipole.

Can equipotential surfaces intersect? – discuss.

45–6   Sketch graphs to show the mutual electric potential energy E_p as a function of their separation r for

(*a*) point charges $+ Q$ and $+ Q$
(*b*) point charges $- Q$ and $+ Q$.

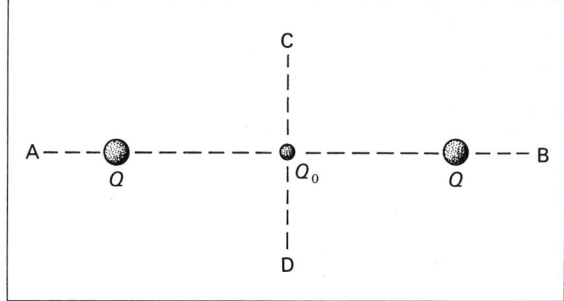

For Qu. **45–7**

45–7   Refer to the diagram. The two fixed charges Q are both positive.

(*a*) When the test charge Q_0 is displaced along the line AB and released, it undergoes periodic motion.
  (*i*) What is the sign of Q_0?
  (*ii*) What would happen if it were given a similar displacement along CD?
  (*iii*) What would happen for any other direction?
(*b*) Suppose that Q_0 is negative. Discuss how its electric potential energy varies over the area of the diagram, and sketch the shape of surface which would result in similar variations of gravitational potential energy.

45–8   Consider points distant r from a long wire of positive linear charge density λ. Sketch graphs which show the variation with r of

(*a*) the electric field strength E, and
(*b*) the electric potential V.

45–9   Distribution of charge on a conductor of varying curvature. Consider two spheres A and B of radii a and b ($b > a$), some distance apart, but joined by a wire. They carry charges Q_A and Q_B.

(a) Write down, in terms of the charges and radii, the condition that the spheres are at the same potential.

(b) Calculate the ratio of their charge densities.

(c) Write down the ratio of the electric field at the surface of A to that of B.

Suppose that instead of being two spheres joined by a wire, the conducting system consisted of a single metallic surface of variable curvature. What conclusions can be drawn?

45–10   Principle of the Van de Graaff generator. Consider a pair of *concentric* conducting spherical shells of radii b and a ($b > a$) carrying like charges Q_A and Q_B. Discuss, *in terms of the potentials of the two spheres*, what happens to the charges if the spheres are joined by a conducting wire.

45–11 Does the divergence of the leaves of a gold leaf electroscope indicate a charge or a potential difference? If a charge, on what body? If a p.d., between which two conductors?

*45–12 By considering the stability of equilibrium of the moving plate of an attracted-disc electrometer, explain what happens when the p.d. between the plates is gradually increased. (*Note that the stability criterion makes it necessary, in one design, to use a spring of relatively large force constant, and that this reduces the sensitivity of the instrument.*)

Quantitative Problems

Plane Equipotential Surfaces

†45–13 A battery of e.m.f. 500 V is connected across a pair of conducting plates of large area. Calculate the electric field strength between the plates for the following situations:

(a) the plates separated by 5.0 mm

(b) the plate separation x increased to 10 mm, the battery remaining connected

(c) $x = 5.0$ mm, and an identical battery connected in parallel with the first

(d) $x = 5.0$ mm, and an identical battery connected in series with the first.

$$[(a)\ 0.10\ \text{MV m}^{-1}\qquad (b)\ 50\ \text{kV m}^{-1}]$$

45–14 In an oscilloscope a potential difference of 20 V between the Y-plates gives an electron-beam displacement of 30 mm. What size potential difference would be needed for a displacement of 45 mm? (*Hint: refer to Qu.* **43–18** *to see how to relate the electron displacement to the electric field between the plates.*)

45–15 An earthed plate of area 0.10 m² acquires a charge -8.85 nC when placed 2.0 mm away from a similar plate of charge $+8.85$ nC. Calculate the value of

(a) E between the plates

(b) V at a point mid-way between the plates.

Assume ε_0 $[(a)\ 10\ \text{kV m}^{-1}\qquad (b)\ +10\ \text{V}]$

*45–16   Attracted-disc electrometer. These are the dimensions of a particular attracted-disc electrometer: plate area 1.0×10^{-2} m², plate separation 10 mm. When a p.d. V was applied between the plates, each plate attracted the other with an electric force 0.50 mN. Calculate the value of V.

Assume ε_0 $[1.1\ \text{kV}]$

Spherical Equipotential Surfaces

†45–17 Point charges $+0.90$ nC and $+0.10$ nC are placed 100 mm apart. What is the value of the electric potential at the neutral point?

Assume $1/4\pi\varepsilon_0$ $[0.14\ \text{kV}]$

†45–18 Suppose that the electron of a hydrogen atom is 53 pm from the proton. Calculate

(a) the electric field it experiences

(b) the p.d. necessary between a pair of parallel plates 10 mm apart to achieve the same size field.

Assume $1/4\pi\varepsilon_0$, e $[(a)\ 0.51\ \text{TN C}^{-1}\qquad (b)\ 5.1\ \text{GV}]$

45–19 Small spherical charges of $+2.0$, -2.0, $+3.0$ and -6.0 nC are placed in order at the corners of a square of diagonal 0.20 m.

(a) Calculate the electric potential at the centre.

(b) Calculate the size of the electric field at the centre, and mark its direction on a diagram.

Are there points within the square at which

(i) the electric potential is zero

(ii) the electric field is zero?

Assume $1/4\pi\varepsilon_0$ $[(a)\ -0.27\ \text{kV}\qquad (b)\ 3.7\ \text{kV m}^{-1}]$

45–20 Two small spheres of radius 1.0 mm are given charges ± 9.0 nC, and placed *in vacuo* with their centres separated by a distance 100 mm. Assume that the charge distribution on each sphere is unaffected by the presence of the other, and calculate

(a) the potential at the surface of each sphere established by its own charge

(b) the potential at the surface of each sphere established by the other sphere's charge

(c) the potential difference between the spheres.

How much work would be done by an external agent in transferring 1.0×10^6 electrons from the positive sphere to the negative? What becomes of this energy?

Assume $1/4\pi\varepsilon_0$, e $[(c)\ 0.16\ \text{MV};\ 26\ \text{nJ}]$

45–21 Concentric spheres capacitor. A sphere of radius 0.10 m is given a charge $+5.0$ nC. It is then surrounded by but insulated from a concentric sphere of radius 0.20 m which is earthed. Calculate

(a) the induced charge on the outer sphere
(b) the potential of the inner sphere (i) due to its own charge (ii) due to that of the outer sphere
(c) the potential of the outer sphere (i) due to its own charge (ii) due to that of the inner sphere
(d) the p.d. between the spheres.

What electric charge, placed on the inner sphere, would cause the p.d. between the spheres to become 1.0 volt? (*This is numerically equal to the* **capacitance** *of the system.*)

Assume $1/4\pi\varepsilon_0$ [(d) 0.22 kV; $+22$ pC]

45–22 What charge would have to be given to the 0.10 m sphere of the previous question to cause its potential to become 1.0 volt if it were isolated? (*Notice that because your answer is less than* $+22$ *pC the isolated sphere has a smaller capacitance for electric charge than it has when it is surrounded by an earthed sphere.*)

Assume $1/4\pi\varepsilon_0$ [11 pC]

45–23 The isolated sphere of a *Van de Graaff* machine has a radius 90 mm, and is charged to a potential of $+1.0$ kV. An oxygen molecule 0.21 m from the surface of the sphere loses an electron. Calculate

(a) the charge on the sphere,
(b) the field strength at the location of the ionized air molecule,
(c) the resultant repulsive force on the gas ion, and
(d) its consequent acceleration if its mass is 5.3×10^{-26} kg.

Assume $1/4\pi\varepsilon_0$, e

[(a) $+10$ nC   (b) 1.0 kN C^{-1} (c) 0.16 fN   (d) 3.0 Gm s^{-2}]

45–24 When the potential of the sphere of a *Van de Graaff* generator is 1.0 MV, the moving belt is transferring charge to the inside of the sphere at the rate 30 μA. What power is needed to drive the belt against repulsive forces?

[30 W]

45–25 The sphere of a *Van de Graaff* machine is to be charged in air to a potential of 0.20 MV, and it is known that it will discharge by sparking when the electric field in the air exceeds 3.0 MV m^{-1}.

(a) What minimum radius sphere must be used?
(b) What charge will the sphere acquire when its potential reaches the maximum allowed?

Assume $1/4\pi\varepsilon_0$ [(a) 67 mm   (b) 1.5 μC]

Cylindrical Equipotential Surfaces

*45–26 Coaxial cylinders. (a) A long cylindrical conductor of radius 10 mm is given a linear charge density $+0.50$ nC m^{-1}. What would be the p.d. between the conductor and a similar coaxial cylindrical conductor of radius (i) 20 mm (ii) 27.2 mm?

(b) Suppose that the second conductor is earthed.
  (i) What would be the induced linear charge density?
  (ii) Would the conductor of 20 mm radius or 27.2 mm radius have the greater effect in reducing the potential of the original cylinder?
(c) Suppose we choose $V = 0$ when $r = \infty$ m. What would be the (theoretical) potential of the original cylinder
  (i) isolated
  (ii) when surrounded by the larger cylinder?

Assume $1/4\pi\varepsilon_0$, ln 2 = 0.693

[(a) (i) 6.2 V (ii) 9.0 V   (b) (i) -0.50 nC m^{-1}
(c) (i) ∞ V (ii) $+9.0$ V]

*45–27 Dielectric breakdown in an ionization chamber. A cylindrical ionization chamber has a central wire anode of radius 0.20 mm, and a coaxial cylindrical cathode of radius 10 mm. It is filled with a gas whose dieletric strength is 2.2 MV m^{-1}. What is the largest p.d. that should be applied between anode and cathode if the gas is not to suffer electric breakdown before radiation penetrates the mica window?

[1.7 kV]

*45–28 Calculation of acquired k.e. for the magnetron effect. In one method of measuring the specific charge of the electron, electrons are accelerated from a cylindrical cathode of radius a towards a co-axial anode of radius b through a p.d. of V_{ac}. (*They are deflected by a magnetic field in such a way that they just fail to reach the anode.*) In the absence of the magnetic field, calculate the following for a tube in which $a = 50$ μm, $b = 3.0$ mm and $V_{ac} = 20$ V:

(a) the electrons' k.e. as they hit the anode
(b) the fraction of the final k.e. which they acquire in moving the first 0.15 mm of this path.

Assume $1/4\pi\varepsilon_0$, e [(a) 3.2 aJ   (b) 0.34]

Miscellaneous Energy Problems

45–29 Calculation of electrical p.e. (a) A conducting sphere radius a is given a very small charge δQ while it is at potential V and carrying a charge Q. How much work δW is done (i) in terms of V and δQ (ii) in terms of Q, a and δQ?

(b) How much energy is stored by the sphere in acquiring a final charge Q_0? (Answer in terms of a and Q_0.)

(c) How much work must be done to put a charge -3.0 nC onto a conducting sphere of radius 90 mm?

Assume $1/4\pi\varepsilon_0$ [(b) $Q_0^2/8\pi\varepsilon_0 a$ (c) 0.45 μJ]

45–30 Suppose that two α-particles are placed 0.10 nm apart.

(a) One is held, and the other released. What will its k.e. be when they are relatively far apart?
(b) What would their k.e.'s be if both were released?

Assume $1/4\pi\varepsilon_0$, e [(a) 9.2 aJ]

45–31 Suppose that, in *Geiger* and *Marsden*'s experiment to detect the existence of the atomic nucleus, an α-particle of k.e. 1.0 pJ is heading directly for a massive gold nucleus (which has 79 protons). How close does the α-particle approach?

Assume $1/4\pi\varepsilon_0$, *e*   [36 fm]

45–32 (*a*) What is the electric p.e. of an electric dipole consisting of two charges ± 1.1 nC separated by 90 mm?

(*b*) A third charge 4.0 nC is now brought by an external agent to a point 50 mm from either charge. How much work is done by the agent?

(*c*) What is the final energy of the three-charge system? Is it a *bound* system?

Assume $1/4\pi\varepsilon_0$ [(*a*) -0.12 μJ]

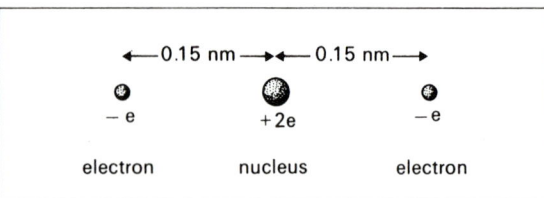

For Qu. **45–33**

45–33 *The electric p.e. of an atom.* Refer to the diagram, which shows a simple model of the instantaneous charge configuration of a helium atom. How much work must be done against electrostatic forces to make the nucleus and electrons infinitely separated?

Assume $1/4\pi\varepsilon_0$, *e*   [5.4 aJ]

*45–34 (*For the following problem the student will find it helpful to consider the gravitational analogy of a satellite in orbit round a very massive body.*) Consider a model of the atom in which an electron orbits a proton in an orbit of radius 53 pm.

(*a*) What is the electron's k.e.?

(*b*) What is the electric p.e. of the electron-proton system?

(*c*) Is the gravitational p.e. worth considering?

(*d*) What is the total energy of the system? (*If you considered an electron at infinity to have zero p.e. – an arbitrary choice – then answer* (*d*) *will be negative. This shows that we are dealing with a* **bound system**.)

(*e*) What is the ionization energy of the hydrogen atom?

Assume e, m_e, $1/4\pi\varepsilon_0$ [(*a*) $+2.2$ aJ   (*b*) -4.4 aJ]

*45–35˙ (*The student will find the previous question helpful.*) Suppose an electron passed close to a proton such that at their least separation distance of 53 pm the electron's k.e. was 4.4 aJ. What were

(*a*) the total energy of the system

(*b*) the binding energy of the system

(*c*) the shape of the electron trajectory?

What would the trajectory have been for (*i*) a slightly greater speed (*ii*) a slightly smaller speed?

46   Capacitors

Questions for Discussion

46–1 What is the radius of a conducting sphere which has a capacitance of 1 F? Name a body of this size.

46–2 Discuss the *physical* reasons for the capacitance *C* of a parallel-plate capacitor being

(*a*) proportional to the plate area *A*, and

(*b*) inversely proportional to the plate separation *d*.

If the plates are given predetermined charges $\pm Q$, does either of these factors influence the electric field between the plates?

46–3 A capacitor has plates of unequal size. If a battery is connected across it, do the plates acquire charges of equal magnitude?

46–4 Two conductors carry like charges of the same magnitude. Can there be a p.d. between them?

46–5 A pear-shaped conductor is charged, and acquires the same potential at all points. A proof plane touching either a flat surface or a curved surface of the pear acquires charges of different size – why?

*46–6 *Fringing effects.* In practice parallel-plate capacitors are non-ideal – that is their electric field lines are *not* equally spaced and parallel. Consider how the charge on the plates is distributed in practice, and use the expressions $E = \sigma/\varepsilon_0$ and $V = Ed$ to decide whether the capacitance of a real capacitor is greater or less than $\varepsilon_0 A/d$.

46–7 Suppose a capacitance C is charged to a potential V at a frequency p, and then discharged by a contact breaker. Show that the system behaves like a resistance $1/pC$. (*The value of C can be measured using the Maxwell bridge, an adaptation of the Wheatstone bridge.*)

46–8 A capacitor is made of two concentric spheres of radii a and b such that $(b - a) \ll a$. By considering this system to approximate to a parallel plate capacitor, find a value for its capacitance. (*Note that because the radial field lines are only approximately parallel, the effective plate area A is such that $4\pi a^2 < A < 4\pi b^2$.*)

46–9 O-M Of what order of magnitude is the capacitance of (*a*) a smoothing capacitor (*b*) a reservoir capacitor (*c*) a wireless tuning capacitor (*d*) the sphere of a laboratory *Van de Graaff* generator?

46–10 O-M Compare the k.e. of rotation of the Earth with the energy stored in its electrostatic field. The value of E near the surface is 0.1 kV m^{-1}.

Quantitative Problems

Calculation of Capacitance

46–11 (*a*) A parallel-plate capacitor whose plates have an area 1.0 m^2 and which are separated by 2.0 mm is connected across the terminals of a 100 V battery. Calculate
  (*i*) the electric field between the plates
  (*ii*) the magnitude of the charge density on the plates
  (*iii*) the charge on either plate
  (*iv*) the capacitance of the system. (*Use $C = Q/V$.*)

  (*b*) A conducting plate 1.0 mm thick is now inserted between the plates, the battery being still connected. Calculate
  (*i*) the electric field inside the conducting plate
  (*ii*) the electric field in the remaining air space
  (*iii*) the induced surface charge densities on the faces of the plate
  (*iv*) the capacitance of the new system.

Assume ε_0
  [(*a*) (*iii*) $\pm 0.44 \text{ μC}$ (*b*) (*ii*) 0.10 MV m^{-1} (*iv*) 8.9 nF]

46–12 *A measurement of* ε_0. Refer to the diagram, in which the circuit shows a method of measuring ε_0. Calculate its value from the following information. The capacitor is a parallel-plate air capacitor of area 0.10 m^2 and plate separation 3.0 mm. (*The field lines can be considered parallel at all points.*) When the switch performs 50 complete vibrations in each second, and the voltmeter reads 0.10 kV, then the sensitive ammeter reads 1.5 μA.

  [9.0 pF m^{-1}]

46–13 The Earth is a conducting sphere of radius 6.4 Mm.
  (*a*) What is its capacitance?
  (*b*) How many electrons must be removed from each m^2 of

surface (*i*) to raise its potential by 1 V (*ii*) to produce a field of size 1 V m^{-1}?

Estimate the number of electrons in a volume of earth of area 1 m^2 and depth 10 mm, and compare this with your answers.
Assume ε_0, e   [(*b*) (*i*) 9 (*ii*) 5.5×10^7]

*46–14 A coaxial cable consists of a pair of conducting cylinders of radii 1.0 and 2.0 mm separated by air. What is the capacitance of each metre length of cable?
Assume $\ln 2 = 0.693$, $1/4\pi\varepsilon_0$ [80 pF m^{-1}]

*46–15 A coaxial cable is made of two conducting cylinders of radii 1.0 mm and 2.72 mm. What length of cable would have a capacitance 1.0 F? (*The enormous answer demonstrates that the farad is a very large unit.*)
Assume $1/4\pi\varepsilon_0$ [18 Gm]

Connecting Capacitors

46–16 *Earthing.* A metal sphere of radius 100 mm is given a charge 6.0 nC, and is then **earthed** (put into electrical contact with the Earth). By considering the Earth to be a conductor of radius 6.4 Mm, and initially uncharged, show that this process effectively discharges the small sphere (i.e. calculate its final charge).

  [94 aC]

46–17 (*a*) Two capacitors of capacitance 2.0 μF and 4.0 μF are each connected in turn to a 250 V d.c. supply. What size charge is produced on the capacitors' plates?

  (*b*) The capacitors are now connected in parallel, such that plates of unlike sign are joined. Suppose that the capacitor plates acquire charges of size Q_2 and Q_4 respectively. Write down (*i*) the value of $(Q_2 + Q_4)$ (*ii*) the condition that the p.d. across each capacitor is now the same.

  (*c*) Calculate the values of Q_2, Q_4 and the resultant p.d.
  [(*c*) 0.17 mC, 0.33 mC, 83 V]

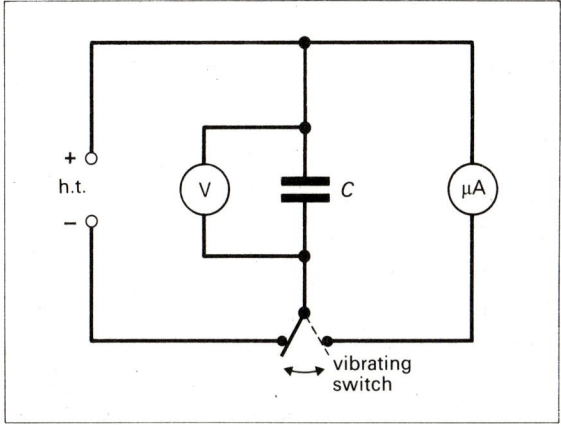

For Qu. **46–12**

46–18 Two capacitors of capacitance 4.0 μF and 6.0 μF are each connected to batteries of e.m.f. 100 V and 200 V respectively, and they each become fully charged. They are then connected in parallel, with connections being made between plates of the same sign.

(a) What, after connection, is the size of the separated positive and negative charges?
(b) What is the final p.d. across the capacitors' plates?

(*Try to solve* (b) *from first principles if you can, rather than using* $C = C_1 + C_2$. *Think in terms of the proportion of the charge taken by the respective capacitors.*)

[(a) ± 1.6 mC   (b) 0.16 kV]

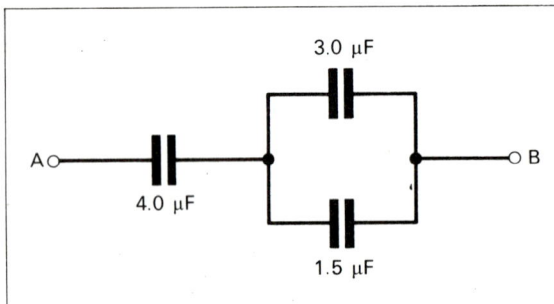

For Qu. **46–19**

46–19 Refer to the diagram.
(a) What is the equivalent capacitance of the system as it stands?
(b) It is found that as V_{AB} is increased steadily, the 4.0 μF capacitor suffers electrical breakdown (its dielectric material becomes conducting). What would be the resulting equivalent capacitance after breakdown?
(c) Suppose that the 3.0 μF (and *not* the 4.0 μF) capacitor is the first to break down. What then would be the equivalent capacitance?

[(a) 2.1 μF]

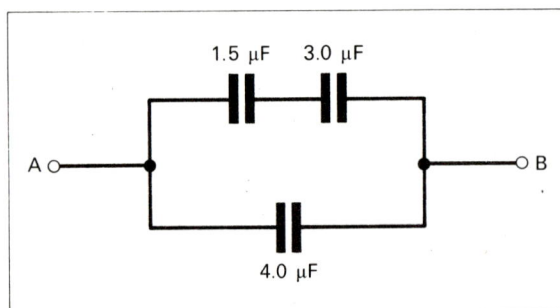

For Qu. **46–20**

46–20 Refer to the diagram, and answer the questions asked in Qu. **46–19**.

Energy in Capacitors

†**46–21** Suppose 1000 4.0 μF capacitors were connected in parallel, and then charged to a p.d. of 25 kV. For how long would the stored energy keep a 100 W lamp burning at normal brightness?

[12 ks]

46–22 A battery of e.m.f. 100 V is used to charge fully a 10 μF capacitor. Calculate

(a) the energy stored in the capacitor's electric field
(b) the energy converted by the seat of e.m.f.
(c) the change of stored energy (a) – (b).

Account for (c). (*This is always equal to* − (a) − *this is not a special case. Refer to Qu.* **5–29** *for a similar situation in which the energy dissipated is half that converted.*)

[(a) 50 mJ   (b) 0.10 J]

46–23 A 2.0 μF capacitor is charged to 0.10 kV, and a 1.0 μF capacitor is charged to 0.20 kV.

(a) What energy is stored in the electric field of each? The two capacitors are then connected (in parallel) with like charges together. Calculate
(b) the equivalent capacitance of the system
(c) the size of the charges that the capacitor keeps separate
(d) the new p.d. between the capacitor's plates
(e) the energy now stored in the system's electric field.

Account for the disparity between (e) and the total for (a).

[(a) 10 and 20 mJ   (e) 27 mJ]

46–24 Drops of impure water of radius 1.0 mm fall from a metal tube maintained at a potential of 50 V, and they fill a spherical conducting vessel of negligible thickness and radius 10 mm. Calculate

(a) the charge on each drop
(b) the total charge on the vessel when full
(c) the electrostatic potential energy carried by each drop
(d) the electrostatic p.e. of the full vessel.

Where does most of the energy in (d) come from? (*Why is the answer to* (d) *not* 1000 × *that of* (c)?)

Assume ε_0 [(a) 5.6 pC   (b) 5.6 nC   (c) 0.14 nJ   (d) 14 μJ]

46–25 *Energy density.* A parallel-plate capacitor consists of a pair of very large parallel plates separated by 1.0 mm. When the capacitor is charged the surface charge density near the middle of the plates becomes 8.85 nC m^{-2}. Calculate

(a) the electric field between the plates,
(b) the p.d. between the plates,
(c) the capacitance of 1.0 m² of plate area near the centre of the system,
(d) the energy stored in the electric field between this area of plates, and
(e) the **energy density** of the electric field (the energy stored in unit volume).

(*The energy density can be calculated from* $\frac{1}{2}\varepsilon_0 E^2$. *Does your answer agree?*)

Assume ε_0 [(e) 4.4 μJ m^{-3}]

*46–26 *Calculation of the mutual attraction of a pair of charged plates.* A pair of parallel plates, each of area 1.0 m², are initially separated by 4.0 mm, and then the plates are given charges of ± 2.0 nC respectively. Calculate

(a) the capacitance of the system,
(b) the sizes of the separated charges, and hence
(c) the energy stored in the electric field between the plates.

The plates now have their separation increased to 8.0 mm by

an external agent who does work against the mutual attraction of the unlike charges.

(d) By what factor is the p.d. between the plates changed?
(e) Use your answer to (c) to evaluate the new electric potential energy.
(f) How much work was done by the external agent?
(g) Use the fact that the agent applies a force of constant magnitude to find its value.

Assume ε_0 [(c) 0.90 nJ   (e) 1.8 nJ   (g) 0.23 μN]

47   Dielectric Materials

Questions for Discussion

47–1   Explain similarities and differences between explanations for these two phenomena.

(a) A permanent magnet can attract *any* (non-magnetized) sample of ferromagnetic material.
(b) A charged cellulose acetate rod can attract small pieces of *uncharged* insulator (such as human hair).

47–2   Two capacitors have the same plate areas, and the same capacitances. One is filled with air ($\varepsilon_r = 1.0$) and the other with oil ($\varepsilon_r = 2.0$). They are connected in turn to the same battery. Find the ratio of the energy densities in the electric fields between their plates.

47–3   *Capacitor with fixed charge.* A capacitor is given a fixed charge Q, and then the air between its plates is replaced by a dielectric oil for which $\varepsilon_r = 2$. By what factors do the following change?

(a) The electric field strength.
(b) The p.d. between the plates.
(c) The capacitance of the system.
(d) The energy stored in the electric field.

If the energy stored by the system decreases, what can you say about the work done on the system by the external agent that introduced the oil? Was the oil, while filling the space, attracted into it, or repelled from it?

47–4   *Capacitor with constant p.d.* A capacitor is left permanently connected to a battery of fixed e.m.f., and then the air between its plates is replaced by a dielectric oil for which $\varepsilon_r = 2$. By what factors do the following change?

(a) The p.d. between the plates.
(b) The electric field strength.
(c) The surface charge density.
(d) The charge on either plate.
(e) The capacitance of the system.
(f) The energy stored in the electric field.

Was the oil, while filling the space, attracted or repelled by the capacitor plates? Did the external agent that introduced the oil do positive or negative work? How do you reconcile your answers to (f) and this last question?

47–5   Can induced polarization charges contribute to an electric current?

*47–6   Polarization of molecules by distortion of electron clouds shows a resonance effect which accounts for optical dispersion. The bond stretching or bending mechanism shows resonance which accounts for infra-red dispersion. Estimate the frequencies of the electric field for both types of dispersion.

*47–7   Gases may be polarized by an electric field in three ways: (i) *distortion* of electron clouds in all molecules, (ii) *stretching* or *bending* of bonds in polar molecules, and (iii) *orientation* of polar molecules.

(a) Discuss the relative importance of each effect.
(b) What effect would temperature changes have on the relative permittivities of polar and non-polar molecules?
(c) Suggest what effect the frequency of the electric field would have on each type of polarization.
(d) How would the polarization behaviour of liquid and solid molecules be likely to differ from that of gas molecules?

Quantitative Problems

47–8   A capacitor is made by putting 4.0 mm of a plastic of relative permittivity 4.0 between a pair of metal plates of area 1.0 m^2 and separated by 5.0 mm. The capacitor is charged by connecting a 2.0 kV h.t. supply across the plates. Calculate

(a) the ratio E_d (dielectric): E_a (air)
(b) the ratio $V_d : V_a$
(c) the field in the air
(d) the surface charge density on the metal plates
(e) the induced surface charge density on the dielectric faces
(f) the charge on the metal plates
(g) the capacitance of the system, using $C = Q/V$.

Assume ε_0

[(c) 1.0 MV m^{-1} (d) 8.9 μC m^{-2} (e) 6.6 μC m^{-2} (g) 4.4 nF]

47–9   A capacitor consists of a pair of parallel plates of area 0.20 m^2 separated by 1.0 mm. Half this area is filled with paraffined paper 1.0 mm thick ($\varepsilon_r = 2.0$), and the other half with air ($\varepsilon_r = 1.0$). Calculate its equivalent capacitance. (*Hint: to an approximation the arrangement is equivalent to two simple capacitors connected in parallel.*)

Assume ε_0   [2.7 nF]

47–10   A capacitor consists of a pair of parallel plates of area 0.10 m^2 separated by 2.0 mm, of which 1.0 mm is air ($\varepsilon_r = 1.0$) and 1.0 mm is polystyrene ($\varepsilon_r = 2.5$). Adopt the following procedure to find its capacitance. Consider a charge ± 0.885 pC (for convenience) to be placed on the metal plates, and by finding the electric field strengths in the air and polystyrene respectively, calculate the p.d. between the plates. Then use the defining equation $C = Q/V$.

Assume ε_0   [0.63 nF]

47–11   In one design of capacitor thin sheets of metal of area 80 mm × 80 mm sandwich between them a piece of paper whose thickness is 40 μm. The relative permittivity of the paper is 4.0, and its dielectric strength is 20 MV m^{-1}. Calculate

(a) its capacitance
(b) the maximum p.d. which should be applied to it.

Assume ε_0   [(a) 5.7 nF   (b) 0.80 kV]

47–12   An insulator, which has $\varepsilon_r = 6.0$ and a dielectric strength $E_{max} = 4.0$ MV m^{-1}, is used as the dielectric material in a parallel-plate capacitor of capacitance 5.0 nF. If the capacitor is to be able to withstand a p.d. of 3.0 kV, what is the minimum area of the plates?

Assume ε_0   [7.1 × 10^{-2} m^2]

X Current Electricity

48 Current and Resistance

Questions for Discussion

48–1 What experimental evidence is there for believing that the phenomena of electrostatics and current electricity are both caused by identical electric charges?

48–2 The conventional direction of an electric current is opposite to the drift velocity of electrons, and yet most electric currents consist of drifting electrons. Is there any advantage in changing the convention?

48–3 An electric current is carried through a metal by the movement of free electrons. When they 'collide' with lattice ions they give up momentum. Is there a net force on the metallic lattice while the current passes?

48–4 A metallic resistor increases its internal energy by taking k.e. from the electrons moving through the lattice. Discuss the nature of the collision between the electron and the lattice ion.

48–5 *Johnson noise,* which is the electrical counterpart of *Brownian* motion in fluids, is a small randomly fluctuating alternating voltage which exists across the terminals of any resistor. What causes it? How can the effect be reduced?

48–6 Express the following units in terms of the units of the primary quantities, m, kg, s and A:
(*a*) coulomb (*b*) volt (*c*) ohm.

48–7 Derive an equation which shows how the flow of charge caused by a potential difference is analogous to the heat flow equation

$$\frac{Q}{t} = \lambda A \left(\frac{\theta_2 - \theta_1}{x} \right)$$

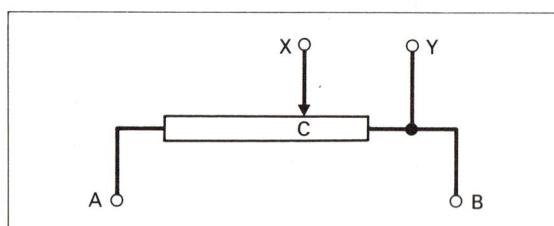

For Qu. **48–8**

48–8 Refer to the diagram. It is often said that
$$V_{XY}/V_{AB} = R_{BC}/R_{AB}$$
Under what conditions is this true?

48–9 How would you join a number of ohmic resistors to give (*a*) maximum resistance, and (*b*) minimum resistance?

How would you alter the p.d. across your combination in order to double the total current flowing through it? What difference would this make to the power dissipated?

48–10 *Variable resistors.* Describe briefly some possible designs of variable resistors. Explain how one of your designs can be used as (*a*) a rheostat, and (*b*) a potential divider.

48–11 Describe how you would test whether the filament of a light bulb obeys *Ohm's Law.* Would the temperature of the filament have any effect on your results?

48–12 When a metal rod is heated, its resistance, length and cross-sectional area all change. If you were investigating the variation in resistivity ρ with temperature, how would you deal with all three factors?

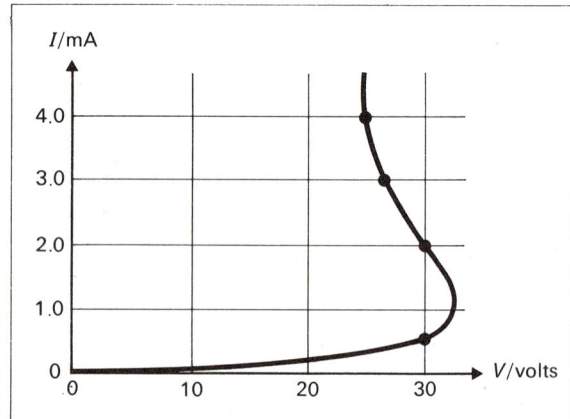

For Qu. **48–13**

48–13 *Thermistor.* The diagram shows the characteristic curve for a typical thermistor.
(*a*) Explain the shape of the curve.
(*b*) Sketch a graph to show variation of resistance of the thermistor with applied potential difference.
(*c*) Discuss some uses of a thermistor.

48–14 What is a *superconductor*? What form does the energy take when an electric current flows through a superconductor? What happens if the temperature increases slightly, and just causes the phenomenon to disappear?

48–15 O-M Estimate the average distance travelled between collisions for free electrons in copper. How does this compare with a typical atomic diameter?

Quantitative Problems

Current and Conductors

†48–16 Van de Graaff generator. The belt of a generator is 0.80 m wide and travels with a speed of 28 m s^{-1}. Charge is carried into the sphere at a rate corresponding to 0.20 mA. Calculate the average surface density of charge on the belt.

[8.9 μC m^{-2}]

†48–17 Derivation of $I = nAev_d$. In a particular wire electrons move with an average drift speed 0.50 mm s^{-1} through a cross-section of area 4.0 mm^2. There are 1.0×10^{29} electrons in each m^3 of wire. Consider a cylinder of wire of length 12 mm.

(a) How long will it take all the electrons within that cylinder to drift through one end face? (Ignore random motion.)
(b) How many electrons are there within the cylinder?
(c) What total charge is carried by the electrons?
(d) By considering the rate at which charge leaves the cylinder, calculate the size of the current.

Assume e   [(d) 32 A]

48–18 Electron drift speed: copper. Copper has a density 8.9×10^3 kg m^{-3}, and a relative atomic mass 64. A particular copper wire carrying a current 2.0 A has a cross-sectional area 0.10 mm^2. Calculate

(a) the number density of copper atoms
(b) the number density of conduction electrons (assuming each atom contributes one free electron)
(c) the electron drift speed for this current.

The thermal speed of the electrons is ~ 1 Mm s^{-1}. How will the collision frequency between the electrons and the crystal lattice change when the current is switched off?

Assume N_A, e   [(b) 8.4×10^{28} m^{-3} (c) 1.5 mm s^{-1}]

48–19 Electron drift speed: germanium. A sample wafer of n-type germanium, of thickness 0.50 mm and width 4.0 mm, has a number density of electrons 6.0×10^{20} m^{-3}. Suppose, for simplicity, that any current through it is carried entirely by electrons. What would be their average drift speed when the current is 4.8 mA? (Compare your answer with that for copper: the different orders of magnitude are relevant in a discussion of the Hall Effect (p. 182).)

Assume e   [25 m s^{-1}]

48–20 Electron beam. The electron beam in a cathode ray tube travels 0.30 m and the beam current is 4.0 mA. If the electron speed is 30 Mm s^{-1}, calculate the number of electrons in the beam at any given time. What is the power dissipated at the screen?

Assume e, m_e [2.5×10^8, 10 W]

48–21 Electrolysis ammeter. A current can be recorded by this instrument by measuring the mass m deposited on an electrode in an electrolyte when a current I flows for a time t. The value of m is given by M_mIt/evN_A, where M_m is the molar mass of the deposited material and v is the number of elementary charges per ion.

(a) Calculate the current if 1.2 g of copper ($M_m = 64 \times 10^{-3}$ kg mol^{-1}) are deposited in 1.0 ks.
(b) What mass of silver ($M_m = 108 \times 10^{-3}$ kg mol^{-1}) would be deposited by the same charge?

Assume e, N_A [(a) 3.6 A   (b) 4.0 g]

Resistance and Resistivity

†48–22 Resistivity and resistance. The resistivities in this problem are all measured at 293 K. Calculate the resistance between

(a) the ends of a copper block of cross-sectional area 12×10^{-4} m^2, length 0.60 m and resistivity 17 nΩ m
(b) the ends of a similar carbon block of resistivity 35 μΩ m
(c) the faces of a wafer of pure silicon of thickness 5.0 mm and area 8.0×10^{-4} m^2 if its resistivity is 60 Ω m.

[(a) 8.5 μΩ   (b) 18 mΩ   (c) 0.38 kΩ]

†48–23 Potential divider. A variable resistor used to control the p.d. across a device has maximum resistance 15 Ω. What range of p.d.'s is available when a constant current of 50 mA is made to flow through it? What proportion of the maximum resistance would be used for a p.d. of 25 mV?

[0 to 0.75 V, 1/30]

48–24 Stranded cables. A copper cable consists of seven strands each of which have area 5.0 mm^2 and resistivity 17 nΩ m. Calculate

(a) the resistance of 1.0 km of cable,
(b) the length of single strand cable having the same resistance as in (a), and
(c) the number of strands so that the cable has a resistance less than 0.30 Ω.

[(a) 0.49 Ω   (b) 0.14 km   (c) 12]

48–25 Current density. The electric current density J in a conductor is given by $J = I/A$, where I is the uniform current through a cross-sectional area A.

(a) Calculate the value of J in a wire of cross-sectional area 5.0×10^{-7} m^2 when a current of 20 mA is flowing.
(b) Show that the applied electric field within a cylindrical conductor is given by $E = \rho J$ where ρ is the resistivity of the material of the conductor.
(c) Show also that p, the power dissipated per unit volume is given by $p = J^2\rho$.

*48–26 Microscopic explanation of Ohm's Law. By considering simple classical ideas, it can be shown why metals obey Ohm's Law.

(a) Show how the drift speed v_d of electrons is related to the current density J.
(b) Find an expression for J in terms of the acceleration of

an electron starting from rest and τ the time between collisions.

(c) Use your two equations to show that the electrical conductivity of the metal is given by $\sigma = \frac{1}{2}(e^2 n/m_e)\tau$.

(d) Explain to what extent this shows that metals obey *Ohm's Law*.

(e) Calculate τ for copper using $\sigma = 6.0 \times 10^7$ S m^{-1} and $n = 8.4 \times 10^{28}$ m^{-3}.

Assume m_e, e   [(e) 51 fs]

*48–27 Resistance by integration. A thick piece of wire of circular cross-section tapers uniformly from a radius r_1 to a radius r_2 in a length l. If the resistivity of the material of the wire is ρ, then what is the wire's resistance? (*Hint: apply* $\delta R = \rho \delta l / A$ *to an element of wire.*)

[$l\rho/\pi r_1 r_2$]

Temperature Variation of Resistance

48–28 The field windings of an electric motor are made of copper of temperature coefficient of resistance 3.90×10^{-3} K^{-1}. If the resistance of the windings is 50.0 Ω at 300 K, calculate

(a) the resistance of the windings at 273 K, and

(b) the temperature of the windings when their resistance has risen to 60.0 Ω.

[(a) 45.2 Ω (b) 357 K]

48–29 Working temperature of a lamp. A tungsten filament rated at 250 W, 230 V has a resistance of 20 Ω at 293 K. Its mean temperature coefficient of resistance when it heats up is 5.0×10^{-3} K^{-1}. What is its working temperature?

[2.4×10^3 K]

49   Energy Transfer in the Circuit

Questions for Discussion

49–1 Two students are asked whether one should use a high or low resistance to obtain a large heating effect. One uses $P = V^2/R$, and says *low*; the other uses $P = I^2 R$, and says *high*. Comment.

49–2 Discuss the energy changes which take place in the following electrical devices when an electric current flows through them: (a) a battery (both charging and discharging) (b) a light bulb (c) an electric motor (d) a non-inductively wound resistance coil.

49–3 Several wires of different resistivities but identical dimensions are connected in turn between two points across which is maintained a fixed potential difference. In which wire will the rate of *Joule* heating be greatest?

49–4 Why are a.c. systems of electric power transmission often preferable to d.c. systems? Which industrial and domestic applications require *either* a.c. *or* d.c.?

49–5 How *efficient* is an arrangement in which a source of e.m.f. supplies maximum power to an external circuit?

49–6 What properties would you consider in selecting a wire for (a) heating, and (b) a fuse?

49–7 O-M What time interval elapses before an overloaded fuse melts?

Quantitative Problems

Energy Conversion

†49–8 Life of a battery. The charge that passes through a particular 12 V battery during charging is 0.50 MC. Assuming that the p.d. across the terminals remains constant while the battery discharges, calculate the time for which it can supply 0.45 kW.

[13 ks]

†49–9 Immersion heater. A 5.0 kW immersion heater takes an hour to raise the temperature of 45 kg of water from 293 K to 373 K. Calculate and account for the average power loss.

Assume c_{H_2O}   [0.81 kW]

†49–10 Paying for electrical energy. A mains p.d. of 240 V is connected across a 30 Ω resistor for 5.0 hours. How much would an Electricity Board charge for this at the rate of $\frac{1}{2}$ p for 1.0 MJ?

[$17\frac{1}{2}$ p]

†49–11 Fuse. Calculate how many 1.0 kW electric fires could be safely connected to a 0.24 kV mains circuit fitted with a 13 A fuse.

[3]

†49–12 A beam of protons, carrying a current of 20 μA, strikes a copper block. If the energy of each proton is 1.8 pJ, calculate the power dissipated in the copper block.

Assume e   [0.22 kW]

†49–13 A bulb marked 12 V 36 W carried a direct current for 0.30 ks.

(a) What was the *probable* current?
(b) What charge passed a given cross-section of the filament?
(c) How many electrons carried the charge?
(d) How does this charge compare with that on each plate of a 1.0 μF capacitor charged to 0.40 kV? Comment.

Why *probable* current?

Assume e   [(d) 2.2 × 10⁶ times]

49–14 A train of mass 1.0 × 10⁵ kg operating from a 25 kV supply reaches a speed of 20 m s⁻¹ on the level in 50 s. Ignore energy losses, and calculate the average effective current that it uses.

[16 A]

49–15 The power of a heater is 0.80 kW when the applied p.d. is 0.24 kV and the temperature of the wire is 1.2 × 10³ K. If the p.d. is kept constant and a cooling liquid maintains the wire temperature at 5.6 × 10² K, what would be the new value for the power dissipated by the heater? The mean temperature coefficient of resistance for the wire material is 5.0 × 10⁻⁴ K⁻¹.

[1.0 kW]

49–16 *Maximum power.* The lighting circuit of a car has a total resistance of 8.0 Ω and is connected permanently across a 12 V battery which should be taken as having an internal resistance of 1.0 Ω. A further resistor of resistance *R* is connected across the terminals of the battery so that maximum power will be dissipated in the external circuit. Calculate

(a) the value of *R*,
(b) the power dissipated in *R*, and
(c) the percentage loss of power in the lighting circuit caused by this connection.

[(a) 1.1 Ω   (b) 32 W   (c) 68 %]

Transmission of Electrical Power

49–17 A consumer who requires to use 5.0 kW at 0.24 kV connects his electrical device to the local power station by mains leads which have resistance 2.0 Ω. Calculate

(a) the current flowing in the mains leads,
(b) the p.d. between the leads at the power station, and
(c) the percentage loss of power in the leads.

[(a) 21 A   (b) 0.28 kV   (c) 15 %]

49–18 The power output at a power station is 500 MW at a p.d. of 132 kV. If the resistance of the complete power line is 4.0 Ω, calculate

(a) the current flowing in the line,
(b) the drop in potential down the line, and
(c) the percentage loss of power in transmission.

[(a) 3.79 kA   (b) 15.2 kV   (c) 11.(5)%]

49–19 The transmission cable from a power station to the transformer of a small town has resistance 4.0 Ω. The transformer is 90 % efficient and steps down the p.d. to 240 V. During maximum demand the effective resistance of the town is 60 mΩ. Calculate

(a) the current used by the town,
(b) the power supplied to the transformer,
(c) the minimum grid p.d. at the town end if the power loss in the cable must not exceed 2.0 kW, and
(d) the potential drop across the transmission cable.

[(a) 4.0 kA   (b) 1.1 MW   (c) 48 kV   (d) 89 V]

50   E.M.F. and Circuits

Questions for Discussion

50–1 What is the difference between an e.m.f. and a potential difference? Under what conditions is the terminal p.d. of a battery greater than its e.m.f.?

50–2 Does the direction of a current flowing through a battery enable us to predict the direction of the e.m.f. of that battery? What determines the magnitude of the e.m.f.?

50–3 What is the cause of the *internal resistance* of a cell? How does it depend upon the size of the cell? What changes occur to the internal resistance and the e.m.f. of a cell as it gets older?

50–4 Discuss carefully the energy changes within a battery, and the directions of the forces acting on the charge-carriers, when the battery is (a) discharging, and (b) being charged.

50–5 What difference does the solder joining the two metals which form a thermocouple make to the net e.m.f.?

50–6 *Thermoelectric thermometers.* If one junction is kept at a fixed temperature, the e.m.f. $\mathscr{E}$ when the temperature of the other junction is greater by $\Delta\theta$ is given by $\mathscr{E} = a\Delta\theta + b\Delta\theta^2$, where a and b are constants. Sketch a graph to indicate how $\mathscr{E}$ varies with $\Delta\theta$. Mark on the graph

(a) the *neutral* temperature,
(b) the range suitable for temperature measurement, and
(c) typical scales.

What factors would decide which materials were most suitable for a thermoelectric thermometer?

50–7 Express the units of R and C in terms of the primary units kg, m, s and A, and hence find the unit of RC.

50–8 Is the time constant of an RC circuit changed when one uses a battery of large e.m.f.?

50–9   O-M How much energy is stored in a new 1.5 V U2 cell?

Quantitative Problems

The Circuit Equation

†**50–10** A cell is connected in series with an 8.0 Ω resistor and a switch. A high resistance voltmeter is connected across the cell and reads 3.6 V when the switch is open and 3.2 V when the switch is closed. Calculate (a) the e.m.f. of the cell, and (b) its internal resistance.

†**50–11** A generator of e.m.f. 20 V and internal resistance 0.50 Ω is used to charge a car battery of e.m.f. 12 V and internal resistance 0.10 Ω. They are connected in series together with a resistance R whose value is adjusted to give a charging current of 2.0 A. Calculate

(a) the value of R
(b) the power of the generator
(c) the total rate of dissipation of electrical energy
(d) the rate at which the battery stores chemical energy
(e) the efficiency of the whole operation.

[(e) 60 %]

†**50–12** A battery drives a current of 3.0 A round a circuit consisting of two 2.0 Ω resistors connected in parallel. When these resistors are connected in series the current is 1.2 A. Calculate

(a) the e.m.f. of the battery,
(b) the internal resistance of the battery, and
(c) the power dissipated in a resistor in each case.

[(a) 6.0 V   (b) 1.0 Ω   (c) 4.5 W, 2.9 W]

50–13 A current of 4.0 A is driven through a resistor by two cells of different e.m.f.'s and internal resistances which are connected in series. When the terminals of one cell are reversed the current falls to 0.80 A. Calculate the ratio of the e.m.f.'s of the two cells.

[1.5 : 1]

50–14 A tetrahedron is formed from six straight wires of equal length and each of resistance 5.0 Ω. A cell of e.m.f. 4.0 V and internal resistance 2.0 Ω is connected across two of the corners. Calculate the current flowing through the cell.

[0.89 A]

Kirchhoff's Rules

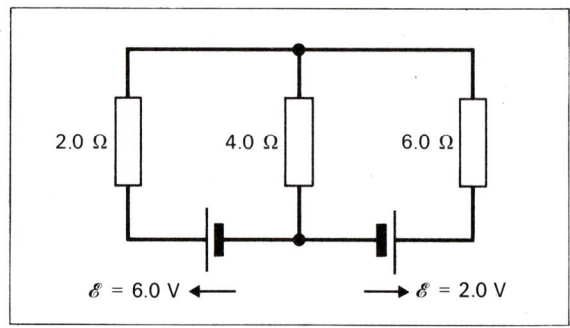

For Qu. **50–15**

50–15 Refer to the diagram. Two cells of negligible internal resistance are connected to three resistors as shown. Calculate

(a) the currents through the three resistors,
(b) the total power dissipated in the resistors, and
(c) the rates of energy conversion in each cell.

Discuss your answers to (b) and (c).

[(a) 1.2 A, 0.91 A, 0.27 A   (b) 6.5 W   (c) 7.1 W, 0.55 W]

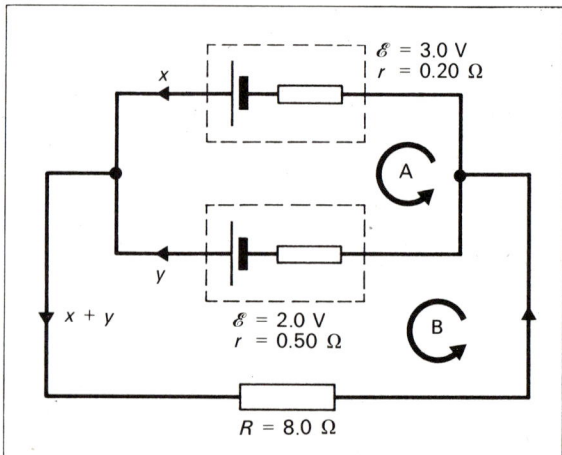

For Qu. **50–16**

50–16 Refer to the circuit diagram.

(*a*) Apply *Kirchhoff's* second rule to the two loops marked A and B, and from the two equations calculate the values of *x* and *y*.

(*b*) Calculate the rate of conversion of electrical energy into internal energy in the 8.0 Ω resistor.

(*c*) Describe what is taking place in the 2.0 V cell.

[(*a*) $x = 1.7$ A, $y = -1.3$ A   (*b*) 0.9 W]

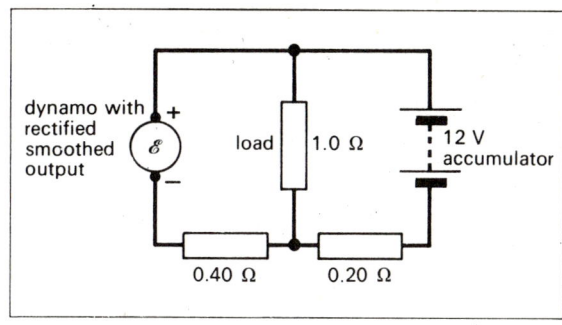

For Qu. **50–17**

50–17 Refer to the diagram opposite.

(*a*) At what value of $\mathscr{E}$ would the accumulator first begin to charge?

(*b*) What is the efficiency of the system when the charging current is zero? (*Hint: the purpose of the system is to energize the load.*)

(*c*) What is the charging current when $\mathscr{E} = 20$ V?

[(*a*) 17 V   (*b*) 71%   (*c*) 4.7 A]

The RC Circuit

*50–18 *Capacitative time constant.* Calculate the number of time constants that it would take for the following to reach 80% of their equilibrium values:

(*a*) the charges on the plates of a capacitor being charged, and

(*b*) the energy stored by the capacitor.

[(*a*) 1.6   (*b*) 2.2]

*50–19 A simple series circuit consists of a 2.0 MΩ resistor and a 1.0 μF capacitor with a 5.0 V source of e.m.f. Calculate the following after the circuit has been connected for 1.0 s:

(*a*) the capacitative time constant,

(*b*) the rate of charging the capacitor,

(*c*) the rate of storing energy in the capacitor,

(*d*) the power dissipated in the resistor, and

(*e*) the power of the source of e.m.f.

[(*a*) 2.0 s   (*b*) 1.5 μA   (*c*) 3.0 μW   (*d*) 4.6 μW   (*e*) 7.6 μW]

*50–20 Sketch graphs showing the variation of *Q* and *I* with time for a capacitor being (*a*) charged, and (*b*) discharged. Mark in the scale for a 2.0 μF capacitor in a circuit of resistance 3.0 kΩ if the e.m.f. used is 20 V, showing the maximum values in each case.

51 Principles of Electrical Measurements

Questions for Discussion

51–1 Give some examples of experiments you have done which involve a *null* adjustment. What is the advantage of such methods?

51–2 What factors contribute to the *accuracy* of a potentiometer? How may one control the *sensitivity* of a wire?

51–3 What advantage is gained by arranging for the balance point to be near

(*a*) the end of a *potentiometer* wire, but

(*b*) the centre of a metre *bridge* wire?

51–4 *Potentiometer faults.* What action would you take if you observed the following? The deflection of the galavanometer

(*a*) is permanently zero

(*b*) has the same size for all jockey positions

(*c*) flickers randomly

(*d*) varies steadily as the jockey is moved, but is always in the same direction (two possibilities).

51–5 In a particular potentiometer measurement there was a substantial contact resistance between the jockey and the slide-wire. What effect did it have on the balance point?

51-6 Devise a means of measuring a p.d. of 20 V using a potentiometer across whose resistance there is a maximum p.d. of 2.0 V.

51-7 Does it affect the balance point of a potentiometer experiment if the slide-wire heats up?

51-8 Describe qualitatively the effect of a potentiometer wire whose diameter tapers steadily over its whole length.

51-9 Why is a *Wheatstone* bridge unsuitable for comparing very low resistances?

51-10 The four resistances available for a *Wheatstone* bridge circuit are about 100 Ω, 100 Ω, 4 Ω and an unknown, thought to be close to 4 Ω. How would you arrange them, and where would you connect the galvanometer?

51-11 What are the advantages and disadvantages of a thermistor compared with a metal resistance thermometer?

Quantitative Problems

Ammeters and Voltmeters

51-12 *Galvanometer adaptation.* The maximum current that can be passed through the coil of a measuring instrument is 2.0 mA. If it has resistance of 5.0 Ω, how can it be adapted to measure

(a) the current in an electric cable carrying up to 18 A, and
(b) the p.d. between two points across which there may be up to 100 V?

[(a) shunt 0.56 mΩ (b) bobbin 50 kΩ]

51-13 *Ammeter-voltmeter measurement of resistance.* (a) A resistance R of *true* value 30 Ω was determined by putting a high resistance ammeter in series with it, and using a voltmeter of (low) resistance 120 Ω to measure the p.d. across R *alone*. What value for R was calculated from the relation

$R = $ (voltmeter reading)/(ammeter reading)?

(b) The method was now modified so that the voltmeter was connected across both R and the ammeter *together*, and the size of the error of the value calculated for R was the same. Find the resistance of the ammeter.

(*In practice neither method is accurate, but the size of the meter resistance will determine which is preferable.*)

[(a) 24 Ω (b) 6.0 Ω]

51-14 *A multirange ammeter.* Refer to the diagram, in which the moving-coil galvanometer has an f.s.d. of 10.0 mA. It has been converted to a dual-range ammeter by the shunt resistors shown. Calculate their resistances.

[$R_1 = 3.3$ mΩ, $R_2 = 30$ mΩ]

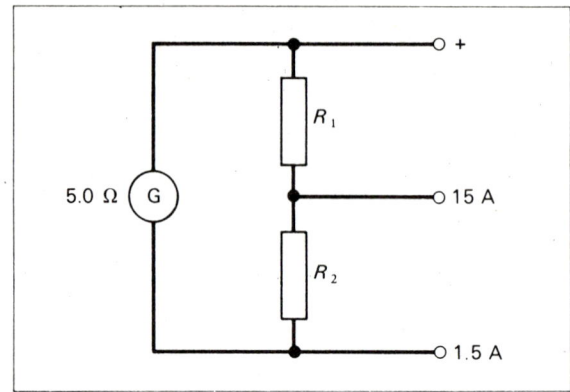

For Qu. **51-14**

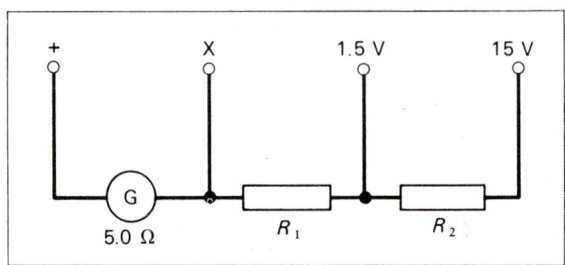

For Qu. **51-15**

51-15 *A multirange voltmeter.* Refer to the diagram, in which the moving-coil galvanometer has an f.s.d. of 10.0 mA. It has been converted to a triple-range voltmeter by the series resistors shown.

(a) What should be marked on terminal X?
(b) Calculate R_1 and R_2.

[(b) $R_1 = 145$ Ω, $R_2 = 1.35$ kΩ]

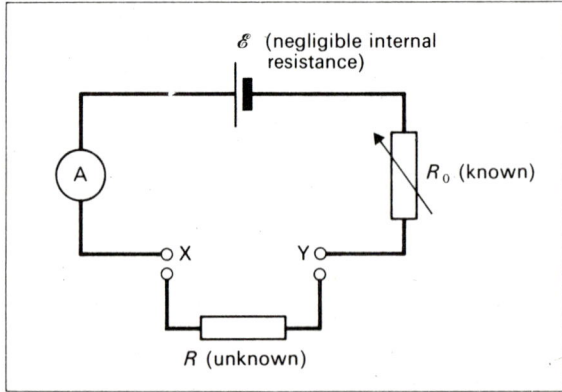

For Qu. **51-16**

51–16 *The ohmmeter.* Refer to the circuit diagram. When X and Y are shorted, the current in the circuit is $\mathscr{E}/R_0 = I_0$. When R is inserted the current becomes αI_0.

(a) Express the unknown R in terms of R_0 and α.
(b) For what values of α will the instrument give reliable values for R?
(c) How may its useful range be extended?

[(a) $R = R_0(1 - \alpha)/\alpha$]

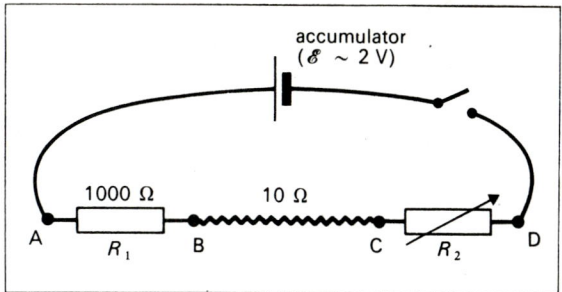

For Qu. **51–22**

The Potentiometer

†51–17 *The potentiometer principle.* An accumulator drives a steady unknown current through a straight uniform resistance wire 1.0 m long. The e.m.f. 1.018 V of a standard cell is balanced by connecting it across the points marked as 200 mm and 709 mm on a scale laid beside the wire. Calculate

(a) the potential drop per mm of wire
(b) the maximum p.d. which this instrument could measure.

†51–18 *Measurement of p.d.* The terminals of a torch bulb carrying a current are found to give a balance point when connected across a length 654 mm of the wire of Qu. **51–17**. What is the p.d. across its terminals?

[1.31 V]

†51–19 *Comparison of e.m.f.'s.* Using the potentiometer of Qu. **51–17**, a dry cell gave a balance length of 750 mm, but no such balance could be found for a battery consisting of two *Nife* cells in series.

(a) Comment on the e.m.f.'s of the dry cell and the battery.
(b) Would it be possible to modify the circuit to make either give a balance length close to 0.95 m?

†51–20 *Measurement of current.* The potentiometer of Qu. **51–17** gave a balance length of 867 mm when connected across a standard resistance of 1.50 Ω.

(a) What current flowed through the resistance?
(b) If the resistance had been measured by an absolute method (such as the *Belham* a.c. method – Qu. **56–18**), would this constitute an *absolute* measurement of current?

[(a) 1.16 A]

51–21 *Measurement of internal resistance.* A cell on open circuit is balanced by 600 mm of a potentiometer wire. When the same cell is passing a current through a 5.0 Ω resistor, the balance length for the cell terminals is reduced to 500 mm. Calculate the internal resistance of the cell.

In what way is a potentiometer superior to a voltmeter for this measurement?

[1.0 Ω]

51–22 *Measurement of a very small e.m.f.* Refer to the diagram. R_2 is adjusted until a 1.01 V standard cell just gives a balance point at C (BC = 1.00 m).

(a) What is the p.d. across 1.0 mm of slide wire?
(b) What approximate length of identical straight wire would have had to be used to attain this sensitivity in the absence of R_1 and R_2?

(c) The e.m.f. of a thermocouple is balanced by 490 mm of the wire between B and C. Calculate its value.

[(a) 10 μV mm^{-1} (b) 0.2 km   (c) 4.9 mV]

51–23 *Comparison of small resistances.* Two resistors connected in series carry a current of 0.40 A. One has a resistance 15 mΩ, and that of the other is unknown.

(a) What is the balance length for the p.d. across the 15 mΩ resistor using (i) Qu. **51–22** (ii) Qu. **51–17**?
(b) The p.d. across the unknown resistor gives a balance length of 840 mm on the more sensitive wire. What is its resistance?

[(a) (i) 600 mm (ii) 3.0 mm   (b) 21 mΩ]

Bridge Circuits

51–24 *Measurement of temperature coefficient.* A standard resistance is fitted into one arm of a *Wheatstone* bridge, and a coil of wire into the other. At 273 K their resistances are in the ratio 1 : 1.01, and when the coil is heated to 373 K the ratio becomes 1 : 1.39. Calculate the temperature coefficient of resistance of the coil of wire.

[3.8 × 10^{-3} K^{-1}]

51–25 *A copper resistance thermometer.* A copper coil was put into one arm of a metre bridge, and a fixed resistance of 100 Ω into the other. When the temperature T of the copper coil was increased the balance point moved as follows:

T/K	273	323	373
balance point position/mm	500	550	589

(a) Calculate the resistance R corresponding to the values of T.
(b) Plot a graph of R against T.
(c) From your graph estimate values for (i) the temperature coefficient of resistance (ii) the temperature of the coil corresponding to a balance point of 528 mm.

[(c) (i) 4.3 × 10^{-3} K^{-1} (ii) 30(0) K]

51 26   Refer to the diagram.

(*a*) Calculate the p.d. between X and Y.
(*b*) A galvanometer of 0.10 kΩ resistance is now connected between X and Y. Comment on (without calculating exactly) the size and direction of the current through it.
(*c*) What shunt resistance placed across the 4.0 Ω resistor would balance the bridge?

[(*a*) 1.0 V   (*c*) 2.2 Ω]

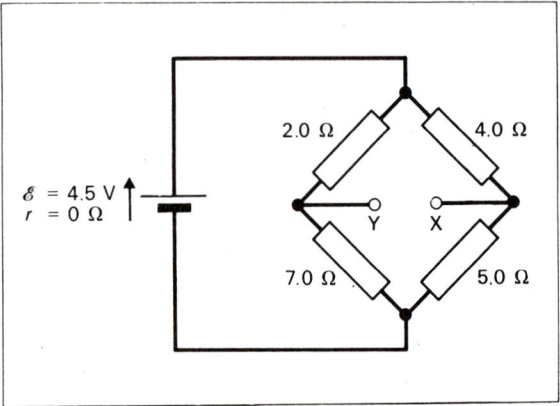

For Qu. **51–26**

51–27   *End-corrections.* The uniform wire of a metre bridge has a resistance 10.0 Ω, and is soldered at both ends. Known resistances of 4.00 Ω and 6.00 Ω are fitted into the arms, and give a balance point 390 mm from one end. When the resistors are interchanged the new balance point is 610 mm from the same end.

(*a*) Calculate the resistances of the *soldered contacts* at the ends of the wire. (Assume there are no other errors.)
(*b*) If the experimenter had been measuring the ratio of the

(known) resistances, and had used the arithmetic mean of his two values, what would have been his percentage error?

[(*a*) both 0.50 Ω   (*b*) 4%]

52   The Magnetic Field Defined from its Effect

Questions for Discussion

52–1   The size of a magnetic force F_m may be written
$$F_m = \lambda B'Qv.$$
To what extent, if at all, are we free to choose the numerical value and unit of the constant λ?

52–2   If a stationary positive charge experiences an electric force F_e, then the direction of F_e is that of the *E*-field. Why do we not follow a similar procedure for the definition of the *B*-field?

52–3   The SI unit for magnetic flux density *B* is the tesla (T). Write the alternative forms of this unit (*a*) N A^{-1} m^{-1} and (*b*) Wb m^{-2} in terms of the fundamental units m, kg, s and A. (*Note that* N A^{-1} m^{-1} *is convenient when we consider forces, and* Wb m^{-2} *when we consider magnetic flux.*)

52–4   Show that the units J T^{-1} and A m^2 are dimensionally equivalent.

52–5   Does magnetic flux Φ have a direction? Is Φ a vector? What do you understand by a negative value for Φ?

52–6   Can there be a magnetic flux through a closed surface?

52–7   **Cosmic rays** are charged particles incident on the Earth from outer space. It is observed that when they arrive at the equator more are travelling from West to East than vice versa. What is the sign of the charge that they carry?

52–8   Can the k.e. of a charged particle be changed by its movement through a *B*-field?

52–9   Can a charged particle be accelerated by magnetic forces if it moves in a straight line through a magnetic field?

52–10   Show that, for particles carrying a given charge *Q* and moving at speed *v* perpendicular to a given *B*-field, the linear momentum *p* is proportional to the radius *r* of the circular orbit described.

52–11   Show tnat the k.e. of a given ion describing circular orbits of radius *r* in a *B*-field is proportional to r^2B^2. (*This is why in a cyclotron r and B are made as big as possible – about 1.5 m and 2 T respectively.*)

52–12 A coil carries a clockwise current when viewed from a particular side. A *B*-field is now directed through the coil from that side. Would this cause the coil to be in a state of tension or compression? What would happen if (*a*) either *I or B*, or (*b*) both *I and B*, were reversed?

*52–13 Consider the true vector quantities *F* and *v*, and the pseudovector quantities *B* and *A*. The product $Q\mathbf{v} \times \mathbf{B}$ gives a vector quantity (force *F*), yet the product $\mathbf{B} \cdot \mathbf{A}$ produces a scalar (magnetic flux Φ). Discuss the distinction between these two products, with particular reference to the relative orientation of the (pseudo-)vector quantities.

*52–14 Suppose that *B* had been defined by the equation $\mathbf{F} = Q\mathbf{B} \times \mathbf{v}$ (instead of $\mathbf{F} = Q\mathbf{v} \times \mathbf{B}$). What rules or laws in electromagnetic theory would need revising?

*52–15 A frictional force is velocity-dependent and **non-conservative**. Is a magnetic field of force conservative? Can we define for it a magnetic p.e., in the same way as we define an electric and a gravitational p.e.?

*52–16 Say which of the following interactions can be described in terms of (*a*) a **monopole**, and (*b*) a **dipole**: (*i*) magnetic, (*ii*) electric and (*iii*) gravitational.

*52–17 Indicate *two* ways in which an electron can have an electromagnetic moment.

*52–18 Can you suggest why a cyclotron is more suitable for accelerating protons to a given high energy than it is for electrons?

52–19 **O-M** What is the electromagnetic moment of an electron orbiting the proton in a hydrogen atom in a state with an angular momentum $h/2\pi$? (*This quantity is referred to as a* **Bohr magneton**. *Its size is convenient for comparison purposes in atomic physics.*)

52–20 **O-M** What is the Earth's electromagnetic moment?

Quantitative Problems

(*Further questions on the motion of charged particles in E- and B-fields will be found in* Chapters **61** *and* **63**, *and under the heading* Hall Effect *in* Chapter **60**. *Questions on the moving-coil galvanometer are on* p. 165.)

Flux Density and Flux

†52–21 A particle of charge size 0.16 aC was moving at a speed 10 Mm s^{-1} when it entered a magnetic field whose direction made an angle $\pi/4$ rad with the direction of *v*. The field caused the particle to experience a force of 20 fN: what was its size?

[18 mT]

†52–22 Calculate the magnetic flux through the areas described below.

(*a*) The circular cross-section of a solenoid of area 0.010 m^2 normal to a uniform *B*-field of 70 mWb m^{-2}.

(*b*) An Earth inductor of area 0.25 m^2 whose normal is inclined at an angle $\pi/3$ rad to the Earth's magnetic field where $B = 70$ μWb m^{-2}.

(*c*) A spherical surface of area 2.0 m^2 at whose centre is a circular coil of radius 25 mm carrying a current 3.0 A.

[(*a*) 0.70 mWb   (*b*) 8.8 μWb]

The Motion of Charged Particles

52–23 Near the equator the *B*-field of the Earth has a horizontal resolved part of about 30 μT. A cosmic-ray proton arrives with a vertical velocity of 0.28 Gm s^{-1}. Compare the sizes of the forces exerted on it by the Earth through the magnetic and gravitational interactions.

Assume e, m_p, g_0   [$F_m/F_g = 1 \times 10^{11}$]

52–24 A proton of k.e. 4.0 pJ moves at right angles to a magnetic field of 0.25 T in a bubble chamber. What is the radius of the circular arc that it describes?

Assume e, m_p   [2.9 m]

52–25 Calculate the *cyclotron frequency* for (*a*) electrons, and (*b*) protons, moving through a magnetic flux density of 0.50 T.

Assume e/m_e, e, m_p   [(*a*) 14 GHz   (*b*) 7.6 MHz]

52–26 *The cyclotron.* In the cyclotron ions are made to pass periodically between one *dee* (a specially shaped hollow metal conductor) and another. An alternating p.d. is applied between the dees so that the k.e. of the ions is increased twice per revolution. They are caused to describe their spiral orbits by a *B*-field at right angles to their plane of movement, and the frequency of the applied p.d. is synchronized with that of the orbital motion.

(*a*) If the p.d. between the dees at the critical instant is 160 kV, by how much would the k.e. of α-particles increase during 100 revolutions? (*Ignore relativistic effects.*)

(*b*) The α-particles have different speeds and orbital radii: how is it that the frequency of the applied p.d. can be synchronized for all of them?

(*c*) How would the frequency change if protons were substituted for α-particles?

(*d*) What changes occur in the orbits during the 100 revolutions?

Assume e   [(*a*) 10 pJ]

*52–27 *Cyclotron frequency.* In general a charged particle moving through a magnetic field is made to describe a helical trajectory, the frequency *f* of which depends upon the particle's specific charge *q* and the flux density *B* of the field. No other variables (such as helix radius or particle speed) are involved.

(*a*) Use the method of dimensions to find how *f* is related to *q* and *B*. (*The dimensionless constant is* $1/2\pi$.)

(b) The cyclotron frequency for electrons spiralling in the ionosphere is of the order 1.0 MHz. What is the magnetic flux density?

Assume e/m_e [(b) 36 μT]

*52–28   *Helical trajectory.* An electron moving through a vacuum at a speed 2.0 Mm s^{-1} enters a uniform magnetic field of flux density 0.50 T along a path making an angle π/6 rad with the magnetic field direction.

(a) With what speed does the electron drift along the direction of the magnetic field?
(b) With what speed does it describe a circular path in a plane perpendicular to the field (*the motion being superimposed on that of (a)*)?
(c) What is the time period of this circular motion?
(d) Calculate the separation, along the field direction, of successive turns in the helical trajectory.

Assume e/m_e [(b) 1.0 Mm s^{-1} (c) 71 ps   (d) 0.12 mm]

Forces on Conductors

†52–29   A straight conductor of length 50 mm carries a current of 1.5 A, and experiences a force of 4.5 mN when placed in a uniform magnetic field of flux density 90 mN A^{-1} m^{-1}. What is the angle between the direction of the magnetic field and that of the conductor?

[42°]

52–30   *Proof of F = BIl.* A wire of cross-sectional area 2.0 mm^2 is placed perpendicular to a uniform magnetic field of flux density 0.50 N A^{-1} m^{-1}. The metal has a number density of conduction electrons 1.0 × 10^{28} m^{-3}, and these electrons have a steady average drift speed 0.40 mm s^{-1}. Calculate

(a) the average magnetic force acting on one conduction electron
(b) the number of conduction electrons in a 30 mm length of wire
(c) the total force F experienced by these electrons, and hence by that length of wire
(d) the current through the wire
(e) the size of F, using $F = BIl$.

Note that your answers to (c) and (e) should be the same.

Assume e [(c) 19 mN   (d) 1.3 A]

52–31   A horizontal electric cable of density 8.0 × 10^3 kg m^{-3} carries a current of 545 A directed from West to East. The horizontal component of the Earth's magnetic flux density is 18 μT, and causes a force on the wire of the same size as the gravitational pull of the Earth, but oppositely directed. Calculate

(a) the cable's linear weight density
(b) the cable's linear mass density
(c) the cable's cross-sectional area
(d) the **current density** (the current per unit cross-sectional area).

Find out (from a reference book if necessary) whether such a current density is realistic.

Assume g_0 [(d) 4.4 GA m^{-2}]

52–32   A circular coil in a moving-coil loudspeaker is immersed in a radial **B**-field of constant size 0.40 T. It has 50 turns of average length 50 mm each. What is the instantaneous force that the coil experiences when it is at rest, and carrying a current of 0.15 A?

[0.15 N]

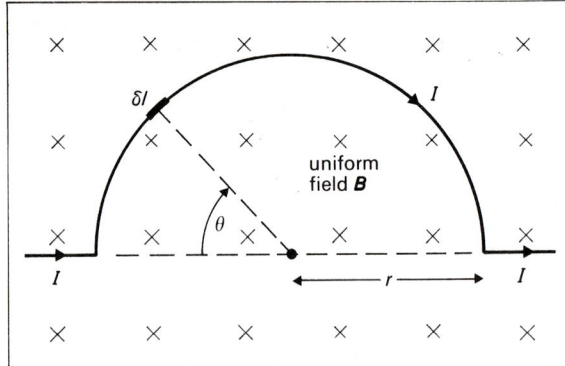

For Qu. **52–33**

*52–33   Refer to the diagram. Calculate the size of the magnetic force acting on the semicircular arc of wire shown. (*Hint: you will find the variable θ a suitable one to use for the integration.*)

[BI(2r)]

Electromagnetic Moment

†52–34   *Electromagnetic moment.* A rigid square coil of side 80 mm and carrying a current 2.0 A was placed in a vertical plane so that its normal made an angle θ with a horizontal magnetic field of **B** = 30 mT.

(a) Calculate the magnitudes and directions of the magnetic forces experienced by the four sides.
(b) Mark on a diagram the pair of forces which produce a couple on the coil. What is the moment arm?
(c) Calculate the torque T of the couple.
(d) Calculate the *electromagnetic moment* m from the *Sommerfeld* definition $T_{max}/B = m$.
(e) What would be the effect on m of increasing the number of turns from 1 to 100?

[(d) 13 mN m T^{-1} = 13 mA m^2]

†52–35   A coil of 200 turns and area 5.0 × 10^{-4} m^2 carries a current of 0.30 A.

(a) What is its electromagnetic moment?
(b) What torque would it experience if its axis were placed (i) parallel to, and (ii) perpendicular to a **B**-field of 4.0 mT?

[(a) 30 mA m^2 (b) (ii) 0.12 mN m]

52–36   A small permanent magnetic dipole had a time period of oscillation of 0.80 s in a uniform field for which B = 9.0 mT. What is the value of B in a field for which the time period of the same magnet is 1.0 s?

[5.8 mT]

*52–37 *Potential energy of a magnetic dipole.* The p.e. $E_{p,e}$ of an electric dipole can be found from an expression of the form

$$E_{p,e} = pE (1 - \cos \theta_0),$$

in which θ_0 is the angle between p and E.

(a) Use the method of dimensions to discover whether the p.e. $E_{p,m}$ of a **m**agnetic dipole can be expressed by

$$E_{p,m} = mB (1 - \cos \theta_0).$$

(b) In what situation would a dipole have (i) zero, and (ii) maximum p.e., if this equation were used? Could any other situation have been used for the zero? If so suggest alternatives.

(c) How much work would have to be done to rotate the following dipoles from a position where m and B are parallel to one in which they are anti-parallel?

(i) A coil of $m = 20$ A m^2 in a field $B = 15$ mT.

(ii) An atom of $m = 1.0 \times 10^{-23}$ J T^{-1} (roughly equal to the quantity *Bohr magneton*) in a field $B = 0.30$ T.

(*This last situation has important implications in spectroscopy.*)

[(c) (i) 0.60 J (ii) 6.0×10^{-24} J]

53 The Magnetic Field Related to its Cause

Questions for Discussion

53–1 The *Biot-Savart Law* can be expressed by the equation

$$B = \left(\frac{\mu_0}{4\pi}\right) \frac{Qv \sin \theta}{r^2}.$$

To what extent is this equation a summary of experimental results, and to what extent (if at all) is it a definition?

53–2 Sketch a graph of the variation of B with

(a) perpendicular distance r from a moving charge

(b) perpendicular distance a from an infinitely long wire carrying a current.

What are the analogous situations in electrostatics?

53–3 Refer to the diagram, in which X and Y are each positive charges, and use non-relativistic theory to answer the following.

(a) Evaluate qualitatively the field B established by X at the site of Y, and hence the magnetic force F_{XY} exerted by X on Y.

(b) Repeat to find F_{YX}.

(c) Are your results consistent with *Newton's Third Law*?

(*The law of conservation of linear momentum is not violated if one takes into account the linear momentum carried by the electromagnetic fields involved.*)

53–4 In electrostatics *Gauss's Law* $\varepsilon_0 \psi_E = \Sigma Q$ relates the field flux ψ_E and the total charge enclosed ΣQ for any closed surface. What is the corresponding statement of *Gauss's Law* in electromagnetism? (*In vacuo the quantity $\varepsilon_0 \psi_E$ has its analogy in the magnetic field flux Φ.*)

53–5 What methods exist for investigating the shape and extent of magnetic fields?

53–6 Electric fields can be represented on paper by maps of equipotential surfaces. Can we do the same for magnetic fields?

53–7 Does $B = \mu_0 nI$ apply to a solenoid whose cross-section is not circular? If so, what conditions *must* be observed before the equation can be applied?

53–8 How does the magnetic flux through the end of a very long cylindrical solenoid compare with its value through a cross-section near the middle? What *accurate* statement does this enable you to make about the relative flux densities?

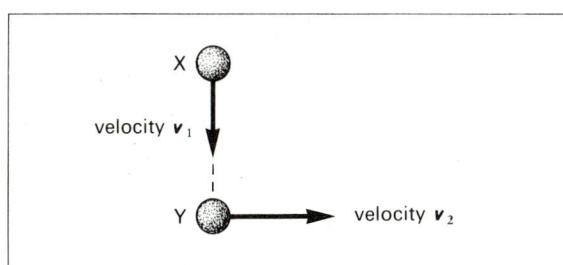

X

velocity v_1

Y

velocity v_2

For Qu. **53–3**

53–9 An electric current is passed through a loosely bound helical solenoid. Will it contract or expand? Try to justify your conclusion by an energy argument.

53–10 The magnetic field lines associated with a steady current in a long straight wire are circles concentric with the wire, and for a given size of field the lines lie on the surface of a cylinder. Describe the field lines that you associate with an *isolated* moving charge.

*53–11 A long solenoid has four layers of windings of relatively thick wire. Will we have to use a modified form of $B = \mu_0 nI$? (*Hint:* use *Ampère's Law.*)

*53–12 Use *Ampère's Law* to show the following:

(*a*) For all points outside an ideal toroid the magnetic flux density is zero.

(*b*) A pair of conductors carrying equal but oppositely directed d.c. can be twisted together to reduce the external field at distant points.

Quantitative Problems

The Biot-Savart Law and Charges

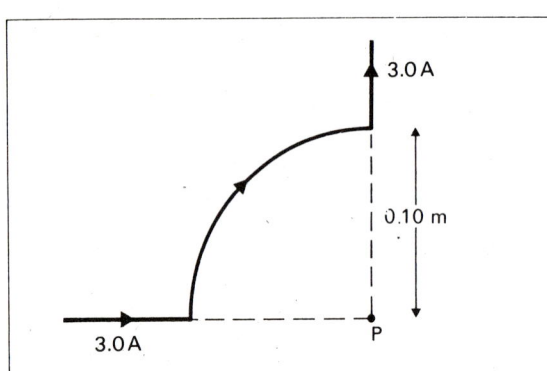

For Qu. **53–13**

†53–13 Refer to the diagram.

(*a*) What contribution to the field at P is made by the straight wires?

(*b*) Calculate the total field at P which results from this arrangement.

Assume μ_0 [(*b*) 4.7 µT]

†53–14 Refer to the diagram. Calculate the contribution ΔB of the element Δl to the field at P for (*a*) $\Delta l = 1.0$ mm, and (*b*) $\Delta l = 2.0$ mm. Suggest what the field might be for $\Delta l = 20$ mm. Are your answers exact?

Assume μ_0 [(*a*) 17 nT]

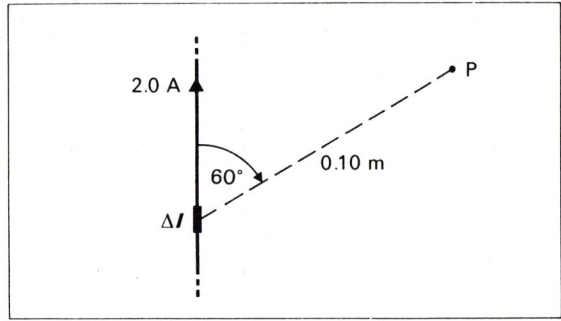

For Qu. **53–14**

53–15 In the ground state of the *Bohr* hydrogen atom we can picture the electron to describe a circular orbit of radius 53 pm at a speed 2.2 Mm s^{-1}.

(*a*) What is the magnitude of the average electric current that is the electron movement?

(*b*) What is the value of B established by the electron at the nucleus? (*You need not use* (*a*).)

Assume μ_0, e [(*a*) 1.1 mA   (*b*) 13 T]

53–16 (*a*) What is the size of the greatest magnetic field created at a point 0.10 m away from an electron moving at 50 Mm s^{-1}?

(*b*) How many electrons moving together at this speed would be needed to create a maximum field (at the same distance) of 60 µT (about the value of the Earth's field)?

Assume μ_0, e [(*a*) 8.0×10^{-17} T   (*b*) 7.5×10^{11}]

53–17 In the *Rowland* experiment to investigate the *B*-field established by moving (electrostatic) charges, the size of B was ~ 0.1 µT. If you tried to detect this field by the oscillation of a permanent magnetic dipole of electromagnetic moment 2 A m^2 and moment of inertia 2×10^{-5} kg m^2, then what time period would you observe?

[*c.* 60 s]

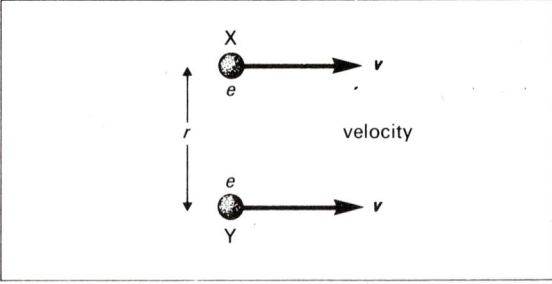

For Qu. **53–18**

*53–18 Refer to the diagram, in which X and Y are protons.

(*a*) Calculate the size and direction of the **magnetic** forces F_m exerted by X on Y and Y on X.

(*b*) Calculate similarly the electric forces F_e.

(c) Using the fact that $1/\varepsilon_0\mu_0 = c^2$, write the ratio F_m/F_e in terms of v and c only.

(d) For what value of v would $F_m = F_e$?

(*Note that classical physics has here led us to an anomaly – the resultant force exerted by X on Y and vice versa would seem to depend on the observer's frame of reference. There are difficulties which can be resolved only by the* **Theory of Relativity**.)

$[(c)\ (v/c)^2]$

The Fields of Circuits

53–19 *Circular coil.* A flat circular coil is wound with 20 turns, each of effective radius 75 mm. A current is passed through it so that the field established by the coil is exactly opposite to that of the Earth (52 μT).

(a) How is the coil oriented?

(b) In which sense does the current flow?

(c) What is the size of the current?

Assume μ_0 $[(c)\ 0.31\ \text{A}]$

53–20 A circular coil of radius 30 mm is found to produce a field of 25 μT at a point 40 mm along its axis.

(a) What is its electromagnetic moment $m\ (= NIA)$?

(b) What maximum torque could it experience in a uniform magnetic field of 6.0 mT?

Assume μ_0 $[(a)\ 16\ \text{mA m}^2\quad (b)\ 94\ \mu\text{N m}]$

53–21 *The solenoid.* An air-cored solenoid has 2.0×10^3 turns in its length of 0.80 m, and carries a current 2.5 A. It has a diameter 80 mm. Calculate the flux density along the axis

(a) at the centre

(b) at one end

(c) 60 mm from one end, inside the solenoid

(d) 60 mm from one end, outside the solenoid.

Repeat calculation (a) for an infinitely long solenoid of the same number of turns per unit length. What would be the percentage error if we treated this solenoid as being long?

Assume μ_0 $[(b)\ 3.9\ \text{mT}\quad (c)\ 7.2\ \text{mT}\quad (d)\ 0.66\ \text{mT}]$

53–22 *Straight wire.* A long vertical wire passes through a perspex sheet on which rests a plotting compass. The plotting compass indicates a neutral point 60 mm due East of the wire. The horizontal component of the Earth's *B*-field is 18 μT. What are the size and direction of the current in the wire?

Assume μ_0 $[5.4\ \text{A}]$

*53–23 Show that for points along the transverse axis distant r from the centre of a coil of electromagnetic moment m, the field B falls off according to

$$B = \left(\frac{\mu_0}{2\pi}\right)\cdot\frac{m}{r^3} \quad \text{for } r \gg a.$$

(*This result should be compared to the corresponding situation for an electric dipole, for which $E = \left(\dfrac{1}{4\pi\varepsilon_0}\right)\cdot\dfrac{p}{r^3}$.)*

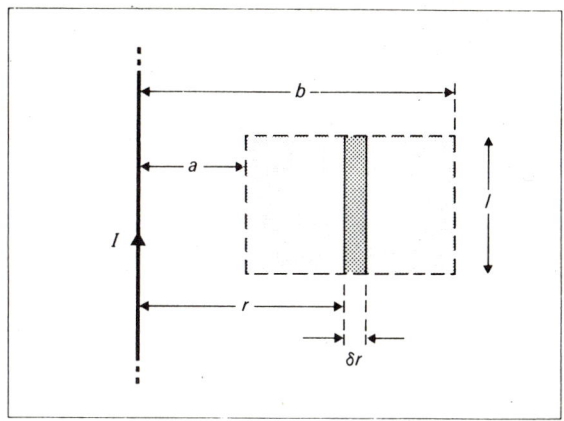

For Qu. 53–24 and 55–19

*53–24 Refer to the diagram, in which the straight wire is very long. What is the magnetic flux through the larger closed rectangle? (*A hint is implied in the figure.*)

$[(\mu_0 Il/2\pi)\ \ln\ (b/a)]$

*53–25 Calculate the value of *B* at the centre of a *square* loop of side 0.10 m in which each of the 50 turns carries a current of 1.5 A.

Assume μ_0 $[0.85\ \text{mT}]$

Forces Between Conductors

†53–26 *Choice of a value for* μ_0. Two infinitely long straight parallel wires carry currents I_1 and I_2. They are placed in a vacuum, and separated by a distance a. Theory indicates that each exerts a force F on a length l of the other, where

$$F = \left(\frac{\mu_0}{2\pi}\right)\cdot\frac{I_1 I_2}{a}\cdot l, \text{ and } \mu_0 \text{ is a constant.}$$

By *definition*, when $I_1 = I_2 = 1$ A (exactly), and $a = l = 1$ m (exactly), then $F = 2 \times 10^{-7}$ N.

(a) Calculate the value of μ_0 (the permeability constant) that must result from this definition of the ampere.

(b) What mass experiences an Earth-pull of 2×10^{-7} N?

(c) Can we put this definition into practice *directly*?

$[(a)\ 4\pi \times 10^{-7}\ \text{N A}^{-2}]$

53–27 Two long straight conductors placed 90 mm apart carry currents of 2.0 A and 4.0 A respectively in the same direction.

(a) Find the position of the line of neutral points in their resultant magnetic field.

(b) What forces do the conductors exert on 0.20 m lengths of each other?

Assume μ_0 $[(a)\ 30\ \text{mm from 2.0 A wire}\quad (b)\ 3.6\ \mu\text{N}]$

Ampère's Law

*53–28 Refer to the diagram, which shows part of a long straight solenoid. **Ampère's Law** states that

$$\oint \boldsymbol{B} \cdot \mathrm{d}\boldsymbol{l} = \mu_0 \left(\begin{array}{c} \text{total current} \\ \text{enclosed} \end{array} \right).$$

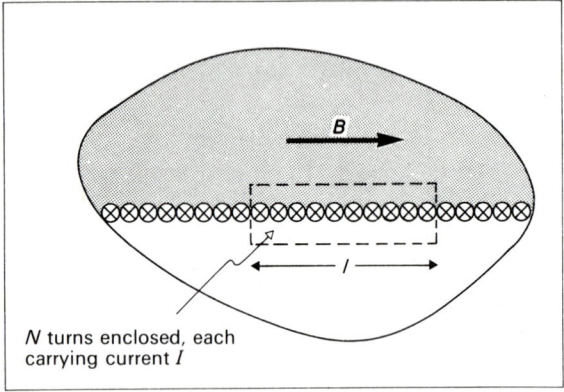

N turns enclosed, each carrying current *I*

For Qu. **53–28**

Apply the law to the dotted path shown, and hence

(*a*) find the value of **B** inside the solenoid, and

(*b*) demonstrate that the size of **B** is the same for points at all distances from the solenoid axis.

What assumptions do you make in your analysis?

*53–29 Use *Ampère's Law* to find how the field **B** at a point P inside a toroid depends upon *r*, the distance of P from the axis of symmetry (the centre of the doughnut). Hence deduce a condition for the field to have approximately the same size at all points within the toroid.

*53–30 A large sheet of width *l* is covered with *N* parallel straight wires, each of which carries a current *I* in the same direction. If each wire is to be treated as long, what is the value of **B** close to the sheet?

$[\frac{1}{2}\mu_0 I(N/l)]$

54   Electrical Measuring Devices

Questions for Discussion

54–1 The coil of a pivoted-coil galvanometer rotates in a radial field, and is deflected from its zero-current position ($\theta = 0$) to its reading position ($\theta = \theta_0$) by a steady current. Plot sketch-graphs on the same axes which show the dependence upon θ of

(*a*) the magnetic deflecting torque,

(*b*) the restoring elastic torque, and

(*c*) the resultant torque.

54–2 Why is a moving-coil galvanometer not an *absolute* instrument? Could it be made so?

54–3 What controls the *accuracy* of a moving-coil galvanometer?

54–4 What connection is there between *Brownian motion* and the sensitivity of a suspension galvanometer?

54–5 How do the sensitivity and moment of inertia (and hence the period of swing) of a *D'Arsonval* suspension galvanometer depend upon

(*a*) the area *A* of the coil,

(*b*) the number of turns *N* of the coil, and

(*c*) the torsion constant *c* of the suspension system?

Discuss the disadvantages of making such a galvanometer more sensitive.

54–6 *The Kelvin measurement of galvanometer resistance.* Refer to the diagram, in which *S* is the resistance to be measured. No other galvanoscope is available. How would you know whether the balance condition for the bridge had been achieved?

*54–7 **O-M** What is the size of the charge sensitivity of a typical moving-coil galvanometer used ballistically? Base your estimate on known values of the galvanometer variables, and compare your answer with that of Qu. **54–12**.

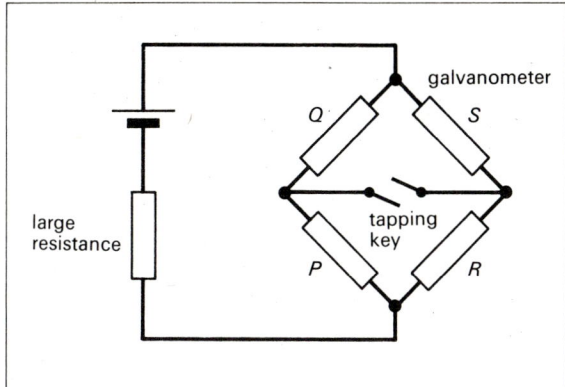

For Qu. **54-6**

54-11 A very sensitive galvanometer gives a deflection of 1.0 rad when the power of its coil is 0.10 aW. (*See Qu.* **54-10**.)

(*a*) To adapt it for use as (*i*) a sensitive ammeter, or (*ii*) a sensitive voltmeter, would you design the coil to have a high or a low resistance?

(*b*) What would be the current sensitivity if the coil resistance were made equal to 1.0 kΩ?

[(*b*) 10 pA rad⁻¹]

The Galvanometer used Ballistically

†**54-12** *Calibration of a ballistic galvanometer.* A standard 1.0 μF capacitor was charged to a p.d. of 12 V, and then discharged through a ballistic galvanometer. The first throw of the reflected light spot was 40 mm on a screen 1.0 m away.

(*a*) What was the galvanometer **charge sensitivity** (θ/Q)? (Ignore damping.)

(*b*) For what purpose(s) could this galvanometer now be used?

(*c*) What charge passed through the galvanometer would produce a first throw of 1.0 rad? (*This is sometimes called the* **galvanometer constant**.)

[(*a*) 1.7 rad mC⁻¹   (*c*) 0.60 mC]

Quantitative Problems

The Moving-Coil Galvanometer

†**54-8** *The moving-coil galvanometer principle.* A pivoted coil of 100 turns each of area 5.0×10^{-4} m² swings in a radial field of 16 mT. The torsion constant *c* of the hairsprings is 2.0 μN m rad⁻¹. A current passing through the coil causes a steady deflection of 0.80 rad. Calculate

(*a*) the magnetic deflecting torque (in terms of *I*),

(*b*) the elastic restoring torque,

(*c*) the size of the current, and hence

(*d*) the **current sensitivity** *S*   (*S* = θ/I).

Quote the value of *S* using a current unit appropriate to the design of this galvanometer.

[(*c*) 2.0 mA]

*54-13 *Theory of the galvanometer used ballistically.* A galvanometer has a coil of 100 turns of mean area 2.0×10^{-4} m². It hangs from a suspension of torsion constant 1.0 μN m rad⁻¹ in a radial field for which *B* = 0.25 T. The moment of inertia of the suspended system is 2.0×10^{-9} kg m². Calculate

(*a*) the angular impulse given to the coil when a charge 1.0 μC is passed through it in a time much less than the period,

(*b*) the intial k.e. of the coil that results from this angular impulse,

(*c*) the elastic p.e. of the suspension when it has been twisted through an angle θ_{m} to the first instantaneous rest position, and hence

(*d*) the value of θ_{m}.

Ignore damping effects.

[(*a*) 5.0 nN m s   (*b*) 6.2 nJ   (*d*) 0.11 rad]

*54-14 A ballistic galvanometer has a current sensitivity of 0.050 rad μA⁻¹. Its swing is effectively undamped, and of period π seconds. What first throw would it give for the rapid movement of 1.0 μC through the coil?

[0.10 rad]

54-9 A galvanometer has 50 turns of wire each of area 6.0×10^{-4} m², and the coil is suspended by a bronze strip of torsion constant 0.30 μN m rad⁻¹. It is observed that when a current of 15 μA is passed through the coil a spot of light reflected from a mirror attached rigidly to the coil is deflected 20 mm on a screen placed 1.0 m away. What is the size of the radial field in which the coil rotates?

[6.7 mT]

54-10 A galvanometer is to be made by winding *N* turns of wire onto a frame whose size is fixed in such a way that if each turn has a cross-sectional area *a*, then the product *Na* is constant.

(*a*) Show that the resistance *R* of the whole coil is proportional to N^2.

(*b*) Hence show that if the radial field *B* has a constant size, and if the area *A* (length × breadth) of each coil is the same, then the *deflection* θ is proportional to the square root of the *power* of the coil.

Other Instruments

†**54-15** *The principle of a current balance.* A long solenoid has 2000 turns m⁻¹ and carries a current 1.5 A. A length of wire is inserted into the solenoid so that a straight part of length 25 mm is perpendicular to the solenoid axis, and the remainder is parallel. The wire carries the same current as the solenoid. What force is exerted on the 25 mm of wire?

Assume μ_0 [0.14 mN]

54–16   *A current balance.* A wire is bent into a rectangular U-shape, and the cross-piece of length 25 mm is inserted into a solenoid so that it lies in a horizontal plane perpendicular to the axis. The solenoid is 0.40 m long, 30 mm in diameter, and has 1000 turns. It is found that when the solenoid and wire are connected in series, and a current is passed, the magnetic interaction between the solenoid and wire causes the latter to experience a force of 0.20 mN. (*This force is measured by balancing a delicate rider on a pivoted beam.*)
What current flows in the circuit?

(*Note that no electrical quantities (other than* μ_0) *are specified in the problem. Since* μ_0 *has a defined value, this constitutes an* **absolute measurement** *of current.*)

Assume μ_0 [1.6 A]

54–17   *The electrodynamometer.* A large circular coil of 100 turns and diameter 0.30 m is mounted with its axis vertical. A small circular coil having 200 turns and radius 25 mm is pivoted about a horizontal axis at the centre of the large coil, and a method is provided for measuring any torque acting on the smaller coil. The axis of the smaller coil is horizontal. The coils are connected in series, and a current I passed through them. Ignore the Earth's field, and calculate

(*a*) (in terms of I) the magnetic flux density established by the larger coil at the location of the smaller coil
(*b*) the electromagnetic moment of the smaller coil
(*c*) the torque acting on the smaller coil
(*d*) the current corresponding to a torque of 0.32 mN m.

Can the coils be arranged so that the neglect of the Earth's field is justified? (*The remarks of Qu.* **54–16** *apply here equally.*)

Assume μ_0 [(*d*) 1.4 A]

55   Principles of Electromagnetic Induction

Questions for Discussion

55–1   Suppose that a bar of length l moves transversely through a magnetic field B at a steady speed v, and as a result charges $\pm Q$ are forced to accumulate at the ends. Let the resulting p.d. between the ends be V.
(*a*) What is the size of the (inward) electric force F_e acting on each charge Q? (*Hint:* $E = V/l$.)
(*b*) What is the size of the (outward) magnetic force F_m on each charge?
(*c*) Equate F_e and F_m for the equilibrium situation to find V in terms of l, B and v.

55–2   To what extent are the various laws for predicting the directions of induced e.m.f.'s independent? Can they all, for example, be summarized by a single statement?

55–3   Can an induced current ever establish a field whose B is in the same direction as that of the agent which caused the induced current?

55–4   Under what conditions will a coil respond to a change of flux-linkage by (*a*) a state of tension, and (*b*) a state of compression?

55–5   The current induced in a coil is viewed from one side and is clockwise. What statements can be made about the directions of (*a*) the magnetic flux density, and (*b*) the change

of flux density that caused it? Suggest *three different ways* in which the change in flux-linkage could have been caused.

55–6   Refer to the diagram. The clockwise current in coil A is increasing.

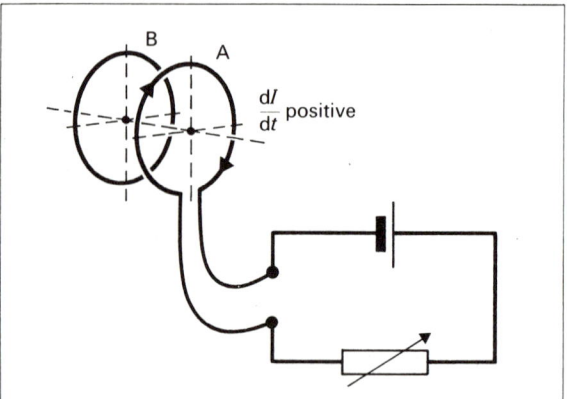

For Qu. **55–6**

(*a*) Suggest *two* different *mechanical* actions which coil B could take so as to oppose the flux change through itself.
(*b*) What *electrical* response by B could bring about *both* the actions you suggested in (*a*)?
(*c*) Are the *magnetic* effects produced by your answer in (*b*) in agreement with *Lenz's Law*?

55–7 A pair of anchor rings of the same size are made of copper and of rope. The magnetic flux-linkage through each is changed at the same rate. Compare (*a*) the induced e.m.f.'s, and (*b*) the induced currents, in the rings.

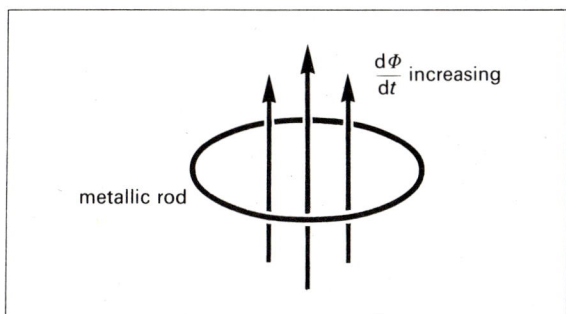

For Qu. **55–8**

55–8 *The ideas behind the betatron.* Refer to the diagram, which shows a metallic loop through which the magnetic field is increasing in the direction shown.

(*a*) In which sense should a current flow round the loop if its flux is to oppose the increase of flux-linkage?

(*b*) What kind of electric field (tangential or radial, in what sense etc.) could have generated this conventional current? Would it exist in the absence of the metal?

(*c*) What would happen to *free* electrons placed at the site of the conductor if they were subjected to the same changing field?

(*Note that in the* **betatron** *the electrons are accelerated by a changing magnetic field, and held in circular orbit by their movement through the same B-field. Electrons cannot be accelerated to high energies in a cyclotron because of relativistic problems which do not arise here.*)

55–9 Sketch a graph of induced e.m.f. against time for a coil made to rotate in a *radial* field (like that of a moving-coil galvanometer).

55–10 A conducting coil is held so that it is linked by the flux between the poles of a magnet, and is then removed in a time Δt. How does the internal energy developed in the coil depend upon Δt?

Quantitative Problems

E.M.F. Induced by Flux-Cutting

†**55–11** Through what size *B*-field must a conductor 0.50 m long be moved at 2.0 m s^{-1} to induce an e.m.f. of 1.0 volt? Is such a field available over large regions in your laboratory?

[1.0 T]

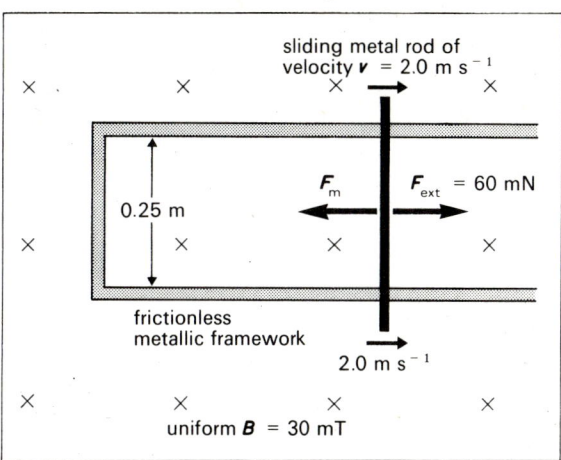

For Qu. **55–12**

55–12 *Calculation of a motional e.m.f.* Refer to the diagram, which shows a sliding rod being pulled at a steady speed through a magnetic field by an external agent who exerts a force F_{ext}. Calculate

(*a*) the size of the magnetic force F_m experienced by the sliding rod,

(*b*) the size and sense of the induced current I in the loop,

(*c*) the rate of working of the external agent,

(*d*) (in terms of the induced e.m.f. $\mathcal{E}$ and I) the power of the seat of e.m.f.

Are other energy sources involved? Assume not, and equate (*c*) and (*d*) to calculate the value of $\mathcal{E}$.

(*e*) What is the resistance of this circuit? To what extent is it an *internal* resistance?

(*Compare this question with Qu.* **56–20** *most carefully.*)

[(*b*) 8.0 A   (*c*) 0.12 W; $\mathcal{E} = 15$ mV   (*e*) 1.9 mΩ]

55–13 A conducting fluid flowing at 0.25 m s^{-1} through a cylindrical tube of diameter 40 mm is subjected to a transverse *B*-field of 60 mT. What size e.m.f. will appear across the width of the tube? (*This idea has been applied in medicine, using an alternating B, to measure the rate of blood flow through the arteries.*)

[0.60 mV]

55–14 *A rotating disc.* A copper disc of radius 50 mm is rotated at 5.0 revolutions s^{-1} about its axis which is parallel to a uniform field of 40 mT.

(*a*) What is the area swept out per revolution by any radius of the disc?

(*b*) How much flux does that radius cut per revolution?

(*c*) What induced e.m.f. exists between the centre of the disc and a point on the rim?

(*In the* **Lorenz** *disc method for measuring R, such an e.m.f. is balanced against the p.d. developed across a resistor carrying the current responsible for establishing the B-field.*)

[(*c*) 1.6 mV]

55–15 Consider a conducting rod 1.0 m long which rotates about a vertical axis through one end in the horizontal plane clockwise when viewed from above. A magnetic field of 50 mT is directed vertically downwards, and the rotational frequency of the rod is 1.0 revolutions s^{-1}.

(a) Calculate the centripetal force that acts on the electrons at the end of the rod.

(b) What magnetic force (size and direction) acts on these electrons?

(c) Does the centripetal force make a significant contribution to the p.d. across the ends of the rod?

(d) Calculate the size of the induced e.m.f.

In which direction does it act? Which end of the rod becomes postive?

Assume e, m$_e$ [(b) 0.050 aN   (d) 0.16 V]

55–16 *Generation of a sinusoidal e.m.f.* A coil of 80 turns, each of area 1.0×10^{-3} m^2, is rotated at a constant angular speed ω in a uniform magnetic field of 30 mT.

(a) What is the magnetic flux linking each turn when the normal to the coil makes an angle ωt with the magnetic field direction?

(b) At what rate is this flux changing?

(c) What is the induced e.m.f. across (i) each turn, and (ii) the whole coil, if $\omega = 100\pi$ rad s^{-1}?

(d) What is the relative orientation of the coil and the field when the peak e.m.f. is reached, and what is its size?

[(d) 0.75 V]

E.M.F. Induced by Flux-Linking

55–17 The flux density through a coil changes with time at the steady average rate of 20 mT s^{-1}. The coil has 500 turns, each of area 2.0×10^{-2} m^2, and has a total resistance 40 Ω.

(a) How long does it take to produce 10 mJ of internal energy in the coil?

(b) By how much has the magnetic field increased during this time?

[(a) 10 s]

55–18 A single coil of area 3.0×10^{-2} m^2 is placed with its axis normal to a uniform magnetic field of 40 mT, and then its area is reduced steadily to 1.0×10^{-2} m^2 over a period of 2.0 s. Calculate

(a) the change of flux-linkage, and

(b) the size of the e.m.f. induced in the coil.

[(b) 0.40 mV]

*55–19 Refer to the diagram for Qu. 53–24 (p. 163). If I is sinusoidal, and takes the form

$$I = (2.0 \text{ A}) \sin (100\pi t/s)$$

then what will be the peak e.m.f. induced in the closed rectangle shown?

[$100 \mu_0 I \ln (b/a)$ A s^{-1}]

Induced Charge

†55–20 *Calculation of induced charge.* A coil of 20 turns and area 5.0×10^{-4} m^2 is part of a circuit whose resistance is 25 Ω. It rests with its plane at right angles to a uniform *B*-field of 40 mT, and is then rapidly removed to a field-free region.

(a) What are the initial and final values of the flux-linkage through the coil?

(b) How much charge is made to pass any cross-section of the circuit?

Would your answer have been different if the movement had been a slow one?

[(b) 16 μC]

55–21 In an Earth inductor experiment the induced charges were compared using a ballistic galvanometer. When the coil was rotated so as to cut the horizontal component of the Earth's *B*-field (18 μT) then the b.g. showed a first throw of 30 divisions. When the vertical component was cut in the same way the throw was 69 divisions. Calculate, for that place,

(a) the angle of dip, and

(b) the size of the Earth's total magnetic field.

[(a) 67°   (b) 45 μT]

55–22 An iron-cored solenoid has 1000 turns of fine insulated copper wire. The *B*-field through the iron is changed at a uniform rate over a 5.0 s time interval from + 0.60 T (in one direction) to − 0.30 T (in the opposite direction). The effective area of cross-section of the copper coils (roughly that of the iron core) is 2.5×10^{-3} m^2, and they are connected into a circuit of resistance 9.0 Ω. Calculate

(a) the charge displaced past a cross-section of the circuit

(b) the size of the average current

(c) the size of the instantaneous current when $B = 0$.

Does the current flow change direction during the operation?

[(a) 0.25 C   (b) 50 mA]

55–23 *Measurement of flux density by b.g.* A search coil of 50 turns and mean area 1.0×10^{-4} m^2 is to be used in conjunction with a ballistic galvanometer to measure a flux density. The complete galvanometer circuit has a resistance of 0.60 kΩ, and with this arrangement the charge sensitivity is found to be 10 divisions μC^{-1}. When the search coil is held correctly between the poles of an electromagnet, and rapidly removed, the galvanometer's first throw is 80 divisions. Calculate the size of the induced charge, the change of flux-linkage for the coil, and hence the flux density between the poles of the electromagnet.

[0.96 T]

56 Applications of Electromagnetic Induction

Questions for Discussion

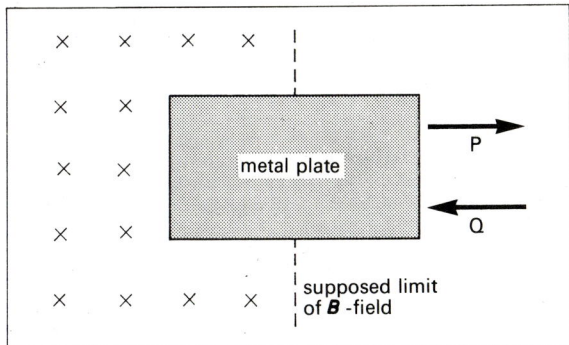

For Qu. **56–1**

56–1 Refer to the diagram. Draw the complete path of an eddy-current loop induced in the metal plate when the plate is forced to undergo (*a*) movement P, and (*b*) movement Q.

56–2 *Electromagnetic damping.* Use *Fleming*'s left-hand motor rule to discover what effect the eddy currents of Qu. **56–1** have on the relative motion of the plate and the source of the *B*-field.

56–3 How could the plate of Qu. **56–1** be slotted to reduce the effects that you analysed in Qu. **56–2**?

56–4 A permanent magnet is held so that the direction of its electromagnetic moment points toward, and is perpendicular to, a conducting sheet. It is then moved *parallel* to the sheet. Mark on a diagram the paths of the eddy currents caused within the sheet.

56–5 A small bar magnet placed vertically above a bowl made of superconducting lead needs no contact forces to support it. Explain.

56–6 If electromagnetic damping is used to speed the use of (say) a highly sensitive analytical balance, does it affect the accuracy?

56–7 What can be learned from the reading on a car ammeter?

56–8 A fixed length *l* of wire is to be wound into *N* circular turns which will become the armature of a generator. What value of *N* would produce the largest e.m.f.?

56–9 Discuss whether you would use a series-wound, shunt-wound or compound-wound d.c. motor in each of the following applications: (*a*) a lathe, (*b*) an electric train, (*c*) a crane, (*d*) a record player, and (*e*) a car starter motor.

56–10 Why do shunt-wound dynamos find more applications than do series-wound dynamos?

56–11 **O-M** Estimate the current used by a car starter motor when starting the engine.

Quantitative Problems

The Transformer

†**56–12** An ideal transformer has 600 turns on the primary, and 120 on the secondary. A resistance of 4.0 Ω is connected across the secondary, and then an alternating p.d. of r.m.s. value 30 V is applied to the primary. Calculate

(*a*) the r.m.s. p.d. across the secondary,
(*b*) the r.m.s. current through the resistor,
(*c*) the power of the secondary circuit, and
(*d*) the r.m.s. current taken by the primary.

How does the primary dispose of the energy that it takes from the source?

[(*c*) 9.0 W   (*d*) 0.30 A]

56–13 *Theory of the ideal transformer.* The primary of a transformer has 50 turns, and is activated by an r.m.s. applied p.d. of 20 V. The secondary has 100 turns, and is on open circuit.

(*a*) What property of the primary determines the (small) *magnetizing current* through it?
(*b*) Why does the primary not remove energy from the source?
(*c*) What current flows in the secondary?
(*d*) What contribution does the secondary make to the flux through the iron core?
(*e*) What is the r.m.s. p.d. across the secondary?

[(*e*) 40 V]

56–14 *Secondary with a resistive load connected.* A load of 40 Ω is now connected across the secondary of Qu. **56–13**.

(a) What current now flows in the secondary?
(b) What contribution does the secondary make to the flux through the core?
(c) What effect would this have on the back e.m.f. in the primary coil if the latter did not respond?
(d) How does it respond (to ensure that the loop equation can still be applied to the primary)?
(e) What power is now extracted from the energy source?
(f) How does the peak flux in the core in this question compare with that in Qu. **56–13**?

|(e) 40 W|

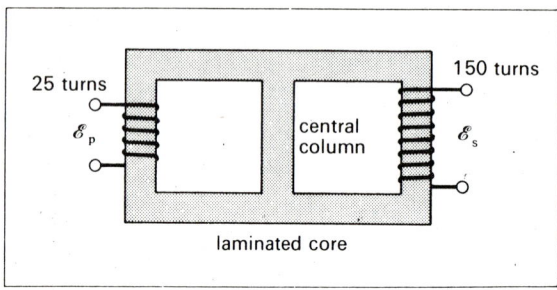

For Qu. **56–15**

56–15 Refer to the diagram. This method of winding the transformer has resulted in two thirds of the flux that links the primary passing through the central column. Find the ratio $\mathscr{E}_s/\mathscr{E}_p$ if there is no flux-leakage from the core. How would *you* have wound the transformer?

*56–16 *Impedance matching.* Maximum power transfer between an energy source and a load resistance occurs when the source internal resistance equals that of the load. It is required to couple an amplifier of internal resistance 2.5 kΩ to a loudspeaker of resistance 4.0 Ω using a transformer of **turns ratio** (N_p/N_s).

(a) If the loudspeaker is connected in the secondary circuit, what is its apparent (or **reflected**) resistance in the primary circuit?
(b) What turns ratio would be suitable?

(*Note two points: (i) the self-inductance of the primary windings is not evident because of the mutually cancelling fluxes, and (ii) the two circuits are not in electrical contact.*)

|(b) 25|

Absolute Measurement of Resistance

56–17 *The Lorenz rotating-disc method.* A circular conducting disc of cross-sectional area 1.0×10^{-3} m² is placed coaxially at the centre of a long solenoid. The solenoid carries a current I, and is wound with 1000 turns m⁻¹. When the disc is rotated at 20 rad s⁻¹ the induced e.m.f. V between its rim and centre is equal to the p.d. across a resistance R which is in series with the solenoid. Calculate

(a) in terms of I, the flux density inside the solenoid, and hence the value of V, and
(b) the value of R.

(*This is referred to as an **absolute measurement** of R, because none of the information given is electrical. μ_0 has a defined value. The simple school laboratory version is neither accurate nor sensitive.*)

Assume μ_0 |(b) 4.0 μΩ|

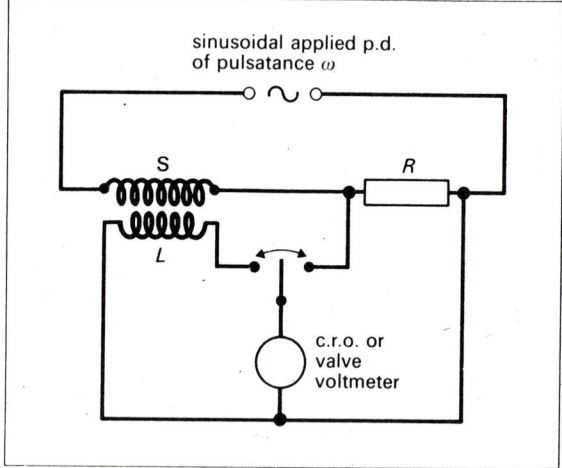

For Qu. **56–18**

*56–18 *The Belham a.c. method.* Refer to the diagram. The long solenoid S has 1000 turns m⁻¹ of mean cross-sectional area 2.0×10^{-3} m², and carries a current $(0.50 \text{ A}) \sin \omega t$. The same current flows through an unknown resistor R. L is a search coil of 500 turns which is wound round the centre of the solenoid.

(a) What is the instantaneous value of B inside the solenoid?
(b) What is the maximum rate at which B changes?
(c) What is the maximum e.m.f. induced across L?
(d) When ω is adjusted to 400π s⁻¹ (four times that of the mains), the maximum p.d. across L equals that across R. Calculate R.

How would you confirm that R was a non-inductive resistor? (*Like the Lorenz method, this is also an absolute method, since the current amplitude cancels, and need not be known. It has the advantage that the induced e.m.f. is far greater, and this enables a larger resistance to be used.*)

Assume μ_0 |(d) 1.6 Ω|

The Motor and Dynamo

†56–19 A motor of 80% efficiency is required to do mechanical work at the rate of 0.64 kW. What current will it take from a 200 V d.c. supply?

[4.0 A]

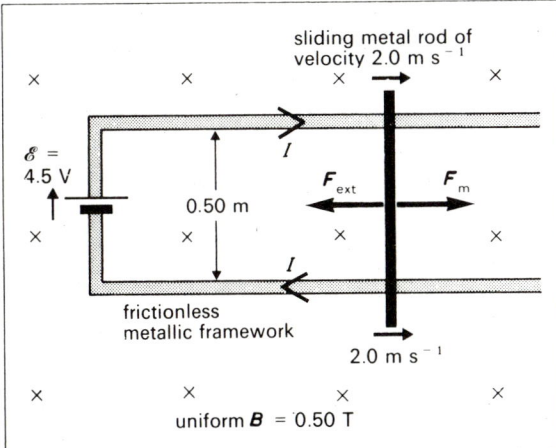

For Qu. **56–20**

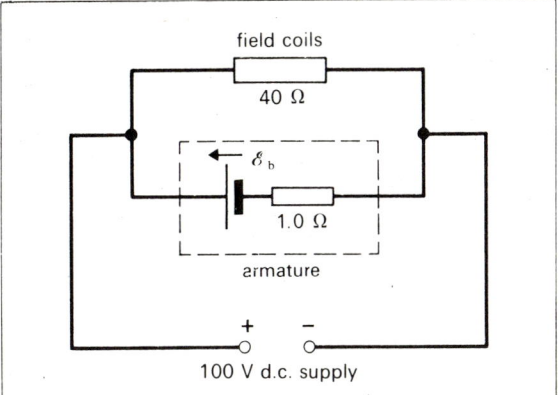

For Qu. **56–21**

(*You can check your answer by making two independent calculations of the power wasted by Joule heating.*)

[(*c*) 92 V   (*d*) 0.69 kW   (*e*) 69%]

56–20 *The fundamental motor principle.* Refer to the diagram. The rod is sliding at a steady speed because it is doing work by exerting a force on an external body. The total circuit resistance (assumed constant) is 0.40 Ω.

(*a*) What is the sense and size of the e.m.f. induced in the completed circuit?

(*b*) Find the current *I* by applying the loop equation.

(*c*) At what rate does the battery supply energy?

(*d*) At what rate does the circuit dissipate energy by *Joule* heating?

(*e*) What is the useful mechanical power output from the motor?

(*f*) Check your answer to (*e*) by evaluating the magnetic force F_m (= *BIl*) acting on the rod, and so calculating *Fv*.

(*The student should compare this question with Qu.* **55–12** *most carefully.*)

[(*e*) 5.0 W   (*f*) F_m = 2.5 N]

56–21 *The shunt-wound motor.* Refer to the diagram, which is a schematic arrangement of a shunt-wound motor. $\mathscr{E}_b$ represents the back e.m.f. developed in the armature. The net power taken from the source is 1.0 kW. Calculate

(*a*) the current taken from the supply,

(*b*) the current through the field coils, and hence that through the armature,

(*c*) the back e.m.f. in the armature,

(*d*) the mechanical power output, and

(*e*) the efficiency of the motor.

56–22 *The torque exerted by an armature.* A shunt-wound motor whose armature has a resistance of 2.0 Ω operates from 110 V d.c. mains. When it is working at an angular speed of 0.30 krad s⁻¹ the armature is taking a current of 5.0 A. Calculate

(*a*) the back e.m.f. in the motor,

(*b*) its rate of doing useful work, and

(*c*) the average torque *T* that it exerts. (*Hint: W = Tθ*)

[(*b*) 0.50 kW   (*c*) 1.7 N m]

56–23 *Motor with zero load.* The armature of a motor has 100 turns of mean area 2.0×10^{-2} m², and rotates in a magnetic field whose value may be taken as constant at 0.30 T. No load is attached, and all frictional effects are negligible. What will be its equilibrium rotational speed if a p.d. of 24 V is applied?

[6.4 revolutions s⁻¹]

56–24 *The need for a starting resistance.* A shunt-wound motor operates from a 25 V d.c. supply. When the armature current is 5.0 A it provides 0.10 kW of useful mechanical energy. Calculate

(*a*) the back e.m.f. and hence the armature resistance, and

(*b*) the current that would flow through the motor if it were started from rest without the use of a starting series resistor.

(*A current of this size would probably burn out the armature.*)

[(*a*) 1.0 Ω   (*b*) 25 A]

*56–25   *Maximum power for a series-wound motor*. Refer to the diagram, which is a schematic representation of the circuit of a series-wound motor.

Write down the loop equation for this circuit, and use it to express the mechanical power supplied by the motor in terms of the single variable I. Hence show that the motor provides maximum power when

$$\mathscr{E}_b = 0.5V.$$

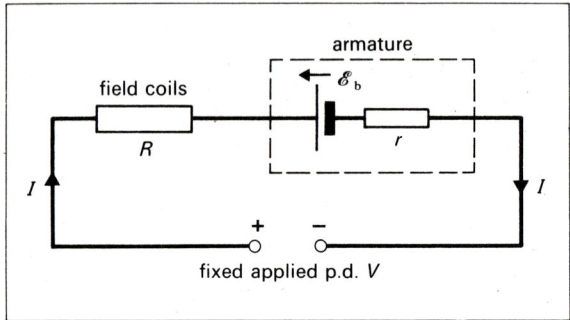

For Qu. **56–25**

57   Inductance

Questions for Discussion

57–1   By expressing each in terms of kg, m, s and A, show that $1\,\text{V s A}^{-1} \equiv 1\,\text{Wb A}^{-1}$ (each being one **henry**.)

57–2   Current I is not a vector quantity. What then does the minus sign imply if we write $\mathrm{d}I/\mathrm{d}t = -10\,\text{A s}^{-1}$? (*You might find it helpful to give examples of other situations in which scalar quantities have negative signs.*)

57–3   An inductor with a changing current is a source of (back) e.m.f. What are the relative directions of the current and the *back* e.m.f. when the inductor is (*a*) absorbing energy, and (*b*) releasing it to an external device? What about the applied p.d.? (*Compare your answers with the charge and discharge, respectively, of an accumulator.*)

57–4   Can the back e.m.f. in an inductor be in the same *sense* as the e.m.f. of the source which gave the inductor its magnetic energy?

57–5   Does a long straight wire have a self-inductance?

57–6   Discuss how, *in principle*, you would calculate the self-inductance of a single circular coil of wire of radius r.

57–7   Two long solenoids of the same length, same number of turns, and carrying the same current produce the same field B along their axes, regardless of their diameters. What about their self-inductances?

57–8   Two inductors L_1 and L_2 are joined in series. What principles enable you to calculate their equivalent self-inductance, and what is its value? Ignore their mutual inductance.

57–9   A p.d. is applied across a circuit of inductance L and total resistance R, and causes the current to build up exponentially to a final value I_0. Use the method of dimensions to find how the *time* taken to reach a fixed fraction of I_0 depends (if at all) upon I_0, R and L. (*Note that the method of dimensions does not enable us to relate I, I_0, R and L directly.*)

57–10   An external agent has to do positive work to separate two parallel straight wires carrying like currents. What effect does this action have on the magnetic energy density in the region between the wires? Explain your answer in terms of the changes in B that he causes.

*57–11   What *electromagnetic* quantities do you need to know to be able to calculate the mutual inductance of two circuits? Consider, as a simple example, a solenoid and a coil. What is the significance of your conclusion?

*57–12   O-M   A parallel circuit has two branches. One contains a torch bulb and a non-inductive resistor, while the other has an identical torch bulb and a coil of self-inductance L. What value must L have if the coil is to cause an observable time-lag in the lighting of the bulb?

*57–13   O-M   What are the magnitudes of the largest E- and B-fields easily obtained in your laboratory? Which field has the greater energy density? (*Qu.* **46–25** *and* **57–21** *will be helpful.*)

Quantitative Problems

Self-Inductance

†57–14 A coil has a flux-linkage of 8.0 μWb turns when it carries a current of 2.0 A. What are

(a) the self-inductance, and

(b) the back e.m.f. in the coil at the instant when the current is changing at the rate 5.0 A s^{-1}?

[(a) 4.0 μH   (b) 20 μV]

†57–15 An inductor of 0.50 H has negligible resistance. What rate of change of current through it will enable an applied p.d. of 10 V to be maintained across its terminals?

[20 A s^{-1}]

57–16 *Calculation of self-inductance.* An air-cored solenoid is 0.80 m long, and its 2.0×10^3 turns have a mean area of 2.0×10^{-3} m^2. It carries a current 0.50 A. Use $B = \mu_0 nI$, and calculate

(a) the value of **B** over the cross-section

(b) the flux-linkage for 200 turns near the centre

(c) their flux-linkage per unit current

(d) the self-inductance per unit volume for regions near the centre.

How would your answers to (c) and (d) change if the current were doubled?

Assume μ_0 [(d) 7.9 H m^{-3}]

57–17 Use the method of Qu. **57–16** to find the self-inductance of an air-filled toroid of mean length 0.50 m, each of whose 1000 turns has a cross-sectional area 4.0×10^{-4} m^2.

Assume μ_0 [1.0 mH]

57–18 A large electromagnet of self-inductance 4.0 H is energized by a current of 10 A. When it is switched off the current falls to 5.0 A in the first 20 ms.

(a) What is the *average* back e.m.f. during this time?

(b) Discuss the relevance of your answer to the design of switches.

(c) How could you demonstrate experimentally that your predicted e.m.f. has the right order of magnitude?

[(a) 1.0 kV]

*57–19 *Energy stored in magnetic field.* What is the instantaneous power of an inductor of self-inductance L when its instantaneous current I is growing at the rate dI/dt? How much electrical energy W will it have converted to stored magnetic energy when the instantaneous current has reached I_0? Find the value of W for an iron-cored toroid whose inductance is 0.50 H when $I_0 = 2.0$ A.

[1.0 J]

*57–20 The coil of a particular superconducting electromagnet used for **nuclear magnetic resonance** (NMR) research has a self-inductance 0.15 kH, and carries a current of 30 A. It is surrounded by liquid helium, whose molar latent heat of vaporization is 85 J mol^{-1}.

(a) How much energy is associated with the magnetic field of the coil?

(b) The superconductor is *quenched,* and suddenly develops a finite resistance. How much helium is boiled off?

(*This magnet can produce fields of* 8 T *which are homogeneous to* 1 *part in* 10^9.)

[(a) 68 kJ   (b) 0.79 kmol]

*57–21 *Energy density of magnetic field.* Use the standard results $L = \mu_0 n^2 V$ and $B = \mu_0 nI$ together with your result from Qu. **57–19** to find the energy density W/V of the magnetic field inside an air-filled toroid. (*The result you should obtain actually holds for all vacuum situations.*) What is the energy density inside a toroid for which $B = 2.0$ mT?

Assume μ_0 [1.6 J m^{-3}]

LR Circuits

*57–22 In an *LR* d.c. series circuit the source has an e.m.f. of 2.0 V, $L = 0.20$ H and $R = 10$ Ω. Write down the loop equation for the situation when the circuit current is growing, and by putting $I = 0$ find the initial rate of current growth when the circuit is first completed. Check that your answer is equal to I_∞/τ_L, where τ_L is the inductive time constant, and I_∞ is the final steady current.

[10 A s^{-1}]

*57–23 An inductor of resistance 4.0 Ω was placed in series with a resistor of resistance 6.0 Ω, and an e.m.f. of 10 V was applied across them. After 10 ms the current had risen to an instantaneous value 0.63 A.

(a) What was the inductor's inductance?

(b) How long would the current take to fall from its steady value to 0.37 A if the source were disconnected, and the inductor and resistor were simultaneously shorted?

(c) What would be the time constant of the circuit if the series resistor were removed?

Assume 1/e = 0.37   [(a) 0.10 H   (b) 10 ms   (c) 25 ms]

*57–24 A source of negligible internal resistance is used to apply a p.d. of 12 V across a resistive inductor, and causes the current to change at the rate 10 A s^{-1} when the instantaneous current is 2.0 A. When the source is disconnected and the coil is simultaneously shorted, the previously steady current again changes at 10 A s^{-1} when the instantaneous value is 2.0 A. Calculate

(a) the values of R and L

(b) the inductive time constant of the circuit.

Sketch the graphs of current growth and decay.

[(a) 3.0 Ω, 0.60 H   (b) 0.20 s]

*57–25 An air-cored solenoid of self-inductance 3.0 mH is connected into a circuit of total resistance 10 Ω. How long does it take for the current to reach 99% of its steady-state value?

Assume ln 100 = 4.61   [1.4 ms]

Mutual Inductance

*57–26 *Calculation of mutual inductance.* The value of B at the centre of a long solenoid is 5.0 mT when it carries a current 1.0 A. A search coil of 100 turns is wound round the solenoid, whose mean cross-sectional area is 8.0×10^{-4} m². Calculate

   (*a*) the flux through the centre of the solenoid,
   (*b*) the flux-linkage of the secondary coil, and
   (*c*) their mutual inductance.

What e.m.f. is induced in the search coil when the solenoid current changes uniformly at the rate 4.0 A s⁻¹? Why do you not need to know the cross-sectional area of the *search* coil?

Assume μ_0 [(*c*) 0.40 mH; 1.6 mV]

*57–27 Calculate the approximate mutual inductance of a pair of concentric coaxial coils, each of 50 turns, whose radii are 0.10 m and 10 mm. When the current in the outer coil changes at the rate dI/dt, an e.m.f. of 0.10 mV is induced in the inner. Calculate dI/dt.

Assume μ_0 [4.9 μH, 20 A s⁻¹]

*57–28 An ideal transformer has a mutual inductance of 0.50 H between its primary and secondary coils. A current of r.m.s. value 1.0 A and pulsatance 300 s⁻¹ energizes the primary. Calculate the r.m.s. p.d. developed across the secondary.

                                   [0.15 kV]

58   Alternating Current

Questions for Discussion

General

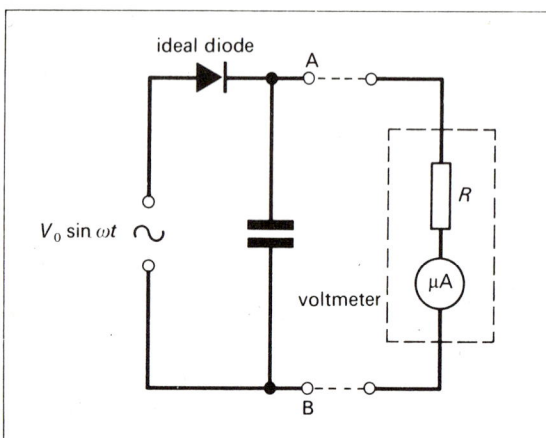

For Qu. **58–1**

58–1 *The diode voltmeter.* Refer to the diagram. (*a*) Suppose the dotted connections are *not* made. Draw graphs to show (*i*) the p.d. across the capacitor, and (*ii*) the current flow to the capacitor, both as functions of time. What impedance does this arrangement eventually present to the source?

(*b*) What would an electrostatic voltmeter indicate if connected across AB?

(*c*) Consider the effect of making the dotted connections. Repeat the graph of (*a*) (*ii*). What value of CR would make the current taken by the voltmeter very small? What value is indicated by the d.c. voltmeter?
(*Note that R can be made ~ megohms, so that the voltmeter as a whole has a very high impedance.*)

58–2 A neon lamp strikes only when the p.d. between its terminals exceeds 180 V, and it is extinguished as soon as the p.d. becomes less than 140 V. Draw a p.d.-time graph for one cycle of the 240 V 50 Hz mains, and shade the areas which correspond to times when the neon flashes. For your discussion assume that a protective resistor is placed in series with the lamp to prevent it burning out.

*58–3 What could cause the a.c. *resistance* of a device to depend upon frequency?

58–4 The build-up of current through an inductor enables it to absorb energy in its magnetic field. Discuss the relationships between the directions of the *applied* p.d., the *back* e.m.f. and the current, when the inductor is in the act of (*a*) storing energy, and (*b*) delivering energy. (*Be careful not to confuse the direction of a current with the sign convention for its having a decreasing or an increasing value.*)

58–5 Charge flow onto the plates of a capacitor builds up a p.d. which eventually opposes further charge flow, and during this time the capacitor absorbs energy in its electric field. Discuss the relative directions of the applied p.d. and current flow when the capacitor is in the act of (*a*) storing energy, and (*b*) delivering energy.

58–6   Can a varying current be said to flow *through* a capacitor? If it cannot, how is it that a capacitor can be used to transmit a varying signal from the first stage of an amplifier to the second? (*To what extent is this analogous to the coupling of a transformer? – see Qu.* **56–16**.)

58–7   The capacitor of an a.c. circuit may be referred to variously as a *blocking, coupling, reservoir* (storage) or *smoothing* capacitor. Compare and contrast these different functions of the same device.

RLC Circuits

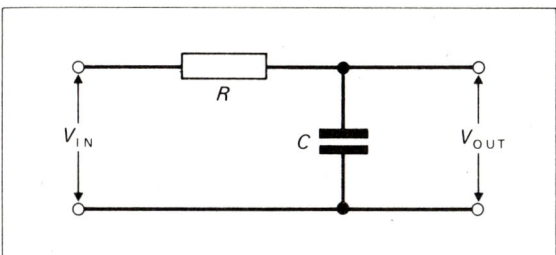

For Qu. **58–o**

58–8   *The RC low-pass filter.* Refer to the diagram. The alternating input p.d. has a fixed amplitude but variable frequency. Sketch a graph of the amplitude of the output p.d. as a function of frequency.

What simple change to this circuit would enable it to act as a **high-pass filter**, in which low-frequency signals are attenuated?

58–9   Using the diagram of Qu. **58–8** as a guide, design a simple *RL* low-pass filter.

*58–10   A charged capacitor is connected into a series circuit containing a resistor, an inductor and a switch. Sketch a graph which plots the charge on either plate of the capacitor as a function of time from the instant that the circuit is completed. (*These temporary currents are referred to as* **transients**.)

*58–11   In a particular *RLC* series circuit the current leads the applied p.d. Which is greater, the inductive or capacitative reactance? What will be the effect on the phase angle, and hence on the current, of gradually decreasing the frequency of the applied p.d.?

*58–12   To investigate the behaviour of an *RL* circuit we can put $C = \infty$ F in the equations that summarize the analysis of an *RLC* circuit, yet if we wish to analyse an *RC* circuit we put $L = 0$ H. Discuss the physical significance of this.

*58–13   Is it possible to introduce an inductor into an *RC* series circuit without altering the size of an existing a.c.?

*58–14   Why should you not read about the r.m.s. *power* of an a.c. circuit?

*58–15   What is the power factor of the primary circuit of an ideal transformer whose secondary is on open circuit?

*58–16   A particular device is to be supplied with a given power by a.c. What is the drawback to having a small power factor in the circuit? (*Hint:* $P = VI \cos \phi$.) How would you attempt to remedy this for an inductive circuit in which the applied p.d. leads the current?

Oscillations and Resonance

58–17   Express the units of the quantities L and C in terms of the fundamental units m, kg, s and A. Hence find the unit $1/\sqrt{LC}$.

58–18   Using the same axes sketch graphs which show values of X_L (positive y-axis) and X_C (negative y-axis) plotted against pulsatance ω. See if you can find evidence for there being more than one frequency at which an *LC* circuit could resonate. (*Think of other resonance situations, such as those of acoustics.*)

*58–19   What electromagnetic quantities can be considered to be analogous to the following mechanical forces: (*a*) an elastic restoring force $F = -kx$, (*b*) a resultant force $F = m(dv/dt)$, and (*c*) a dissipating force such as a viscous force $F_v = -6\pi\eta rv$? (*Qu.* **58–46** *will be found useful*.)

*58–20   The pulsatance ω of the p.d. applied to an *RLC* series circuit is steadily increased from zero to $2/\sqrt{LC}$. Sketch, as functions of ω, the variations in (*a*) the current, (*b*) the phase shift between current and applied p.d., and (*c*) the average dissipation of power in the circuit.

*58–21   The capacitor and inductor of a resonant *RLC* series circuit are adjacent. If $X_C = X_L = 100$ Ω, and $I_{r.m.s.} = 1.4$ A, what would a voltmeter read if it was connected across both devices (together)?

*58–22   *The quality or Q-factor.* In Qu. **11–9** we defined the Q-factor of an oscillating system by

$$Q = \frac{2\pi \times \text{(energy of system)}}{\left(\begin{array}{c}\text{energy supplied to the system} \\ \text{during one oscillation}\end{array}\right)}.$$

(This Q is *not* associated with electric charge.) Consider an *RLC* series circuit at resonance, and write down

(*a*) in terms of I_0 and L, the maximum energy stored
(*b*) in terms of L and C, the time period of oscillation
(*c*) the energy supplied per cycle by a source which maintains a fixed current. (*The source replaces the energy dissipated by the resistor*.)

Hence show that $Q = \sqrt{L/R^2C}$.
(*Q is a measure of the sharpness with which the circuit can be tuned, and is small for large values of R.*)

*58–23   *Q-factor for a coil.* The resistance of a series *RLC* circuit is mostly determined by that of the inductor coil. By writing *Q* in terms of the resonance pulsatance ω_0, *L* and *R* only, show that *Q* equals

(a) $\dfrac{\text{peak p.d. across inductance}}{\text{peak p.d. across resistance}}$

(b) tan ϕ, where ϕ is the phase angle between the current through the coil and the p.d. across it.

(*Expression* (a) *explains the idea of a* **voltage magnification** *during resonance.*)

*58–24   Is it desirable for an oscillatory radio receiver circuit to have a high or a low *Q*-factor?

*58–25   O-M   What values of *L* and *C* could be used in an oscillatory circuit whose natural frequency is equal to that of *Radio 3*?

Quantitative Problems

(*For the generation of a sinusoidal e.m.f. refer to Qu.* **55–16**. *When the value of an alternating p.d. is quoted without qualification, then the r.m.s. value is implied.*)

R, L and C Separately

58–26   *Alternating current and electron behaviour.* A wire of cross-sectional area 0.80 mm^2 is made of material in which the number density of electrons is 1.0×10^{29} m^{-3}. It carries a 50 Hz sinusoidal alternating current of r.m.s. value 4.0 A. Calculate

(a) the r.m.s. drift speed of the electrons

(b) their average drift velocity

(c) their maximum drift speed

(d) their amplitude of periodic oscillation.

Assume that the current is distributed uniformly over the conductor's cross-section.

Assume e   [(a) 0.31 mm s^{-1} (c) 0.44 mm s^{-1} (d) 1.4 μm]

†58–27   If a p.d. of more than 1.0 kV is applied to a particular capacitor then it breaks down. What is the largest r.m.s. p.d. that should be used across it?

†58–28   An ammeter consisting of a moving-coil galvanometer and its shunt resistor are fitted into a full-wave rectifier bridge circuit in order to measure an a.c. It reads 1.00 A. Calculate the values of the peak and r.m.s. currents.

[1.57 A, 1.11 A]

†58–29   A small bulb is rated at 36 W, and lights to its normal brightness when an alternating p.d. of peak value 18 V is applied. What is its resistance? (*You should assume that the resistance of the filament does not vary during a cycle. The fact that it does in practice causes the effective current to differ from the r.m.s. value.*)

[4.5 Ω]

†58–30   What is the self-inductance of an electromagnet which takes a current of 0.50 A from the 250 V 50 Hz mains? Ignore its resistance.

[1.6 H]

†58–31   For what frequency of applied p.d. will a capacitor of 20 μF have a reactance of the same size as that of a 50 mH inductor?

[0.16 kHz]

58–32   *The pure inductance.* An alternating p.d. *V* of pulsatance 300 s^{-1} and r.m.s. p.d. 0.10 kV is applied to an inductor of *L* = 0.50 H. Calculate

(a) the inductive reactance X_L

(b) the r.m.s. current $I_{\text{r.m.s.}}$

(c) the instantaneous power *P*, and hence its maximum value P_0

(d) the maximum energy stored in the magnetic field of the inductor.

Draw a sinusoidal curve to represent the variation of *I* with *t*, and then superimpose on the same sketch the variations of *V* and *P* with *t*. Mark on this figure the instants at which occur (i) V_0 (ii) P_0 (iii) maximum energy storage.

[(a) 0.15 kΩ   (b) 0.67 A   (c) (67 W) sin (600*t*/s)   (d) 0.22 J]

58–33   *The pure capacitance.* An alternating p.d. *V* of pulsatance 300 s^{-1} and r.m.s. p.d. 0.20 kV is applied to a capacitor of *C* = 33.3 μF. Calculate

(a) the capacitative reactance X_C

(b) the r.m.s. current $I_{\text{r.m.s.}}$

(c) the instantaneous power *P*, and hence its maximum value P_0

(d) the maximum energy stored in the electric field of the capacitor.

Draw a sinusoidal curve to represent the variation of *I* with *t*, and then superimpose on the same sketch the variations of *V* and *P* with *t*. Mark on this figure the instants at which occur (i) V_0 (ii) P_0 (iii) maximum energy storage.

[(a) 100 Ω   (b) 2.0 A   (c) (400 W) sin (600 *t*/s)   (d) 1.3 J]

RLC Circuits

*58–34   *Use of phasors.* An ideal inductor (having zero resistance) and a resistor are connected in series into a circuit of pulsatance 300 s^{-1}. The peak p.d. across the inductor is 30 V, and that across the resistor is 40 V. Use the rotating-vector method to find the peak value V_0 of the applied p.d. *V*, and the phase shift between this p.d. and the current in the circuit. Sketch on the same graph the variations with time of the instantaneous p.d. *V*, V_L and V_R.

[$\Delta\phi$ = arctan 0.75]

*58–35   In an *RLC* series circuit the peak values of V_L, V_R and V_C are 120 V, 120 V, and 70 V respectively. Use the rotating-vector method to calculate the peak value of the applied p.d., and the phase shift between this p.d. and the current in the circuit.

[130 V, *V* leads *I* by arctan 0.42]

*58–36 *The inductive resistor.* A particular coil has a resistance 50 Ω and an inductance 0.40 H. An alternating current passed through the coil has an r.m.s. value 1.5 A and pulsatance 300 s^{-1}. An a.c. voltmeter is connected across its terminals. What does it read?

[0.20 kV]

*58–37 When a p.d. of pulsatance 300 s^{-1} was applied to a resistive inductor the impedance was measured to be 10 Ω. When the pulsatance was increased to 750 s^{-1} the impedance increased to only 17 Ω. Calculate the resistance and the inductance of the device.

[8.0 Ω, 20 mH]

*58–38 *The RLC series circuit.* A resistance 80 Ω, capacitance 25 μF and inductance 0.10 H are connected in series to a 200 V supply of pulsatance 400 s^{-1}. Find

(a) the values of X_C and X_L
(b) the total circuit impedance
(c) the r.m.s. current
(d) the phase shift between the instantaneous values of the current and applied p.d.
(e) the power factor
(f) the average rate of transfer of energy from the source to the circuit.

[(b) 0.10 kΩ (c) 2.0 A (d) arctan 0.75 (e) 0.80 (f) 0.32 kW]

*58–39 A 255 V 50 Hz source is connected to an *RLC* series circuit in which $R = 150$ Ω, $X_L = 160$ Ω and $X_C = 80$ Ω.

(a) Calculate the total impedance of the circuit, and hence the r.m.s. current.
(b) What is the r.m.s. p.d. across each of the circuit elements considered separately if they can be treated as being ideal? What is the sum of these r.m.s. values? (*How does it compare with* 255 V?)
(c) What would be the reading of an a.c. voltmeter which has been calibrated for r.m.s. values if it is connected across (i) the resistor alone (ii) the capacitor and the inductor together (iii) all three elements at the same time?

[(a) 170 Ω, 1.5 A   (b) 225 V, 240 V, 120 V
(c) (i) 225 V, (ii) 120 V]

*58–40 The secondary of an ideal transformer is applied to a circuit of power factor 0.80, and to which it delivers a (real) power of 0.20 kW. The transformer has a turns ratio of 10 : 1, and is supplied by a primary applied p.d. of r.m.s. value 200 V. Calculate the currents in the secondary and primary.

[12.(5) A, 1.0 A]

*58–41 *A parallel RLC circuit.* Refer to the diagram. *R* is a purely resistive fluorescent lamp rated at 0.40 kW which is designed to take a current 2.0 A.

(a) What is the resistance of the lamp?
(b) An inductance *L* is now placed in series with the lamp to enable it to be energized by the 240 V a.c. mains. What value should *L* have?

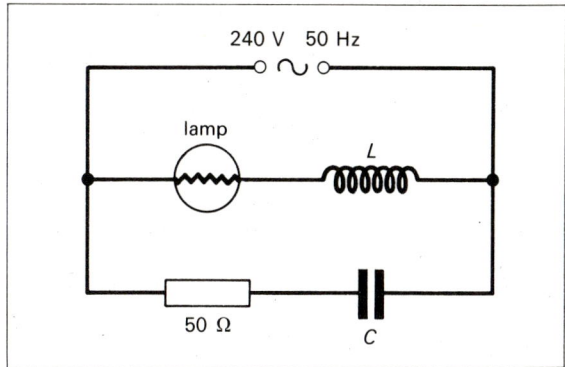

For Qu. **58–41**

(c) A capacitor and resistor are then inserted as shown so that their current lead compensates for the lag introduced by the inductor. Find the value of *C* for this resonant circuit.

(*The whole device now has a power factor of* 1, *since the p.d. applied to it is in phase with the current taken from the supply.*)

[(a) 0.10 kΩ   (b) 0.21 H   (c) 96 μF]

Oscillations and Resonance

*58–42 An ideal inductor of inductance 0.20 H is connected across a capacitance of 0.80 μF which has been charged by a 200 V battery. What is the maximum current through the inductor during the subsequent oscillation?

[0.40 A]

*58–43 An alternating p.d. of 0.10 kV is applied to an *RLC* series circuit of resistance 10 Ω. What p.d. will be developed across the inductor and the capacitor at the resonant frequency, when $X_L = X_C = 0.10$ kΩ?

(*It may be necessary to take precautions to avoid potentially dangerous p.d.'s of this size. On the other hand the tuning of a radio circuit aims at building up a comparatively large response from a small input signal.*)

[1.0 kV]

*58–44 A particular coil has a resistance of 4.0 Ω, and a self-inductance of 0.20 mH. Calculate its *Q*-factor (Qu. **58–22**) at a pulsatance of 0.60 MHz.

[30]

*58–45 A wireless is retuned from *Radio* 4 (0.91 MHz) to *Radio* 3 (0.65 MHz) using a variable capacitor. By what factor is the capacitance changed?

[2.0]

*58–46   *The LC oscillatory circuit.* (*a*) Consider a capacitance C of charge Q_0 at the instant when it is first connected across an inductance L. What is (*i*) the energy associated with the system (*ii*) the instantaneous current?

(*b*) After a time $T/4$ the capacitor is completely discharged. (*i*) What has become of the system's electrical p.e.? Write, in terms of L, an expression for its new form. (*ii*) What, at this instant, prevents the current from becoming zero? (*What is the new source of e.m.f.?*)

(*c*) By considering the analogy of situations (*a*) and (*b*) with: (*a*) a stationary mass held at the end of a stretched spring, and (*b*) the released system at the instant when the spring is at its natural length, write down (*i*) electromagnetic quantities analogous to displacement x, velocity $\dot{x}$, acceleration $\ddot{x}$, mass m and spring constant k, and (*ii*) expressions for quantities analogous to elastic p.e. and k.e.

(*d*) Deduce the natural frequency of oscillation of the circuit.

$$[f = 1/2\pi \sqrt{LC}]$$

59   Magnetic Properties of Matter

Questions for Discussion

Macroscopic Ideas

59–1   To what extent are *Amperian* surface currents real?

59–2   (*a*) Sketch the magnetic flux pattern associated with a short current-carrying solenoid. Now imagine a closed surface to be drawn round the whole of one end of the solenoid. What is the net flux through the closed surface?

(*b*) Repeat part (*a*) for a permanent magnetic dipole. (*Don't forget to include the B-field inside the magnet.*)

(*c*) Try to formulate a statement for electromagnetism which is equivalent to *Gauss's Law* in electrostatics: $\varepsilon_0 \psi_E = \Sigma Q$.

59–3   Two permanent magnets are held close together with their electromagnetic moments parallel, and then released. They acquire kinetic energy. What is the origin of this k.e.?

59–4   Sketch a typical B–B_0 hysteresis curve for a permanent magnetic material. How does your B-axis scale compare with that for the B_0-axis?

59–5   Why can the points on a hysteresis loop *not* be plotted in a random order (as can be, for example, those of the output characteristics of a transistor)?

59–6   In the attached table B_r is a material's remanence, and B_c/μ_0 its coercive force.

Material	B_r/T	$\dfrac{B_c}{\mu_0}$/A m^{-1}
silicon-iron	0.8	30
Vicalloy 11	1.0	3.6×10^4

Which material would you choose for a permanent magnet, and which for a transformer core?

59–7   According to *Lenz's Law* all materials react in an attempt to reduce the effects inside themselves of an external B-field. How would a superconductor respond to a small field? (*Think of an analogous situation in electrostatics.*)

59–8   A moving charged particle entering a magnetic field describes a path which is (in general) a helix. How does the direction of the B-field *caused by this movement* relate to the direction of the original field? What is the relevance of your answer to the magnetic behaviour of materials?

59–9 Why are diamagnetic effects independent of temperature?

59–10 (a) A diamagnetic substance gains energy when placed in an external magnetic field. What is the origin of this energy?
(b) A paramagnetic substance loses energy. If the field is non-uniform then the substance will move. Which way, and why?

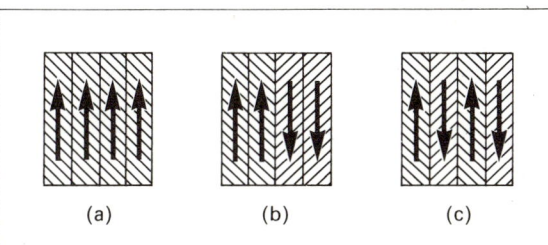

For Qu. **59–11**

59–11 Refer to the diagram, in which are shown three different ways of arranging four magnets. Draw the field lines associated with these arrangements, and by considering the energy associated with each suggest why it is that a ferromagnetic material consists of many small domains rather than a single large one.

59–12 All ferromagnetic minerals become purely paramagnetic at a depth of about 20 km below the Earth's surface. Suggest a reason for this.

*59–13 Describe the external magnetic field pattern created by a negative uniformly charged sphere rotating about a diameter. How are the sense and direction of its electromagnetic moment related to those of its angular momentum? How would such a sphere respond if placed so that its axis made a non-zero angle with the direction of a uniform **B**-field? (*Hint: how does one steer a moving bicycle?*)

Microscopic Ideas

59–14 In what way (if any) are the magnetic properties of an isolated ion of iron different from those of any other isolated ion of a paramagnetic material, such as manganese? How can one account for the special (ferro-) magnetic properties of iron *in bulk*?

59–15 An orbiting electron has an electromagnetic moment. An ion, such as Cu^+, has many orbital electrons: why is its electromagnetic moment usually zero?

59–16 A permanent magnet is brought close to a soft iron nail. Describe the behaviour of the nail's domains during this process, and their external effects. What will they do when the magnet is removed?

59–17 Suggest a simple microscopic connection (in terms of domain behaviour) between (a) the *existence* of permanent magnetism, and (b) the fact that energy must be supplied to take a ferromagnetic material through a hysteresis cycle.

59–18 **O-M** If the linear dimensions of a domain are ~ 0.1 mm, how many iron atoms constitute a typical domain?

*59–19 **O-M** Find values for
(a) the average translational energy $\frac{3}{2}kT$ of a room-temperature gas molecule, and
(b) the energy $2\,mB$ needed to reverse the orientation of the electromagnetic moment **m** of such a molecule from parallel to an external **B**-field to antiparallel. Take the size of **B** to be the largest readily available in your laboratory.

Comment on the significance of your answer.

Quantitative Problems

59–20 *An anchor-ring measurement of B.* (a) What is the (constant) size of **B** inside an air-cored toroid of 2000 turns m^{-1} which carries a current of 3.0 A?
(b) A secondary coil is wound round this toroid and connected to a ballistic galvanometer. When the toroid current is switched off the b.g. gives a first throw of 40 mm. The toroid is now rewound with only 200 turns m^{-1}, and supplied with 0.30 A. Calculate the new size of **B**, and predict the throw that would be given on the same b.g. by switching off.
(c) The toroid in its new form is now given an iron core by winding it onto an anchor ring so that its geometry is unchanged. When the 0.30 A current is switched off the throw of the b.g. is 400 mm. What was the change in the size of **B** inside the iron?
Assume μ_0 [(a) 7.5 mT   (b) 0.40 mm   (c) 75 mT]

59–21 From a ballistic galvanometer experiment of the type described in Qu. **59–20** the following results were calculated:
$$B_{air} = 0.25 \text{ mT, and } B_{iron} = 0.20 \text{ T.}$$
For *this value* of the magnetizing current, calculate
(a) the relative permeability μ_r of the iron
(b) its magnetization (electromagnetic moment per unit volume).
Why the italicized words?
Assume μ_0 [(a) 8.0×10^2 (b) 0.16 MA m^{-1}]

59–22 Calculate the relative permeability of
(a) liquid oxygen at 54 K, at which temperature its magnetic susceptibility χ_m is $+4.0 \times 10^{-3}$
(b) graphite, for which $\chi_m = -4.4 \times 10^{-5}$

*59–23   *Paramagnetism.* A particular sample of material contains one mole of atoms, each of which has an electromagnetic moment 1.0×10^{-23} A m^2.

(*a*) What would be the electromagnetic moment of the specimen if *all* these atoms could be aligned in the same direction?

(*b*) What, in practice, prevents an external field from causing anything like complete alignment?

(*c*) *Estimate* a value for the saturation magnetization **M** (electromagnetic moment per unit volume) of this specimen.

Assume N_A [(*a*) 6.0 A m^2]

*59–24   *Demagnetization and coercive force.* The coercive force of the material of a particular magnet is 5 kA m^{-1}. (**Coercivity** $= B_c/\mu_0$.) What current would have to be passed through a long solenoid of 800 turns m^{-1} to demagnetize it?

[6 A]

*59–25   *The bar magnet.* An iron ion has an electromagnetic moment of 1.8×10^{-23} A m^2, and in solid iron the number density of such ions is 1.0×10^{29} m^{-3}. Consider an iron cylinder 100 mm long and of cross-sectional area 100 mm^2. Calculate, for saturation (complete alignment of all the elementary dipoles)

(*a*) its electromagnetic moment **m**

(*b*) its magnetization **M** $(= \textbf{\textit{m}}/V)$

(*c*) the torque needed to hold it with its axis perpendicular to a uniform external **B**-field of 0.10 T

(*d*) its time period of small oscillation in a field of $B = 18$ μT. Take its moment of inertia to be 6.0×10^{-5} kg m^2.

(*Does your order of magnitude in* (*d*) *agree with your experience?* 18 μT *is the horizontal component of the Earth's field.*)

[(*a*) 18 A m^2 (*b*) 1.8 MA m^{-1} (*c*) 1.8 N m   (*d*) 2.7 s]

XI Electronic, Atomic and Nuclear Physics

60 Semiconduction

Semiconduction

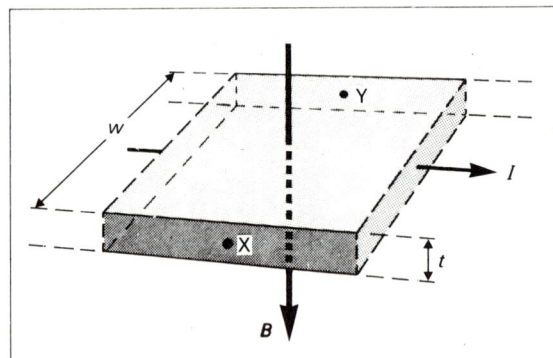

For Qu. 60–1, 60–2, 60–3 and 60–4

60–1 *The Hall effect.* Refer to the diagram, in which the slab of conducting material has a number density n of charge-carriers, each having a charge e.

(a) By considering the magnetic forces on the moving charge-carriers explain why a p.d. is set up between X and Y. Determine the direction of the electric field along the line XY, and hence the polarity of this p.d. for (i) positive, and (ii) negative charge-carriers.

(b) The p.d. between X and Y is called the **Hall p.d.**, and has a size

$$V_H = \frac{BI}{tne}.$$

(i) Why do you think w does *not* appear in this equation?
(ii) What sort of samples (materials and dimensions) are likely to produce measurable p.d.'s?

*60–2 *Derivation of the Hall p.d.* Refer to the diagram. In an experiment using copper a particular wire had $w = 2.0$ mm, $t = 0.50$ mm, and carried a current 16 A. The value of B was 0.10 T. Calculate

(a) the electron drift velocity,
(b) the transverse **magnetic** force F_m on each electron,
(c) the value of the electric field E between X and Y which exactly opposes F_m, and hence
(d) the p.d. Ew between X and Y which results from the *lateral* movement of charge.

Take $n_e = 1.0 \times 10^{29}$ m^{-3}. You can check your answer to (d) by using the expression of Qu. **60–1**.

Assume e [(a) 1.0 mm s^{-1} (b) 1.6×10^{-23} N
(c) 0.10 mN C^{-1} (d) 0.20 μV]

*60–3 *Available conduction electrons per atom.* Refer to the diagram. In a particular experiment with copper the following measurements were made: $B = 1.6$ T, $I = 50$ A, $t = 0.50$ mm and $V_H = 9.1$ μV.

(a) Calculate the number density n_e of conduction *electrons* in copper.
(b) Different considerations suggest that the number density n_a of *atoms* in copper is 8.4×10^{28} m^{-3}. How many conduction electrons are provided on average by each atom in the lattice?

Assume e [(a) 1.1×10^{29} m^{-3} (b) 1.3]

*60–4 *Hall effect for a semiconductor.* Refer to the diagram. In an experiment with a sample wafer of n-type germanium of thickness 0.40 mm and number density of conduction electrons 6.0×10^{20} m^{-3}, the size of the transverse B-field was 0.16 T. What current would have to be passed through the specimen to obtain a *Hall* p.d. of 10 mV? (*Compare these orders of magnitude most carefully with those of Qu. **60–2**. What is the principal cause of the different Hall p.d.'s?*)
Assume e [2.4 mA]

60–5 Discuss the possibility of using the *Hall effect* to measure the size of B in a magnetic field. (*Such a device is called a* **Hall probe** *and can be made very small.*)

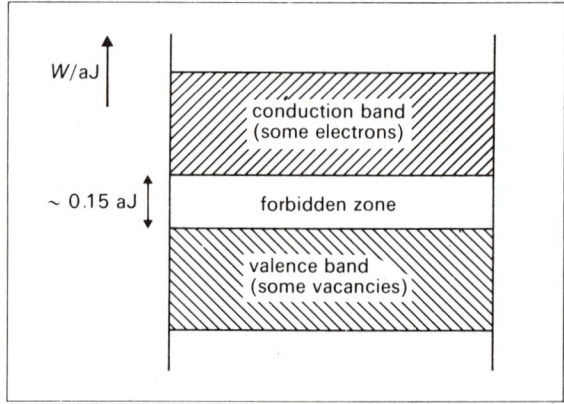

*For Qu. **60–6***

60–6 *Energy band diagrams.* Refer to the diagram which is for a semiconductor at room temperature. When the electrons are found in the energy states available in the **valence band**, the crystal is in the state of lowest possible energy (the **ground state**). The **conduction band** contains the next energy levels in which the electrons are effectively free to move within the lattice. The **forbidden zone** is the energy gap between these two bands. Draw corresponding energy band diagrams for (a) a metal and (b) an insulator. How will the introduction of donor and acceptor impurities affect the forbidden zone of a semiconductor?

60–7 Discuss the electronic configurations of impurity materials suitable for making pure tetravalent silicon into (a) a p-type extrinsic material, and (b) an n-type material. (*In the pure crystal neighbouring atoms can be considered to distribute their electrons in such a way that any one nucleus has a share of 8 orbital electrons.*)

60–8 Suppose a single trivalent atom replaces a tetravalent atom in a crystal. Where will the resulting positive hole be found?

60–9 Does a slab of n-type material carry a *net* negative charge?

60–10 Of the charged particles in a semiconducting crystal, only the electrons are free to move through it. Suggest a mechanism by which the movement of so-called positive charge-carriers (*positive holes*) can take place in such a crystal.

60–11 When a current flows through a p-type material, positive holes move toward the negative terminal of the battery, and combine with electrons at the ohmic electrode connected to the boundary of the crystal. Why does the crystal not become negatively charged?

60–12 Do you think that it is possible, under any circumstances, to obtain perfectly insulating silicon?

60–13 Why is silicon often preferred to germanium for making semiconductor devices?

60–14 **O-M** A sample of very pure germanium has 1 impurity atom to 10^9 of germanium. Estimate the distance between impurity atoms.

The Diode

60–15 The diffusion or *contact potential* across the depletion layer of a p-n junction of thickness 1.5 μm is 0.40 V, and this prevents further diffusion of majority carriers. Assume that the electric field E in the layer is constant, and compare its value with that of a typical parallel-plate capacitor.

60–16 Discuss in terms of both potential barriers and effective resistances the result of connecting a battery across a junction diode using (a) forward, and (b) reverse bias.

60–17 Why is it that when n- and p-type materials are joined as they are in the junction diode, the electrons and positive holes do not *all* recombine, and thus remove the possibility of conduction?

60–18 Why should a resistor be put in series with a diode when its characteristic is being determined?

60–19 What is the cause of the leakage current in a reverse biased p-n junction? Why is this current greater for a germanium diode than for a silicon diode?

60–20 When the maximum peak inverse voltage of a diode is exceeded the reverse current is comparatively large, and almost independent of the applied p.d. Suggest a use for this effect (the **avalanche effect**).

60–21 Discuss on a microscopic level the physical causes of breakdown in a junction diode.

60–22 Why is it not possible to design an extremely small semiconducting diode for power rectification?

60–23 *The varactor diode.* In what respect does a junction diode behave like a capacitor? Which region behaves as the dielectric? How can the capacitance of the device be varied continuously by simple means?

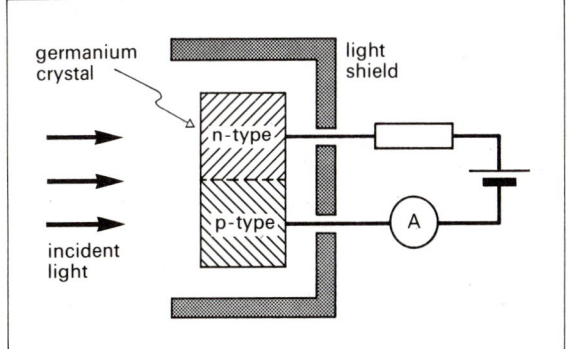

For Qu. **60–24**

60–24 *The p-n junction photodiode.* Refer to the diagram.

(a) In which direction is the p-n junction biased? What, in the absence of incident light, would be the order of magnitude of the current?

(b) What effect does light have on the junction, and hence on the ammeter reading?

(c) How would you adapt this effect for switching?

(*This device is the semiconductor equivalent to the photoelectric cell. In practice the **phototransistor** is often preferred because it can act as its own amplifier.*)

60–25 *The solid state detector.* When ionizing radiation passes near the junction of a reverse-biased diode, a small pulse of current passes through the detector circuit and may be counted by a scaler. Explain why this happens and how particles of different ionizing energy can be distinguished.

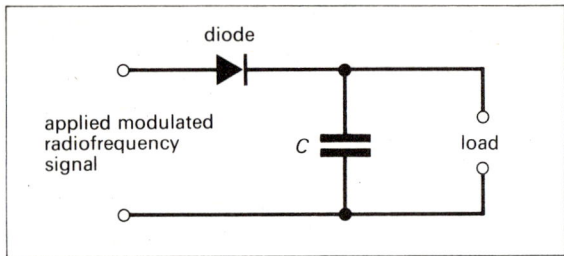

For Qu. **60–26**

60–26   **The demodulator. Refer to the diagram. C has a low reactance to radiofrequency signals. Sketch the following waveforms: that of*

    (a) an applied radiofrequency (carrier) signal whose amplitude has been modulated by an audiofrequency signal (*hint: think of a beat waveform*),
    (b) the signal passing through the diode, and
    (c) the signal applied to the load.

The Transistor

60–27   What is the origin of the word *transistor*?

60–28   In the p-n-p transistor some of the positive holes from the emitter combine with electrons in the base. How, usually, are these electrons replaced? What happens if they are not?

60–29   In the p-n-p transistor most positive holes passing from the p-type emitter into the base fail to combine with the electrons of the n-type base. Suggest two reasons for this.

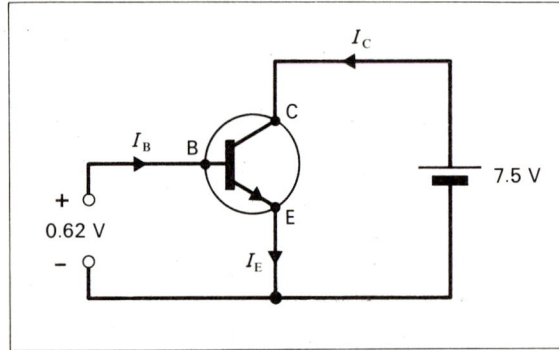

For Qu. **60–30**

60–30   *Action of the silicon n-p-n transistor.* Refer to the diagram.

    (a) Which of the two junctions (E-B and B-C) is forward biased, and which reverse?

(b) Suppose that $I_E = 2.00$ mA, and that 0.50% of the electrons diffusing into the base combine there with holes. Then what would be the values of I_B and I_C?
(c) If the p.d. across the emitter-base junction is increased to 0.64 V then the new value of $I_E = 4.00$ mA. What are the new values of I_B and I_C?
(d) Calculate the value of the ratio ($\Delta I_C / \Delta I_B$), the current gain or amplification. (*This quantity is called the **small signal forward current transfer ratio**, symbol h_{fe}.*)

    [(c) 20 μA, 3.98 mA   (d) 199]

60–31   Give an explanation in terms of the behaviour of charge-carriers as to why the transfer characteristic is a straight line.

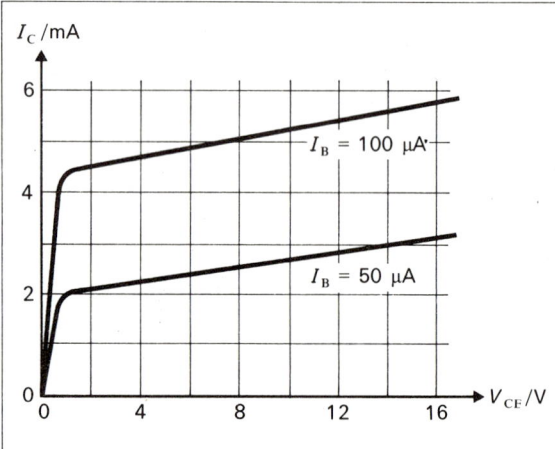

For Qu. **60–32**

60–32   *Output characteristic.* Refer to the diagram which shows how the collector current varies with the collector-emitter p.d. for two different values of the base current.

    (a) Beyond the sharp change of slope the output resistance of the transistor is defined as ($\Delta V_{CE} / \Delta I_C$) for a fixed I_B. Measure the average value of the output resistance from the two slopes on the graph.
    (b) The small signal forward current transfer ratio is defined as ($\Delta I_C / \Delta I_B$) for a fixed V_{CE}. Measure the average value of this ratio from the graph.

    [(b) 51]

60–33   *Voltage amplification.* Refer to the diagram. When V_{BE} is changed from 0.62 V to 0.64 V, then I_C changes from 2.0 mA to 4.0 mA.

    (a) What is the change in the p.d. V_R across the resistor?
    (b) What is the change in the p.d. V_{XY} between the output terminals X and Y?
    (c) Calculate the *voltage gain* ($\Delta V_{XY} / \Delta V_{BE}$).

What is the phase relationship between ΔV_{BE} and ΔV_{XY}?

    [(c) 100]

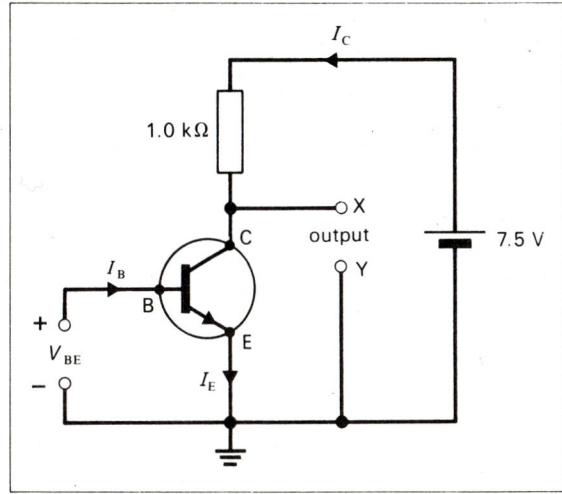

For Qu. **60–33**, **60–35** and **60–36**

*60–34 Why is it necessary to use an input source of high resistance if undistorted (linear) amplification is required from a transistor with a common-emitter circuit?

60–35 *Power amplification.* Using the information given in Qu. **60–33**, together with a value for h_{fe} of 200, calculate
  (a) the change in the input power,
  (b) the change in the power developed by the resistor, and
  (c) the ratio [(b)/(a)].
What is the source of power that enables this amplification to be produced?

[(a) 6.6 μW   (b) 12 mW   (c) 1.8×10^3]

60–36 *Alternating current amplification.* Refer to the diagram. Suppose that an input signal causes V_{BE} to oscillate sinusoidally about a mean value 0.63 V, and with an amplitude 10 mV.
  (a) Draw the graph showing how the potential of X varies with time.
  (b) What would be the effect on V_X of inserting a **blocking capacitor** between X and the collector? Draw a further graph to illustrate your answer.

*60–37 *The multivibrator.* This oscillator is used to generate many different harmonics, the frequencies of which are controlled by the discharge times of two capacitors. Find out a suitable circuit and explain how the waveforms are produced. How can it generate square-waves?

61 The Free Electron

The Electron and Thermionic Emission

61–1 What is an electron? Why do you believe that it exists?

61–2 What evidence is there that the electron is common to *all* matter?

61–3 What is the evidence for believing that cathode rays are *not* electromagnetic waves?

61–4 In an attempt to demonstrate the deflection of cathode rays by an electrostatic field *Hertz* used a partly evacuated tube, but the experiment was unsuccessful. Suggest reasons for this.

61–5 What conclusions can be drawn from the following *group* of observations?
  (a) At standard pressure the mean free path of gas molecules is about 0.1 μm ($\sim 10^3$ molecular diameters).

  (b) At standard pressure cathode rays (accelerated through a p.d. ~ 10 kV) can pass through about 10 mm of air, and still cause a screen to fluoresce.
  (c) e/m_e for cathode rays is about 2×10^3 times the specific charge of the hydrogen ion.

61–6 It is not possible to evacuate glass bulbs to *zero* pressure. Suggest, in terms of electron mean free path, a criterion which could be adopted for the evacuation of a thermionic vacuum tube.

61–7 Propose a simple model for the behaviour of an electron escaping from a metal surface. Use it to predict the shape of graphs plotting (a) the restraining force on the escaping electron, and (b) the p.e. which it has gained, both as functions of its distance r from the metal. Relate sketch (b) to the work function energy.

61–8 Explain why power has to be supplied to a heated filament if it is to continue to give thermionic emission. Discuss the relative importance of the factors you mention.

61–9 Calculate the mass of the electron from the experimental values of N_A, e/m_e and F.

The Specific Charge of the Electron

61–10 Suppose the following are provided: (*i*) known variable magnetic fields *B*, (*ii*) a source of variable known p.d. *V* for increasing the speed of charged particles, and (*iii*) known variable electric fields *E*. Devise *three* ways of causing charged particles to undergo movements from which you could calculate their specific charge *q*.

61–11 What speed do electrons acquire if they are accelerated from rest through a p.d. of 1.0 kV? Does it make any difference if the electric field is non-uniform? Describe qualitatively what would happen with an accelerating p.d. of 400 kV.

Assume e/m_e [19 Mm s⁻¹]

61–12 *Helmholtz coil system.* The magnetic deflection of an e/m_e tube is to be established using a *Helmholtz system* in which each coil has 125 turns of mean radius 75 mm. What magnetizing current will be needed to produce a (reasonably uniform) field of 0.50 mT?

Assume μ_0 [0.33 A]

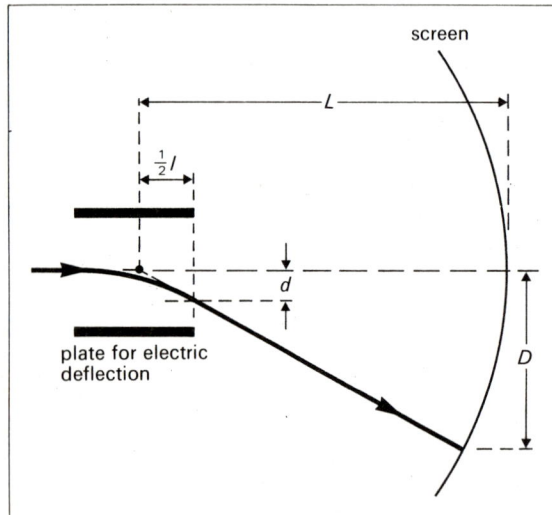

For Qu. **61–13**

61–13 *The geometry of magnetic deflection.* Refer to the diagram. For a particular e/m_e tube a magnetic deflection experiment gave the following measurements: $D = 25$ mm, $l = 40$ mm and $L = 0.20$ m. Calculate
(*a*) the deflection *d*, and so
(*b*) the radius *r* of the circular path.

[(*a*) 2.5 mm   (*b*) 0.32 m]

61–14 *Measurement of electron speed.* In a *Thomson* experiment electrons of speed *v* were first deflected by a magnetic field of 0.50 mT, and were then restored to the undeflected position by an electric field of 12 kV m⁻¹.
(*a*) Write down, in terms of *e* and *v* (*i*) the magnetic force F_m and (*ii*) the electric force F_e.
(*b*) Put $F_m = - F_e$ to calculate the value of *v*.

[(*b*) 24 Mm s⁻¹]

61–15 The following results were obtained from a *Thomson* thermionic e/m_e tube. The size of the deflecting *B*-field from the *Helmholtz* coils was 0.40 mT, and this caused the electrons to be deflected into a circular path of radius 0.25 m. When an electric field of 7.2 kV m⁻¹ was applied, the spot on the screen was restored to its undeflected position. Calculate
(*a*) the electrons' speed, and hence
(*b*) their specific charge.

[(*b*) 0.18 TC kg⁻¹]

61–16 What were the sources of uncertainty in the *Thomson* e/m_e experiment? To what extent were they insuperable using his design of apparatus?

61–17 In a *Thomson* e/m_e experiment does it matter whether the measured deflection is produced by magnetic or electric forces?

61–18 In a *Thomson* e/m_e experiment it is found that when the beam undergoes *magnetic* deflection the previously well-defined spot becomes considerably blurred. What could be the cause?

61–19 In the thermionic version of the *Thomson* e/m_e tube do all the electrons travel at the same speed?

61–20 *Dunnington's method.* Refer to the diagram. The time of flight *T* of electrons describing part of a circular arc of radius *r* is deduced from the frequency *f* of an alternating potential. The circular path is caused by a constant field *B*.
(*a*) Write down the angular speed ω of the electrons in terms of *B* and e/m_e.
(*b*) What, in terms of θ, is the time of flight for the arc shown?
(*c*) The arrangement was such that a collector showed minimum response when $f = 1/T$. Relate f, θ, *B* and e/m_e.

(*This method, similar in principle to that of the cyclotron, was used by Dunnington in 1933. It is more satisfactory than Thomson's, since f, θ and B can all be measured with great accuracy.*)

[(*c*) $e/m_e = f\theta/B$]

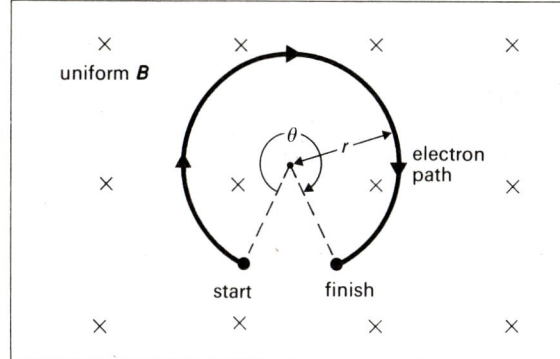

For Qu. **61–20**

61–21 *The magnetron effect.* (*See also Qu.* **45–28**.) Electrons are accelerated radially from a fine wire cathode to a coaxial cylindrical anode of radius 3.0 mm. They are then subjected to an axial magnetic field of 22 mT which just prevents them from reaching the anode. Calculate

(*a*) (in terms of e/m_e) the speed acquired by the electrons during their acceleration from the cathode, and

(*b*) a value for e/m_e, given that their speed is 5.9 Mm s⁻¹.

For (*b*) you may assume that the electrons reach their final speed just after leaving the cathode. (*This is the basis of **Hull's method** for measuring e/m_e.)

[0.18 TC kg⁻¹]

61–22 O-M What errors are introduced in a *Thomson* experiment by neglecting the deflection produced by (*a*) gravitational forces, and (*b*) magnetic forces, exerted by the Earth?

The Millikan Measurement of *e*

61–23 The most precise determinations of *e* come from indirect average measurements (such as $e = F/N_A$). What then is the *true* significance of the *Millikan* experiment?

61–24 Is it possible to demonstrate conclusively that there are not two elementary quanta of charge, a positive one of $\Delta Q_p = +2e$, and a negative one of $\Delta Q_n = -e$?

61–25 Why, in the *Millikan* oil-drop experiment, does one use the same type of oil as is used in vacuum apparatus?

61–26 Are *Millikan* oil drops affected by *Brownian* motion?

61–27 If, in an oil-drop experiment, you were to calculate the following values for the charge carried by a drop, what conclusion would you be entitled to draw about the elementary charge?

Q/aC	0.32	0.96	0.32	0.64	1.28

Is your conclusion consistent with (if not identical to) modern thinking? (*Think of analogous situations with oil film thicknesses and steps in cleaved mica crystals.*)

†**61–28** An oil drop of weight 20 fN acquires two surplus electrons. What p.d. should be applied between a pair of horizontal parallel metallic plates 10 mm apart to hold it in equilibrium? Which plate should be positive?

Assume e [0.62 kV]

61–29 The following questions refer to an oil drop which is observed between a pair of parallel metal plates.

(*a*) When the electric field is 0.30 MV m⁻¹, and the drop has one excess electron, then it remains stationary. What is the weight of the drop?

(*b*) The same drop suddenly starts to fall at a measured terminal speed of 0.20 mm s⁻¹, a speed which remains the same whether or not a p.d. is applied between the plates. Why? If the viscous resistive force is written $F_v = -kv$, then what is the value of *k*?

(*c*) What would be the terminal velocity of this drop if it acquired 3 excess electrons?

Assume e [(*a*) 48 fN (*b*) 0.24 nN s m⁻¹ (*c*) 0.40 mm s⁻¹↑]

61–30 O-M If an oil drop has a radius 1.5 μm and carries a charge $+e$, by what percentage do (*a*) the electric, and (*b*) the gravitational forces on it change when it captures a positively charged air ion?

61–31 O-M How far does an oil drop fall under gravity before it acquires 99% of its terminal velocity? What is the corresponding time interval?

The Cathode Ray Oscilloscope

(*For electron ballistics in the c.r.o. see also Qu.* **43–18**.)

61–32 How is the electron beam in a c.r.o. focused?

61–33 What is the function of a mumetal screen in a c.r.o.?

61–34 Why is it usual to connect one of each pair of deflecting plates to the anode of a c.r.o., and to earth the anode itself?

61–35 What is the purpose of *synchronization* in the c.r.o., and how is it achieved?

61–36 Explain in mechanical terms (using forces and kinematics) why the deflection of a c.r.o. spot is roughly proportional to the p.d. applied across the deflecting plates.

61–37 A signal of amplitude 5 V is to be applied to a c.r.o. whose Y-plates need a p.d. of 100 V to produce f.s.d. How can a useful deflection be achieved?

61–38   A p.d. which considerably exceeds that required for f.s.d. is applied to (*a*) a moving-coil galvanometer, and (*b*) a c.r.o. Compare their responses.

61–39   A linear c.r.o. time-base has a frequency 50 Hz. Sketch the pattern seen on its screen when the alternating p.d. applied to the Y-plates has a frequency (*a*) 25 Hz (*b*) 100 Hz.

61–40   How would you calibrate an audiofrequency oscillator using a c.r.o. and a known frequency of exactly 100 Hz? Assume both waveforms to be sinusoidal.

61–41   Consider the following series a.c. circuits: (*a*) *L* and *R*, (*b*) *C* and *R*, and (*c*) *L* and *C*. What, in each case, would be seen on the screen of a c.r.o. if the p.d.'s across the respective devices were connected to the X- and Y-plates? Assume that the peak values are all equal.

61–42   Draw the circuit diagram of an arrangement by which a c.r.o. can be used to display the current – applied p.d. characteristic of a diode rectifier. How could you display half-wave and full-wave rectification and their smoothed outputs?

61–43   What connections and adjustments would you make to a c.r.o. to observe the waveform of a musical note whose fundamental is thought to have a frequency 210 Hz?

†61–44   The Y-plates of a c.r.o. have a sensitivity 10 V mm^{-1}. What is the r.m.s. p.d. of an a.v. source which causes a peak-to-peak deflection of 71 mm?

[0.25 kV]

†61–45   *Measurement of time.* The time-base of an oscilloscope has a frequency 10 kHz, and gives the spot a horizontal amplitude 0.10 m. A radar pulse and its echo from a distant object are applied to the Y-plates, and give peaks whose centres are separated by 40 mm. How far away is the object?
Assume c   [6.0 km]

61–46   The current through a c.r.o. is 0.16 mA, and the electrons are accelerated through a p.d. 1.0 kV. Calculate

(*a*) the number of electrons that strike the screen during each second,
(*b*) the momentum of each, and
(*c*) the average force experienced by the screen from the electrons.

Assume e, m$_e$
   [(*a*) 1.0×10^{15} s^{-1} (*b*) 1.7×10^{-23} N s   (*c*) 17 nN]

*61–47   Calculate, in V mm^{-1}, the sensitivity of the Y-plates of a c.r.o. of the following specification: the cathode-anode p.d. is 1.2 kV, the deflector plates are 40 mm long and 8.0 mm apart, and their centre is 200 mm from the screen. The specific charge of the electron disappears from the final equation. Try to explain the *physical* reason for this.

[2.4 V·mm^{-1}]

*61–48   What part is played by the c.r. tube in the display of television pictures?

*61–49   **O-M** How long does an electron spend in transit between the plates of a typical c.r.o.? What is the maximum value that the frequency of the signal applied to the X-plates can have if the field between the plates is not to change significantly during this interval? For what frequency would the sensitivity become zero?

62   The Photoelectric Effect

Questions for Discussion

62–1   *An analogy.* Ball-bearings of mass 0.2 kg are resting in hemispherical depressions on a table top. What would you expect to happen if you threw at them (*a*) an intense rain of ping-pong balls (*b*) glass marbles (*c*) a small number of 0.2 kg ball-bearings? (*Consider both momentum and energy.*)

62–2   What is the relationship between the maximum speed of photoelectrons from a metal surface, and the intensity of the electromagnetic radiation incident on it? What, if any, would be the effect of using a contaminated surface?

62–3   Can one distinguish *photoelectrons*, after emission, from *thermoelectrons*?

62–4   Why is it very accurate to say that *all* the energy of an absorbed photon is given to an emitted photoelectron? Why, for example, does the lattice take a negligible proportion?

62–5   What are the wave and photon explanations for the fact that the rate of emission of photoelectrons is proportional to the light intensity? Are they equally plausible?

62–6   It is a fact that photoelectrons are not emitted at all if the electromagnetic wave frequency is less than the threshold value, *however long* the surface is irradiated. Can you explain this, using the wave theory?

62–7   Why is it that a tube used to investigate photoelectric emission is

(*a*) usually fitted with a window made of vitreous silica rather than glass, and
(*b*) evacuated?

62–8   Distinguish carefully between *work function* potential, and *stopping* potential. Which is associated with particular experimental conditions, and which with a particular metal species?

62–9   *Einstein* explained the photoelectric effect using the *Planck* quantum theory. In what way did he extend *Planck*'s proposals?

62–10   Suggest reasons why most photoelectrons have a smaller k.e. than the maximum predicted by the *Einstein* theory.

62–11   Sketch graphs of the *current I* (*y*-axis) in a circuit used for investigating the photoelectric effect against (*x*-axis)

(*a*) the *time t* of exposure of the metal surface (with V and I' constant),
(*b*) the light *intensity I'* (V constant)
(*c*) the retarding *potential V* (I' constant).

Suppose the same value of v to be used throughout.

62–12   Sketch, using the same axes, graphs of *stopping potential* V_s (*y*-axis) against *frequency* v of incident light (*x*-axis) for three metals of different work function energies. What effect would an increase of light intensity have on your sketches? Indicate on your sketches the *work function potential*.

62–13   Discuss the mechanism of the **atomic photoelectric effect** by which X-rays (for example) can ionize gases.

62–14   Draw circuits to help explain how a photocell can be used in the following situations: (*a*) to raise a car park barrier, (*b*) to switch on street lighting, and (*c*) to operate a photographic exposure meter.

62–15   O-M   Suggest what minimum wavelength of electromagnetic radiation could be used to ionize a hydrogen atom. Explain your assumptions carefully.

*****62–16   O-M**   In the **photonuclear effect** a γ-ray may eject a nucleon from an atomic nucleus. Suggest a value for the threshold frequency.

*****62–17   O-M**   What is the speed of a typical ejected photoelectron (for which the stopping potential is a few volts)? Would it be necessary to use relativistic mechanics to verify the *Einstein* photoelectric equation with an uncertainty of 1 %?

Quantitative Problems

†62–18   Blue light of wavelength 460 nm irradiates a clean calcium surface which has a photoelectric work function energy of 0.43 aJ. Does photoelectric emission occur?

Assume h, c

†62–19   A monochromatic light source provides 25 W for light of wavelength 450 nm, and this liberates 3.2×10^{11} photoelectrons per second from a fresh potassium surface. The cut-off wavelength for potassium is 550 nm. Calculate the size of the photoelectric emission current from potassium given by

(*a*) this arrangement
(*b*) an otherwise identical one with a 50 W source
(*c*) a 100 W source operating at 600 nm.

Assume e   [(*a*) 51 nA]

†62–20   A lithium surface for which the work function energy $W = 0.37$ aJ is irradiated with light of frequency 6.3×10^{14} Hz. Loss of photoelectrons causes the metal to acquire a positive potential. What will this potential have become by the time its value prevents further loss of electrons from the surface?

Assume h, e   [0.3 V]

62–21   *The time-factor failure of the wave theory.* The irradiance of a particular zinc surface by a monochromatic u.v. source is 0.30 mW m^{-2}. (*The nature of the quantity irradiance is indicated by its unit.*) Suppose that according to the wave theory a typical surface electron can absorb energy from an area of about 1.0×10^{-18} m^2 (from about 100 neighbouring ions).

(*a*) How long would such an electron take to absorb an amount of energy equal to its work function energy 0.68 aJ?
(*b*) Experiment indicates no measurable time-lag for photoelectric emission. What conclusion can you draw?

[(*a*) 2.3 ks]

62–22   *The verification of the Einstein photoelectric equation.* In an experiment on photoelectric emission from a cleaned caesium surface the following measurements were obtained:

stopping potential V_s/volts	0.60	1.0	1.4	1.8	2.2
light frequency $v/10^{14}$ Hz	6.0	7.0	8.0	9.0	10

Plot a graph from which you can deduce

(*a*) the threshold frequency v_0 and its corresponding colour, and
(*b*) the value of h.

Assume e   [(*a*) 4.5×10^{14} Hz   (*b*) 6.4×10^{-34} J s]

62–23   The table lists results from a photoelectric emission experiment:

maximum k.e. of ejected electrons	$K_{max}/10^{-20}$ J	13	8.0	4.0
wavelength of irradiating light	λ/nm	400	450	500

Given that
$$K_{max} = hc\,(1/\lambda - 1/\lambda_0),$$
plot a graph from which you can estimate the cut-off wavelength λ_0 and the *Planck constant h*. Guess (using tables) what metal might have been used.

Assume c   [0.56 μm, 6.0×10^{-34} J s]

62–24   Monochromatic light of wavelength 380 nm falls with an intensity 6.0 μW m^{-2} onto a metallic surface whose work function energy is 0.32 aJ. Calculate

(a) the number of photoelectrons emitted per second per mm^2

of surface if a photon has a 1 in 10^3 chance of ejecting an electron

(b) the maximum k.e. of these photoelectrons.

Assume h, c   [(a) 11×10^3 (b) 0.20 aJ]

62–25   Use the value of *h* to draw an accurate graph of *cut-off wavelength* λ_0/nm against *work function energy W*/aJ for values of *W* ranging from 0.1 to 1.0 aJ. Estimate from your graph the cut-off wavelengths of platinum ($W = 1.0$ aJ), silver (0.76 aJ), thorium (0.56 aJ) and caesium (0.30 aJ), and name the corresponding radiations.

Assume h, c

62–26   Electromagnetic radiation of frequency 1.0×10^{15} Hz falls on a cleaned magnesium surface for which the work function energy is 0.59 aJ. Calculate

(a) the maximum k.e. of emitted electrons

(b) the potential to which the magnesium must be raised to prevent their escape (the **stopping potential**)

(c) the threshold frequency, and hence the cut-off wavelength for magnesium.

Assume h, c, e   [(a) 0.07 aJ   (b) 0.4(6) V   (c) 0.34 μm]

63   X-rays and the Atom

X-rays

†63–1   Electrons were accelerated in an X-ray tube through a p.d. of 25 kV. The X-radiation emitted from the target showed a well defined minimum *cut-off wavelength* at 48 pm. Calculate

(a) the maximum energy of the X-ray photons

(b) the frequency of the X-rays

(c) a value for the *Planck constant*.

(*This method is capable of great accuracy, and is one of the best ways of measuring h.*)

Assume e, c   [(c) 6.4×10^{-34} J s]

63–2   In what way can one use man-made reflection diffraction gratings to measure X-ray wavelengths? What is the significance of our being able to do this?

63–3   How may one obtain information about the energies possessed by electrons closest to the atomic nucleus?

63–4   *The significance of Moseley's work.* Consider the following.

(a) Simple application of the ideas of the *Bohr* atom predicts

that a particular X-ray spectral line of any element should have a frequency
$$v_{K_x} = (3cR/4)(Z - 1)^2,$$
where *Z* is the atomic number.

(b) *Moseley's* results (1913) were presented on a graph on which he plotted $\sqrt{(v/R)}$ against an arbitrary number allotted to each element according to its chemical properties, and increasing by 1 from one element to the next. This graph was a straight line of *x*-axis intercept ∼ 1.

What conclusions can be drawn? Can you account for the −1? (*Hint: the inner shell of an atom contains two electrons.*)

*63–5   *Bragg planes.* Refer to the diagram, which shows the arrangement of ions in a cubic crystal (such as NaCl). The dashed lines show the intersections of the planes of ions with that of the paper.

(a) How is the spacing of these planes related to the ionic separation r_0?

(b) Draw, on this diagram, *two* other sets of similar planes. Which of the three is likely to give the most intense diffracted beam?

[(a) $r_0/\sqrt{5}$]

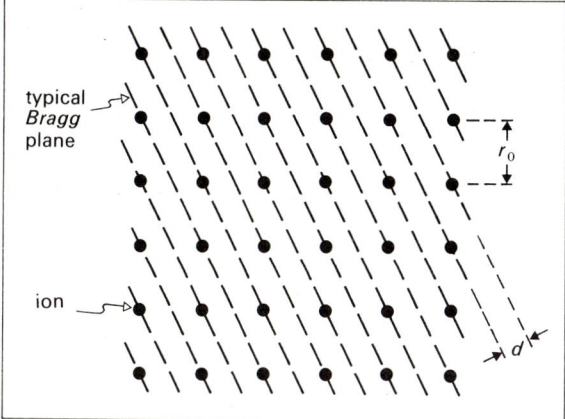

For Qu. **63–5**

*63–6   *Bragg reflection and the Bragg Law.* Refer to the diagram. The incident electromagnetic radiation is scattered (or diffracted) by the electrons in the atoms of the unit cell lying in the *Bragg* planes.

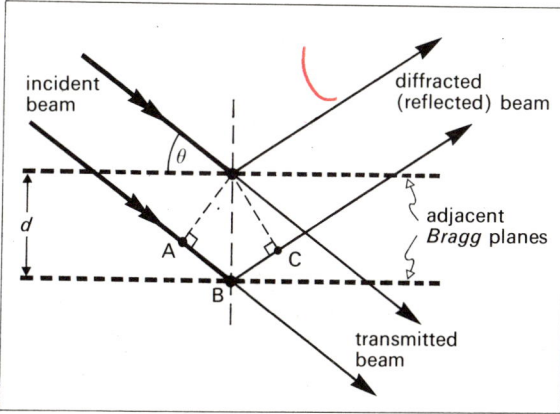

For Qu. **63–6**

(a) Calculate the path difference ABC.
(b) For what value of the **glancing angle** θ will simple theory predict scattered waves to superpose constructively, and hence produce an intense diffracted beam?
(c) Calculate the glancing angle at which NaCl gives an intense second-order beam ($m = 2$) when X-rays of $\lambda = 0.126$ nm are incident on planes for which $d = 0.252$ nm.

[(a) $2d \sin \theta$ (b) $2d \sin \theta = m\lambda, m = 1, 2, 3 \ldots$ (c) $30°$]

63–7   *Measurement of* N_A. The ions in an NaCl crystal are equally spaced at separation 0.281 nm, and the crystal has a density 2.18×10^3 kg m^{-3}. Calculate

(a) the total number of ions in 58.4 g (one mole) of NaCl, and hence
(b) the *Avogadro constant* N_A.

[(a) 1.20×10^{24}]

*63–8   When copper is used as the target of an X-ray tube, an intense monochromatic line is emitted. Its wavelength is measured by obtaining a first order reflection at 14° from an NaCl crystal using *Bragg* planes for which the spacing is 0.28 nm. Calculate

(a) the wavelength of the radiation
(b) the minimum p.d. which can be used between the anode and cathode of the X-ray tube.

Assume e, c, h   [(a) 0.14 nm   (b) 9.1 kV]

*63–9   *Exponential absorption of radiation.* Experiment shows that when monoenergetic electromagnetic radiation passes through an elemental absorbing layer of thickness δx, the change of intensity δI can be written $\delta I = -\mu I \delta x$ where I is the intensity entering that layer, and μ is called the **absorption coefficient**. What is the intensity transmitted by an absorbing plate of thickness x on which the incident radiation has intensity I_0? Can you put a physical interpretation on the quantity $1/\mu$? (*Study its unit.*)

*63–10   The absorption coefficient μ for a particular X-ray frequency from tungsten is 6.9×10^{-2} mm^{-1} when it is absorbed by aluminium, and 4.2 mm^{-1} when the absorber is lead.

(a) By what factor is the intensity reduced by a 4.0 mm thick plate of Al?
(b) What thickness of Pb would halve the intensity?
(*Hint: see the previous question.*)

[(a) 76%   (b) 0.17 mm]

63–11   O–M If X-rays of wavelength 0.1 nm were incident normally on an *optical* diffraction grating, at what angles would one locate the first-order maxima? Could this offer useful information?

The Atom

63–12   Why could *Thomson* not apply the method he used for e/m_e to the measurement of the specific charge of positive ions?

63–13   *Positive ray parabolae.* When *Thomson* used neon in the tube (on its own), he obtained two parabolae, one flatter than the other.

(a) Which species, ^{20}Ne or ^{22}Ne, makes the flatter parabola?
(b) Why does each species form a *semi-parabolic* trace, rather than a *point* trace?
(c) Suggest two *different* interpretations for the fact that a single element gave two traces (based on the information known to *Thomson*).

63–14   Why are molecular and atomic masses compared to that of the ^{12}C atom, rather than (say) ^{16}O?

63–15   What can you deduce from the fact that the relative atomic mass of carbon is usually quoted (correctly) as 12.01?

63–16 *An ionic velocity selector.* A beam of singly ionized chlorine atoms of differing speeds leave an ion source, and enter a sorting chamber through a narrow slit. A transverse **B**-field of 0.20 T is applied to the chamber. What size electric field would enable ions travelling at 0.30 Mm s^{-1} to describe a linear trajectory, and so escape through a defining slit on the opposite side of the chamber? (*Note that this selector is mass-independent.*)

[60 kV m^{-1}]

*63–17 *The Bainbridge mass spectrometer.* The ions of Qu. **63–16** now pass into a deflecting chamber in which a second magnetic field of 0.60 T causes them to describe semicircular paths which finish on a photographic plate. Calculate

(*a*) the radius of the path of the ^{35}Cl ions

(*b*) the linear separation of the traces on the plate given by ^{35}Cl and ^{37}Cl. (*Hint: do not subtract the diameter of one path from the other.*)

Assume $m_u = 1.66 \times 10^{-27}$ kg, e [(*a*) 0.18 m   (*b*) 21 mm]

*63–18 Show that, for a given speed, the *separation* of the traces of singly charged ions on the plate of a *Bainbridge* mass spectrometer depends only upon their mass *difference*, being independent of their masses. What would this separation be for ^{20}Ne and ^{22}Ne used in the instrument of the previous question?

*63–19 *Separation of uranium isotopes.* Protons, and singly charged ions of ^{235}U and ^{238}U are passed in turn through the same velocity selector into the deflecting chamber of a *Bainbridge* mass spectrometer. The protons describe semi-circles of radius 10 mm. What is the *difference* between the radii of the semi-circles described by the uranium ions?

[30 mm]

63–20 Why, in the *Geiger* and *Marsden* experiment, do the α-particles have to travel through a vacuum, although we assume that the gold orbital electrons have a negligible effect on the α-particles?

63–21 *Geiger* and *Marsden* aimed α-particles of mass 6.7×10^{-27} kg and speed 20 Mm s^{-1} at massive gold nuclei for which $Z = 79$.

(*a*) How close could these α-particles approach to the nucleus?

(*b*) How would your answer be modified if an attractive force became operative at small separations?

(*c*) What tentative conclusion can you draw from (*a*), bearing in mind that an atom has a diameter ~ 0.10 nm?

List your assumptions.

Assume $1/4\pi\varepsilon_0$, e [(*a*) 27 fm]

63–22 In the *Geiger* and *Marsden* experiment the distance by which an α-particle would, if undeflected, have missed the centre of the gold nucleus is called the **impact parameter** b. The radius of the gold atom ~ 0.3 nm. What proportion of α-particles incident on an atom will have an impact parameter $\leqslant 30$ fm (an upper limit for the nuclear radius)? (*This is a measure of the probability that an α-particle will be back-scattered by at least a specified angle.*)

[1.0 × 10^{-8}]

63–23 *Atomic models.* For what kind of physical situation would you consider the following atomic models to offer the most satisfactory simple explanation? (*a*) An elastic sphere, (*b*) a sphere with detachable electrons, (*c*) an impenetrable core surrounded by electrons with definite energy levels, and (*d*) as for (*c*), but with the nucleus having a definite structure.

63–24 To what extent can our models of the atom be considered as representing reality?

63–25 Why is the effective radius of a helium atom less than that of a hydrogen atom?

63–26 Why are there roughly half as many electrons in a given mass of any light element (such as carbon) as there are in hydrogen?

63–27 *The Franck and Hertz experiment.* In the original experiment electrons were passed through mercury vapour in a thyratron tube. It was found that the graph plotting the anode current against the p.d. between the grid and the cathode showed a regular periodicity in which all the peaks were spaced at 4.86 V.

(*a*) What is the lowest excitation energy for mercury?

(*b*) What is the wavelength of the radiation emitted by the atoms as they return to their ground state?

(*c*) How would you adapt the apparatus so that this radiation could be detected? (*A thyratron usually has a glass envelope.*)

(*d*) Mercury has another excitation potential at 6.7 V. Why is this not shown by this graph?

(*e*) What is the effect on the k.e. of an electron of an *elastic* collision with a mercury atom?

Assume c, h, e [(*a*) 0.77(8) aJ   (*b*) 255 nm]

63–28 The ionization energies of potassium and neon are 0.7 aJ and 3.5 aJ respectively. What do *these figures* suggest about their chemical activities?

63–29 The molar heat of combustion of coal is about 0.4 MJ mol^{-1}. How much energy is associated with the oxidation of each carbon atom?

Assume N_A [0.7 aJ]

63–30 O-M What would be the mass of an eggshell full of atomic nuclei stripped of their orbital electrons?

63–31 O-M What fraction of its k.e. does an α-particle lose when it makes an elastic collision with a gold nucleus?

The Bohr Atom

Bohr's hypotheses led him to the following value for the total energy E of the hydrogen atom:

$$E = \left(-\frac{m_e e^4}{8\varepsilon_0^2 h^2}\right)\frac{1}{n^2}.$$

$n = 1, 2, 3 \ldots$ *and is called the* **principal quantum number**. *This result will be found useful for the following questions.*

*63-32   What is the physical significance of the minus sign in the equation above?

*63-33   Show, by considering the relation $\Delta E = h\nu$, that the *Planck constant h* has the same unit as angular momentum.

*63-34   Show that if, for an orbital electron in the *Bohr* model of a hydrogen atom, any *one* of its k.e., p.e., speed, rotational frequency, linear momentum, angular momentum or orbital radius is quantized, then *all* must be. (*Bohr merely chose to discuss the angular momentum.*)

*63-35   Calculate the energy change when an electron makes a transition from $n = n_2$ to $n = n_1$ ($n_2 > n_1$). Hence derive a general expression for the frequency of the emitted radiation in the form
$$\nu = cR_H\left(\frac{1}{n_1^2} - \frac{1}{n_2^2}\right).$$
Use known values of m_e, e, c, ε_0 and h to evaluate R_H, the *Rydberg constant*, to two significant figures.

[1.1×10^7 m^{-1}]

*63-36   Combine the equations (*a*) $F = ma$, and (*b*) angular momentum $L = h/2\pi$ (for the ground state) to express the *ground state radius* Q_0 in terms of ε_0, h, m_e and e. Use the known values of these constants to calculate Q_0.   [53 pm]

*63-37   Calculate the *ionization energy* for atomic hydrogen, that is the energy supplied to take an electron from its ground state ($n = 1$) to a position where it is no longer bound to the nucleus ($n = \infty$). Check your answer by comparing it with the value obtained from the measured ionization potential (13.6 V).
Assume m_e, e, ε_0, h [2.2 aJ]

*63-38   It follows from the *Bohr* theory that
$$\frac{1}{\lambda} = (1.10 \times 10^7 \text{ m}^{-1})\left(\frac{1}{n_1^2} - \frac{1}{n_2^2}\right)$$
Use this result to find the wavelengths of the transitions from $n_2 = \infty$ to (*a*) $n_1 = 1$, (*b*) $n_1 = 2$, and (*c*) $n_1 = 3$. Give your answers in μm. These transitions correspond to the **limits** of the **Lyman**, **Balmer** and **Paschen series** respectively. At which end of the series does a limit occur – the long or short wavelength? Name the corresponding part of the electromagnetic spectrum.

[(*a*) 0.091 μm   (*b*) 0.36 μm   (*c*) 0.82 μm]

*63-39   Draw a horizontal spectral diagram for the *Balmer* series of hydrogen. Beneath it draw a similar diagram for singly ionized helium, and comment on the similarity between them.

Mass—Energy Equivalence

*63-40   An electron and a positron may be annihilated in a collision which results in the emission of electromagnetic radiation. If energy and linear momentum are to be conserved, how many quanta must be produced?

*63-41   Calculate the minimum energy required for the production of an electron-positron pair by a γ-ray. What would be the wavelength of such a γ-ray? What would be the minimum energy for the production of a proton-antiproton pair? (*Note that the real threshold value for proton pair-production is substantially higher because of the problem of conserving momentum.*)
Assume h, m_e, c and m_p [0.16 pJ, 1.2 pm, 0.30 nJ]

*63-42   *Atomic binding energy.* The hydrogen atom has a mass 1.67×10^{-27} kg, and an ionization energy 2.2 aJ. Calculate

(*a*) the binding energy of the electron/proton system
(*b*) the mass decrease when the atom is formed from its constituent particles
(*c*) the fractional mass decrease.

With what precision can the masses of the proton, electron and hydrogen atom be measured? Is this mass decrease detectable?
Assume c [(*c*) 1.5×10^{-8}]

*63-43   *Nuclear binding energy.* The relative atomic mass of oxygen-16 is 15.995, that of hydrogen 1.007 8, and that of a neutron 1.008 7.

(*a*) Calculate, in unified atomic mass units, the mass of 8 hydrogen atoms plus 8 neutrons, all at infinite separation at rest.
(*b*) What is the mass decrease when they come together to form an oxygen atom?
(*c*) What fraction is this of the mass of the oxygen atom? Could this be detected by a mass spectrometer whose sensitivity is 1 in 10^5?
(*d*) Calculate the average binding energy per nucleon in the oxygen nucleus.
Assume $m_u = 1.66 \times 10^{-27}$ kg, c [(*d*) 1.3 pJ]

63-44   *The chain reaction.* Suppose a neutron strikes a uranium nucleus and that the fission process which results releases two further neutrons. Suppose now that these two do the same. How many links would there be in this chain if one mole (6×10^{23} atoms) of uranium were to undergo fission?

[79]

*63-45   *Energy from nuclear fusion.* Consider the reaction represented symbolically by
$$^3\text{H } (^2\text{H, n})^4\text{He}$$
(*a*) Use the relative atomic masses listed below to find the energy released when a tritium nucleus fuses with a deuterium nucleus.

$^3\text{H} = 3.016\ 0$ $^4\text{He} = 4.002\ 6$
$^2\text{H} = 2.014\ 1$ $\text{n} = 1.008\ 7$

(*b*) How much energy would become available from the fusion of about 2 grams of deuterium with 3 grams of tritium?
Assume N_A, c, $m_u = 1.66 \times 10^{-27}$ kg   [(*a*) 2.8 pJ (*b*) 1.7 TJ]

*63–46 A **thermonuclear reaction** is one which is caused by the energetic collisions of nuclei in a gas at very high temperatures. What temperature would give the particles a mean translational energy of about 0.2 fJ, and thus enable the reaction to proceed? (*A certain proportion of the particles will have energies far above the mean.*)

Assume k   [10^7 K]

*63–47 O-M In *chemical* reactions the change of binding energy per atom is ~ 1 aJ. Estimate the corresponding change in the total mass of a typical system of 0.2 kg. Do you think this change is detectable? (*In a nuclear reaction the change in binding energy per particle may be* $\sim 10^6 \times$ *as great.*)

*63–48 O-M *Atomic energy.* Determine, by reference or calculation, how much energy (J) is available *from each molecule* in

(*a*) the fission of uranium-235
(*b*) the fusion of tritium and deuterium
(*c*) the oxidation of carbon to CO_2
(*d*) a waterfall of height 100 m.

Compare the results of your investigations. How much energy would be available from a mole of uranium-235?

64 Radioactivity

Questions for Discussion

64–1 Explain in energy terms why small temperature changes have a considerable effect on the rate of a chemical reaction, but *none at all* on a nuclear reaction.

64–2 The distinctive feature of most clocks is the complete *regularity* of some periodic process. A distinctive feature of radioactive decay is that it is completely *random*. How then can we use radioactivity for the measurement of time?

64–3 The fact that emitted α-particles are generally *mono-energetic* (i.e. their energies have a line spectrum) implies that their emission is associated with a *discrete* change of energy. β-particles are emitted with a continuous spectrum of energies. May we conclude that the energy of the nucleus can change by non-quantized amounts?

64–4 It is a theoretical prediction that electrons are not constituents of nuclei. How can one reconcile this with the observed phenomenon of β-decay?

64–5 How can one measure the speed, and hence the k.e., of an α-particle?

64–6 A radioactive emitter is simultaneously giving out two radiations of substantially different ranges. A G.M. tube is set up to receive the radiation through an absorber. Sketch a graph of log (*count rate*) vs. (*absorber thickness*).

64–7 Complete these symbolic representations of nuclear transformations:

(*a*) $^9_4$Be (α, n)...   (*b*) $^{13}_6$C (p, γ)...
(*c*) $^{15}_7$N (p, ...) $^{12}_6$C

64–8 Suggest two different schemes by which $^{214}_{83}$Bi can decay into $^{210}_{84}$Po. Discuss whether you think *both* modes of decay will happen in practice.

64–9 Suggest a scheme of disintegration by which $^{214}_{84}$Po can emit three particles, and thereby become its own isotope.

64–10 *An analogy of nuclear decay.* Imagine 1024 people in an arena. Every ks (about 17 minutes) they toss a coin, and those who throw a head leave the arena.

(*a*) Plot a graph of the expected number N of people in the arena against time, for the first 10 ks. Join your points (unrealistically) by a *smooth* curve.
(*b*) Can you predict when (*i*) a particular individual will leave the arena, or (*ii*) the last person will leave?
(*c*) What is the *half-life* of this process?

64–11 Repeat Qu. **64–10**, plotting your graph on the *same* axes, for the toss of the coin taking place every 4 ks (about every hour). Compare the graphs carefully.

64–12 How is the decay constant λ of a substance related to the probability that a particular *one* of its nuclei will disintegrate?

64–13 The half-life of radium-226 is about one millionth of the age of the Earth. How is it that rocks can contain significant amounts of radium?

64–14 Under radioactive equilibrium conditions the mass ratio of radium-226 to uranium-238 is always $3.3 \times 10^{-7}:1$. Suggest how this fact could be used to measure the half-life of uranium (c. 10^{17} s).

Quantitative Problems

Activity and Decay

†**64–15** The half-life of krypton-92 is 3.00 s. Suppose we have 5.12×10^{20} atoms (just less than a thousandth of a mole).

(a) How many atoms are undecayed after time intervals of (i) 3.00 s (ii) 6.00 s (iii) 12.00 s (iv) one minute? How exact are your answers?

(b) How much time would elapse before the activity was reduced to 2^{-40} of its former value? (*Express* 2^{-40} *as a fraction in terms of powers of 10.*)

[(a) (iii) 3.2×10^{19} (b) 120 s]

†**64–16** (a) What is the decay constant of radium-226, whose half-life is 51 Gs?

(b) What is the half-life of a substance which has a decay constant 6.9×10^{-4} s^{-1}?

[(a) 1.4×10^{-11} s^{-1} (b) 1.0 ks]

64–17 In a large sample of radioactive material there exists a particular nucleus which is known *not* to have disintegrated over the time period of observation, $3\,T_{\frac{1}{2}}$. What is the probability that it will decay in a further interval of (a) $T_{\frac{1}{2}}$ (b) $3T_{\frac{1}{2}}$?

[(b) 0.875]

64–18 *A radioisotope tracer.* Iodine-131 is a radioisotope for which $\lambda = 1.0 \times 10^{-6}$ s^{-1}, and is present in a particular iodine sample in the ratio 1 atom ^{131}I : 10^7 atoms ^{127}I. Iodine-127 is stable.

(a) What is the activity of 0.10 mg of the sample?

(b) A detector responds to 1 disintegration in every 500. What will be its counting rate?

(*Notice the great sensitivity of this method of tracing the behaviour of iodine:* 1 *in* 10^7 *is a very small concentration.*)

Assume N_A [(a) 4.7×10^4 s^{-1} (b) 95 s^{-1}]

64–19 What is the activity of the radon in equilibrium with 1.0 g of radium-226? Radium has a decay constant 1.4×10^{-11} s^{-1}. (*Until recently the size of this activity was often used as a unit, the* **curie**.)

Assume N_A [3.7×10^{10} s^{-1}]

64–20 The half-life of strontium-90, a typical laboratory β-source, is 28 years. What mass of ^{90}Sr will have an activity 3.7×10^4 s^{-1}? (*This activity used to be referred to as a* **microcurie**, *but the term is now obsolete.*)

Assume N_A [7.1×10^{-12} kg]

64–21 *Correction for background count rate.* At time $t = 0$ a detector gave a count rate of 82 counts s^{-1}. After a time of 210 s the count rate had dropped to 19 s^{-1}. When the sample was taken away the average background count rate stayed constant at 10 s^{-1}. What was the half-life of the substance under investigation?

[70 s]

64–22 Radon-222 is a radioactive monatomic gas whose decay constant is 2.1×10^{-6} s^{-1}. What is the initial rate of disintegration of 1.0 mm^3 of pure radon gas at s.t.p.?

Assume R, N_A [5.6×10^{10} s^{-1}]

64–23 Uranium-238 has a half-life of 1.42×10^{17} s. Calculate

(a) the number of atoms in 1.00 mg

(b) the decay constant λ

(c) its **activity**, the number of disintegrations that 1.00 mg undergoes in each second.

How does this activity change over a period of (i) 1 day (ii) 10^6 years?

Assume N_A [(c) 12.3 s^{-1}]

64–24 What is the half-life of a substance whose activity decreases by 1.0% in a time 0.10 Gs (~ 3 years)? (*If you make justifiable approximations in your answer, indicate their physical significance.*)

[6.9 Gs]

*able **64–25** *Measurement of $T_{\frac{1}{2}}$ from a log graph.* The table shows the variation with time of the activity of a sample of radon-220.

activity/arbitrary unit	150	70	33	15
time t/s	0	60	120	180

(a) Plot a graph of lg (*activity*) against t.

(b) Using $N = N_0 e^{-\lambda t}$, show that the graph of ln N against t has a slope $-\lambda$.

(c) Measure the slope of *your* graph to find λ, and hence deduce $T_{\frac{1}{2}}$.

[(c) 5(5) s]

***64–26** *Geological dating.* A particular rock sample contains ^{206}Pb and ^{238}U in the ratio of 1.0 : 5.0 (by weight). Uranium has a half-life 1.4×10^{17} s. Calculate

(a) the number of ^{206}Pb atoms and of ^{238}U atoms in a sample containing (say) 1.0 g of ^{238}U

(b) the original number of ^{238}U atoms in this sample

(c) the age of the rock, using $N = N_0 e^{-\lambda t}$.

You will need to make assumptions – detail them explicitly.

Assume N_A [(a) 5.8×10^{20} (Pb), 2.5×10^{21} (U)
(b) 3.1×10^{21} (c) 4.2×10^{16} s $= 1.3 \times 10^9$ years]

***64–27** *Parent-daughter equilibrium.* When uranium-238 decays, radium-226 is one of a series of disintegration products. A particular rock sample contained N_0 atoms of uranium, decay constant λ_0, and N atoms of radium, decay constant λ.

(a) At *equilibrium*, what is the rate at which

(i) uranium atoms are decaying, and radium atoms are being formed

(ii) radium atoms are decaying?

(b) Express the ratio N/N_0 in terms of the respective half-lives $T_{0,\frac{1}{2}}$ and $T_{\frac{1}{2}}$.

(c) The rock sample had 1.0×10^{23} uranium nuclei. How many radium nuclei were in it? ($T_{0,\frac{1}{2}} = 1.4 \times 10^{17}$ s, and $T_{\frac{1}{2}} = 51$ Gs.)

(d) What mass of radium is this?

Assume N_A　　　[(c) 3.6×10^{16}　(d) 1.4×10^{-8} kg]

*64–28　*Average or mean lifetime* T_{av}. Average lifetime is defined by the equation

$$T_{av} = (sum\ of\ individual\ lifetimes)/N_0 = T_{tot}/N_0.$$

(a) Put $T_{tot} = \int_{t=0}^{t=\infty} t \cdot dN$, and hence show that $T_{av} = 1/\lambda$.

(b) Sketch a graph of N against t, and use it to show the distinction between $T_{\frac{1}{2}}$ and T_{av}.

(c) Calculate T_{av} for radium, for which $T_{\frac{1}{2}} = 51$ Gs.

[(c) 74 Gs]

Energy

64–29　*Energy needed to create an ion-pair*. An α-particle emitted with k.e. 0.67 pJ has a range in air of 26 mm. Analysis of its track in an ionization chamber indicates that collisions have caused an average of 5.0×10^3 ion-pairs per mm of track. How much energy is needed to create a single ion-pair in air?

[5.2 aJ]

64–30　*Radiation dose*. Radiation dose used to be measured in terms of the **röntgen**, the dose received by 1.0×10^{-6} m^3 of air at s.t.p. in which 2.1×10^9 ion-pairs were created. The density of air at s.t.p. is 1.3 kg m^{-3}, and the energy required to produce an ion-pair is 5.1 aJ.

(a) Express the röntgen in terms of the SI unit for dose, the J kg^{-1}. (Your answer will not equal *exactly* the accepted value.)

(b) A permitted *dose rate* for human beings is 1 milliröntgen per week. Express this in W kg^{-1}.

[(a) 8.2 mJ kg^{-1}　(b) 14 pW kg^{-1}]

64–31　*α-emission from uranium-238*. Refer to the diagram, which shows the variation in the p.e. E_p of an α-particle with its distance r from a thorium-234 nucleus.

(a) Explain the form of the curve. How, for example, is the zero of p.e. defined?

(b) What, in classical physics, is the minimum energy needed by an α-particle to escape from the nucleus? What would be its k.e. after escape?

(c) Wave mechanics predicts a 1 in 10^{38} chance for the escape of an α-particle. What would be its speed after emission?

(d) Inside the nucleus a particle has a speed $\sim 20 \cdot$Mm s^{-1}. How many 'collisions' does it make with the 'wall' per second?

(e) How long, *on average*, would such an α-particle take to escape from the nucleus? Explain the significance of the italicized words.

Assume $m_\alpha = 6.7 \times 10^{-27}$ kg, and compare your answer to (e) with the half-life of uranium, 4.5×10^9 years.

[(c) 1(3) Mm s^{-1}　(e) 1×10^{17} s]

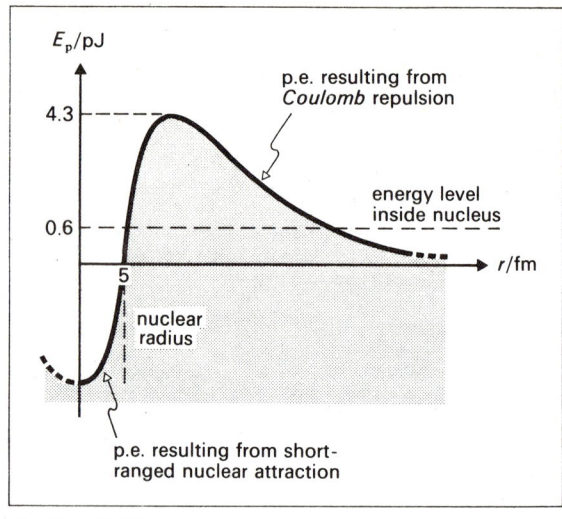

For Qu. **64–31**

Miscellaneous

†64–32　*Range of β-particles*. β-particles of energy 0.2 pJ have a range which can be expressed generally as 6 kg m^{-2}. (*This means that they are stopped by a thickness of superficial density 6 kg m^{-2}.*) Calculate the approximate thicknesses of

(a) air (density 1.3 kg m^{-3}), and

(b) aluminium (2.7×10^3 kg m^{-3})

which would be effective in stopping them.

[(a) 5 m　(b) 2 mm]

†64–33　*Ionization chamber current*. A source emits 3.7×10^7 α-particles in each second, each of which can give rise to 2.0×10^5 ion-pairs. What saturation current could this source produce in an ionization chamber?

Assume e　　　[2.4 µA]

64–34　*The ionization chamber*. The p.d. between the anode and the cathode of an ionization chamber of capacitance 80 pF is 1.0 kV. The battery is disconnected. An α-particle enters the chamber and produces 1.5×10^5 ion-pairs.

(a) What was the original charge on the plates of the chamber?

(b) What neutralizing charge moves to the plates when the α-particle enters?

(c) By what fraction is the charge changed? Could this be observed without amplification?

Assume e　　　[(c) 3.0×10^{-7}]

64–35　*Estimation of the Avogadro constant*. Use the following information to calculate a rough value for N_A. 1.0×10^{-9} kg of radium-226 in equilibrium with its decay products emits 1.5×10^5 α-particles s^{-1}. During one year 1.7×10^{-7} m^3 of gaseous ^{4}He, at a density 0.18 kg m^{-3}, can be collected from 1.0 g of radium.

[6.2×10^{23} mol^{-1}]

XII Essays

65   Essays

Notes on Planning and Writing an Essay

Preparation. Once the topic has been chosen, read widely to sample different viewpoints and alternative approaches. It is a good scheme to make written notes as you read. Possible sources for reading are suggested opposite, and some selected essay titles are given on page 200.

Planning. Identify quite clearly, before you put anything further on paper, the theme which you intend to make the thread of the whole essay. Not only must all your ideas be related to this thread, but it must also be clear *how* they are related.

(1) Now write down, as they occur to you, any further ideas which are relevant to the main theme. These could be main points, developments, qualifications, illustrations, significant orders of magnitude, etc.

(2) Order these random notes by sorting them into natural groups. (For example, one group might deal with experimental work, another with the results of these experiments, and a third with their interpretation.) This grouping process will enable you to visualize the paragraphs of the essay, as you will be associating naturally similar ideas, and each group of ideas will subsequently become a main point.

(3) Write an explicit plan of the final essay along these lines:

Introduction (which can be brief)
First main point
  – subsidiary points, development, qualifications, illustrations, etc.
Other main points
  – subsidiary points etc., following the same lay-out.
Summary and *conclusion*.

The introduction, main points and summary must be seen to be linked by the principal theme.

Writing. The fact that you are writing about a scientific topic makes it all the *more* important to ensure that what you have written is clearly expressed. Only in this way can you hope to communicate efficiently, with, for example, people with a knowledge or vocabulary less specialized than your own. Observe the standard conventions of punctuation. Use *paragraphs* to distinguish your main points, and to make clear the separate steps in your argument: *sentences* deal more briefly with individual ideas or developments. Avoid padding and irrelevance: selective use of clear diagrams will help to achieve this.

When you have finished writing, you can test the continuity of what you have written by trying to read it aloud. Any sentence over which you have to hesitate would probably benefit from rewriting.

Sources of Information

Listed below are several series of books which are available relatively cheaply in paperback form. Some, such as the *Science Study Series,* are written specifically to arouse the interest of pupils in schools, while others, such as the *World of Science Library,* are intended for the more mature non-specialist reader. All provide useful background information of value for the essays suggested later in this chapter.

New titles are being added at a rate that would have made it pointless to list individual titles. Instead we have listed the addresses of publishers, so that they can be asked to supply up-to-date information.

(a) Paperback series

(*i*) **Contemporary Science Paperbacks,** published by
Oliver and Boyd Ltd.
Croythorn House
Ravelston Terrace
Edinburgh EH4 3TJ

(*ii*) **Momentum Books,** published by
Van Nostrand Reinhold Company Ltd.
Molly Millars Lane
Wokingham
Berkshire
These books were written for the *Commission on College Physics.*

(*iii*) **Science Study Series**, published in England by
Heinemann Educational Books Ltd.
48 Charles Street
London W1X 8AH
These books are recommended as being particularly suitable for use in schools, since they were written on behalf of the *Physical Science Study Committee.*

(*iv*) **The World of Science Library,** published by
Thames and Hudson Ltd.
30 Bloomsbury Street
London WC1B 3QP

(*v*) **World University Library**, published by
Weidenfeld and Nicolson Ltd.
11 St. John's Hill
London SW11 1XA

(*vi*) **The Wykeham Science Series** for Schools and Universities, published by
Taylor and Francis Ltd.
Cannon House
10-14 Macklin Street
London WC2B 5NF
The aim of these books is to bridge the gap between school and university. Each book is written by a specialist in collaboration with a schoolteacher.

(b) Other useful sources

(*i*) Details of **Pelicans** in print at any time can be obtained from
Penguin Books Ltd.
Harmondsworth
Middlesex

(*ii*) The periodical **Scientific American** appears monthly, and invariably contains articles by scientists who are authorities in their own fields. Individual articles are often reprinted and sold separately. Details of the **Offprints** that are available can be obtained from
W. H. Freeman and Company Ltd.
58 Kings Road
Reading RG1 3AA
Berkshire
From time to time *Readings from Scientific American* are collected together and made available in book form in the **Scientific American Resource Library.** At present (1971) three volumes of this are devoted to *Physical Sciences and Technology.*

Other *Scientific American* books of interest include **Materials** (1967), **Lasers and Light** (1969), and **Frontiers in Astronomy** (1970).

(*iii*) The following books will be found useful for essays of a biographical or historical nature:
Developments of Concepts of Physics by Arons (*Addison-Wesley*) 1965. This book also contains extensive references for supplementary reading.
Foundations of Modern Physical Science, by Holton and Roller (*Addison-Wesley*) 1958.
Physics for the Inquiring Mind, by Rogers (*Oxford*) 1960.

(*iv*) A reliable encyclopaedia, such as the **Encyclopaedia Britannica**, will contain, in addition to the more obvious biographical studies, appropriate articles dealing with the factual content of physics.

Selected Essay Titles

Historical and Biographical

Models of the atom
The breakdown of classical physics
The development of the quantum theory
The development of a theory of
    gravitation
The relationship between heat and work
The wave and corpuscular theories
    of light
The development of the atomic theory

Biographies: *Newton, Rutherford, J. J.
Thomson, Michelson, Count Rumford,
Ampère, Fermi, Mme Curie, Galileo,
Leonardo da Vinci.*

Mechanics and Gravitation

Tides
Planetary orbits
Artificial satellites
Flight
The atmosphere
The dynamics of athletics
The physics of balance

Thermal Properties of Matter

Very low temperatures
Infra-red radiation
Heat engines
Gas liquefaction
High pressures
Entropy

The Structure of Matter

Intermolecular and interatomic forces
Brownian motion
Lubrication
The structure and properties of metals
Streamlining
Superfluidity
The kinetic theory
Perpetual motion
The structure of molecules

Measurement

Thermometry
The measurement of very high
    temperatures
The *Avogadro* constant
The gravitational constant G
The mass of the Earth
The speed c of electromagnetic radiation
Radioactive half-lives
The elementary charge e
The specific charge of the electron
    and proton
Mass spectroscopy
The radius of the Earth

Wave Motion and Optics

Resonance
Seismology
Water waves
Applications of the *Doppler* effect
Fibre optics
The design of optical instruments
Photography
Photoelasticity
The nature of light
Spectroscopy
Spectroscopy in astronomy
Interferometry
Radar
The production of electromagnetic waves
The detection of electromagnetic waves
The physics of music
Ultrasonics

Electromagnetism

The induction motor
Terrestrial magnetism
The relationship between E and B
The physics of radio
The physics of television
Superconductivity
Magnets and the origin of magnetism
The behaviour of electrons in metals
Particle accelerators

Particle Physics

Cosmic rays
The electron
The proton
The neutron
The neutrino
Plasmas
The particle-wave duality
Elementary particles

Atomic and Nuclear Physics

Radioisotopes
Radioactivity
The detection of radiation from active
    nuclei
High energy physics
Nuclear energy
The structure of atoms
Nuclear fusion
The structure of the nucleus

Other Topics

Fields
Astronomy
Radioastronomy
Geophysics
Crystallography
Holography
Integrated circuits
Computers
The Moon
Meteorology
The physics of the car
Discovery and properties of X-rays
Time
The age of the Earth
Probability in physics
Van Allen radiation belts
Crystals and X-rays
Energy
The transistor amplifier
The production of pure semiconducting
    materials
The laws of physics
Lasers

Three-Figure Tables

LOGARITHMS

N	0	1	2	3	4	5	6	7	8	9
10	0000	0043	0086	0128	0170	0212	0253	0294	0334	0374
11	0414	0453	0492	0531	0569	0607	0645	0682	0719	0755
12	0792	0828	0864	0899	0934	0969	1004	1038	1072	1106
13	1139	1173	1206	1239	1271	1303	1335	1367	1399	1430
14	1461	1492	1523	1553	1584	1614	1644	1673	1703	1732
15	1761	1790	1818	1847	1875	1903	1931	1959	1987	2014
16	2041	2068	2095	2122	2148	2175	2201	2227	2253	2279
17	2304	2330	2355	2380	2405	2430	2455	2480	2504	2529
18	2553	2577	2601	2625	2648	2672	2695	2718	2742	2765
19	2788	2810	2833	2856	2878	2900	2923	2945	2967	2989
20	3010	3032	3054	3075	3096	3118	3139	3160	3181	3201
21	3222	3243	3263	3284	3304	3324	3345	3365	3385	3404
22	3424	3444	3464	3483	3502	3522	3541	3560	3579	3598
23	3617	3636	3655	3674	3692	3711	3729	3747	3766	3784
24	3802	3820	3838	3856	3874	3892	3909	3927	3945	3962
25	3979	3997	4014	4031	4048	4065	4082	4099	4116	4133
26	4150	4166	4183	4200	4216	4232	4249	4265	4281	4298
27	4314	4330	4346	4362	4378	4393	4409	4425	4440	4456
28	4472	4487	4502	4518	4533	4548	4564	4579	4594	4609
29	4624	4639	4654	4669	4683	4698	4713	4728	4742	4757
30	4771	4786	4800	4814	4829	4843	4857	4871	4886	4900
31	4914	4928	4942	4955	4969	4983	4997	5011	5024	5038
32	5052	5065	5079	5092	5105	5119	5132	5145	5159	5172
33	5185	5198	5211	5224	5237	5250	5263	5276	5289	5302
34	5315	5328	5340	5353	5366	5378	5391	5403	5416	5428
35	5441	5453	5465	5478	5490	5502	5514	5527	5539	5551
36	5563	5575	5587	5599	5611	5623	5635	5647	5658	5670
37	5682	5694	5705	5717	5729	5740	5752	5763	5775	5786
38	5798	5809	5821	5832	5843	5855	5866	5877	5888	5899
39	5911	5922	5933	5944	5955	5966	5977	5988	5999	6010
40	6021	6031	6042	6053	6064	6075	6085	6096	6107	6117
41	6128	6138	6149	6159	6170	6180	6191	6201	6212	6222
42	6232	6243	6253	6263	6274	6284	6294	6304	6314	6325
43	6335	6345	6355	6365	6375	6385	6395	6405	6415	6425
44	6435	6444	6454	6464	6474	6484	6493	6503	6513	6522
45	6532	6542	6551	6561	6571	6580	6590	6599	6609	6618
46	6628	6637	6646	6656	6665	6675	6684	6693	6702	6712
47	6721	6730	6739	6749	6758	6767	6776	6785	6794	6803
48	6812	6821	6830	6839	6848	6857	6866	6875	6884	6893
49	6902	6911	6920	6928	6937	6946	6955	6964	6972	6981
50	6990	6998	7007	7016	7024	7033	7042	7050	7059	7067
51	7076	7084	7093	7101	7110	7118	7126	7135	7143	7152
52	7160	7168	7177	7185	7193	7202	7210	7218	7226	7235
53	7243	7251	7259	7267	7275	7284	7292	7300	7308	7316
54	7324	7332	7340	7348	7356	7364	7372	7380	7388	7396

N	0	1	2	3	4	5	6	7	8	9
55	7404	7412	7419	7427	7435	7443	7451	7459	7466	7474
56	7482	7490	7497	7505	7513	7520	7528	7536	7543	7551
57	7559	7566	7574	7582	7589	7597	7604	7612	7619	7627
58	7634	7642	7649	7657	7664	7672	7679	7686	7694	7701
59	7709	7716	7723	7731	7738	7745	7752	7760	7767	7774
60	7782	7789	7796	7803	7810	7818	7825	7832	7839	7846
61	7853	7860	7868	7875	7882	7889	7896	7903	7910	7917
62	7924	7931	7938	7945	7952	7959	7966	7973	7980	7987
63	7993	8000	8007	8014	8021	8028	8035	8041	8048	8055
64	8062	8069	8075	8082	8089	8096	8102	8109	8116	8122
65	8129	8136	8142	8149	8156	8162	8169	8176	8182	8189
66	8195	8202	8209	8215	8222	8228	8235	8241	8248	8254
67	8261	8267	8274	8280	8287	8293	8299	8306	8312	8319
68	8325	8331	8338	8344	8351	8357	8363	8370	8376	8382
69	8388	8395	8401	8407	8414	8420	8426	8432	8439	8445
70	8451	8457	8463	8470	8476	8482	8488	8494	8500	8506
71	8513	8519	8525	8531	8537	8543	8549	8555	8561	8567
72	8573	8579	8585	8591	8597	8603	8609	8615	8621	8627
73	8633	8639	8645	8651	8657	8663	8669	8675	8681	8686
74	8692	8698	8704	8710	8716	8722	8727	8733	8739	8745
75	8751	8756	8762	8768	8774	8779	8785	8791	8797	8802
76	8808	8814	8820	8825	8831	8837	8842	8848	8854	8859
77	8865	8871	8876	8882	8887	8893	8899	8904	8910	8915
78	8921	8927	8932	8938	8943	8949	8954	8960	8965	8971
79	8976	8982	8987	8993	8998	9004	9009	9015	9020	9025
80	9031	9036	9042	9047	9053	9058	9063	9069	9074	9079
81	9085	9090	9096	9101	9106	9112	9117	9122	9128	9133
82	9138	9143	9149	9154	9159	9165	9170	9175	9180	9186
83	9191	9196	9201	9206	9212	9217	9222	9227	9232	9238
84	9243	9248	9253	9258	9263	9269	9274	9279	9284	9289
85	9294	9299	9304	9309	9315	9320	9325	9330	9335	9340
86	9345	9350	9355	9360	9365	9370	9375	9380	9385	9390
87	9395	9400	9405	9410	9415	9420	9425	9430	9435	9440
88	9445	9450	9455	9460	9465	9469	9474	9479	9484	9489
89	9494	9499	9504	9509	9513	9518	9523	9528	9533	9538
90	9542	9547	9552	9557	9562	9566	9571	9576	9581	9586
91	9590	9595	9600	9605	9609	9614	9619	9624	9628	9633
92	9638	9643	9647	9652	9657	9661	9666	9671	9675	9680
93	9685	9689	9694	9699	9703	9708	9713	9717	9722	9727
94	9731	9736	9741	9745	9750	9754	9759	9763	9768	9773
95	9777	9782	9786	9791	9795	9800	9805	9809	9814	9818
96	9823	9827	9832	9836	9841	9845	9850	9854	9859	9863
97	9868	9872	9877	9881	9886	9890	9894	9899	9903	9908
98	9912	9917	9921	9926	9930	9934	9939	9943	9948	9952
99	9956	9961	9965	9969	9974	9978	9983	9987	9991	9996

NATURAL TRIGONOMETRIC FUNCTIONS

Angle		Sine	Cosine	Tangent	Angle		Sine	Cosine	Tangent
Degrees	Radians				Degrees	Radians			
0°	0.000	0.000	1.000	0.000					
1°	0.018	0.018	1.000	0.018	46°	0.803	0.719	0.695	1.036
2°	0.035	0.035	0.999	0.035	47°	0.820	0.731	0.682	1.072
3°	0.052	0.052	0.999	0.052	48°	0.838	0.743	0.669	1.111
4°	0.070	0.070	0.998	0.070	49°	0.855	0.755	0.656	1.150
5°	0.087	0.087	0.996	0.087	50°	0.873	0.766	0.643	1.192
6°	0.105	0.105	0.995	0.105	51°	0.890	0.777	0.629	1.235
7°	0.122	0.122	0.993	0.123	52°	0.908	0.788	0.616	1.280
8°	0.140	0.139	0.990	0.141	53°	0.925	0.799	0.602	1.327
9°	0.157	0.156	0.988	0.158	54°	0.942	0.809	0.588	1.376
10°	0.175	0.174	0.985	0.176	55°	0.960	0.819	0.574	1.428
11°	0.192	0.191	0.982	0.194	56°	0.977	0.829	0.559	1.483
12°	0.209	0.208	0.978	0.213	57°	0.995	0.839	0.545	1.540
13°	0.227	0.225	0.974	0.231	58°	1.012	0.848	0.530	1.600
14°	0.244	0.242	0.970	0.249	59°	1.030	0.857	0.515	1.664
15°	0.262	0.259	0.966	0.268	60°	1.047	0.866	0.500	1.732
16°	0.279	0.276	0.961	0.287	61°	1.065	0.875	0.485	1.804
17°	0.297	0.292	0.956	0.306	62°	1.082	0.883	0.470	1.881
18°	0.314	0.309	0.951	0.325	63°	1.100	0.891	0.454	1.963
19°	0.332	0.326	0.946	0.344	64°	1.117	0.899	0.438	2.050
20°	0.349	0.342	0.940	0.364	65°	1.134	0.906	0.423	2.145
21°	0.367	0.358	0.934	0.384	66°	1.152	0.914	0.407	2.246
22°	0.384	0.375	0.927	0.404	67°	1.169	0.921	0.391	2.356
23°	0.401	0.391	0.921	0.425	68°	1.187	0.927	0.375	2.475
24°	0.419	0.407	0.914	0.445	69°	1.204	0.934	0.358	2.605
25°	0.436	0.423	0.906	0.466	70°	1.222	0.940	0.342	2.747
26°	0.454	0.438	0.899	0.488	71°	1.239	0.946	0.326	2.904
27°	0.471	0.454	0.891	0.510	72°	1.257	0.951	0.309	3.078
28°	0.489	0.470	0.883	0.532	73°	1.274	0.956	0.292	3.271
29°	0.506	0.485	0.875	0.554	74°	1.292	0.961	0.276	3.487
30°	0.524	0.500	0.866	0.577	75°	1.309	0.966	0.259	3.732
31°	0.541	0.515	0.857	0.601	76°	1.327	0.970	0.242	4.011
32°	0.559	0.530	0.848	0.625	77°	1.344	0.974	0.225	4.331
33°	0.576	0.545	0.839	0.649	78°	1.361	0.978	0.208	4.705
34°	0.593	0.559	0.829	0.675	79°	1.379	0.982	0.191	5.145
35°	0.611	0.574	0.819	0.700	80°	1.396	0.985	0.174	5.671
36°	0.628	0.588	0.809	0.727	81°	1.414	0.988	0.156	6.314
37°	0.646	0.602	0.799	0.754	82°	1.431	0.990	0.139	7.115
38°	0.663	0.616	0.788	0.781	83°	1.449	0.993	0.122	8.144
39°	0.681	0.629	0.777	0.810	84°	1.466	0.995	0.105	9.514
40°	0.698	0.643	0.766	0.839	85°	1.484	0.996	0.087	11.43
41°	0.716	0.656	0.755	0.869	86°	1.501	0.998	0.070	14.30
42°	0.733	0.669	0.743	0.900	87°	1.518	0.999	0.052	19.08
43°	0.751	0.682	0.731	0.933	88°	1.536	0.999	0.035	28.64
44°	0.768	0.695	0.719	0.966	89°	1.553	1.000	0.018	57.29
45°	0.785	0.707	0.707	1.000	90°	1.571	1.000	0.000	∞

Index

Selected Physical Constants

c	speed of light *in vacuo*	$3.00 \times 10^8 \, \text{m s}^{-1}$
e	elementary charge	$\pm 1.60 \times 10^{-19} \, \text{C}$
e/m_e	specific charge of electron	$-1.76 \times 10^{11} \, \text{C kg}^{-1}$
ε_0	permittivity constant	$8.85 \times 10^{-12} \, \text{F m}^{-1}$
$\dfrac{1}{4\pi\varepsilon_0}$		$9.00 \times 10^9 \, \text{N m}^2 \, \text{C}^{-2}$
F	*Faraday* constant	$9.65 \times 10^4 \, \text{C mol}^{-1}$
G	gravitational constant	$6.67 \times 10^{-11} \, \text{N m}^2 \, \text{kg}^{-2}$
h	*Planck* constant	$6.63 \times 10^{-34} \, \text{J s}$
k	*Boltzmann* constant	$1.38 \times 10^{-23} \, \text{J K}^{-1}$
m_e	rest mass of electron	$9.11 \times 10^{-31} \, \text{kg}$
m_p	rest mass of proton	$1.67 \times 10^{-27} \, \text{kg}$
μ_0	permeability constant	$4\pi \times 10^{-7} \, \text{H m}^{-1}$ exactly
N_A	*Avogadro* constant	$6.02 \times 10^{23} \, \text{mol}^{-1}$
R	molar ideal-gas constant	$8.31 \, \text{J mol}^{-1} \, \text{K}^{-1}$
σ	*Stefan-Boltzmann* constant	$5.67 \times 10^{-8} \, \text{W m}^{-2} \, \text{K}^{-4}$
V_m	molar volume of ideal gas at s.t.p.	$2.24 \times 10^{-2} \, \text{m}^3 \, \text{mol}^{-1}$
R_∞	*Rydberg* constant	$1.10 \times 10^7 \, \text{m}^{-1}$
T_{tr}	triple point of water	$273.16 \, \text{K}$

Common Physical Properties

p_0	standard atmospheric pressure	0.101 MPa (0.101 MN m^{-2})
ρ_{H_2O}	density of water	1.00×10^3 kg m^{-3}
ρ_{Hg}	density of mercury	1.36×10^4 kg m^{-3}
g_0	standard gravitational field strength	9.81 N kg^{-1}
g_0	standard acceleration due to gravity	9.81 m s^{-2}
T_{ice}	standard temperature $-$ ice point	273 K
T_{st}	steam point	373 K
c_{H_2O}	specific heat capacity of water	4.19 kJ kg^{-1} K^{-1}
	one Earth day	86.4 ks
	one Earth year	31.6 Ms (i.e. about 10π Ms)

Unit Prefixes

Fraction	Prefix	Symbol	Multiple	Prefix	Symbol
10^{-3}	milli	m	10^3	kilo	k
10^{-6}	micro	μ	10^6	mega	M
10^{-9}	nano	n	10^9	giga	G
10^{-12}	pico	p	10^{12}	tera	T
10^{-15}	femto	f	10^{15}	peta	P
10^{-18}	atto	a	10^{18}	exa	E